Lecture Notes in Computer Science 2883
Edited by G. Goos, J. Hartmanis, and J. van Leeuwen

T0239513

**Springer**
*Berlin*
*Heidelberg*
*New York*
*Hong Kong*
*London*
*Milan*
*Paris*
*Tokyo*

Jonathan Schaeffer   Martin Müller
Yngvi Björnsson (Eds.)

# Computers and Games

Third International Conference, CG 2002
Edmonton, Canada, July 25-27, 2002
Revised Papers

 Springer

Series Editors

Gerhard Goos, Karlsruhe University, Germany
Juris Hartmanis, Cornell University, NY, USA
Jan van Leeuwen, Utrecht University, The Netherlands

Volume Editors

Jonathan Schaeffer
Martin Müller
Yngvi Björnsson
University of Alberta
Department of Computing Science
Edmonton, Alberta, Canada T6G 2E8
E-mail: {jonathan;mmueller;yngvi}@cs.ualberta.ca

Cataloging-in-Publication Data applied for

A catalog record for this book is available from the Library of Congress.

Bibliographic information published by Die Deutsche Bibliothek
Die Deutsche Bibliothek lists this publication in the Deutsche Nationalbibliografie;
detailed bibliographic data is available in the Internet at <http://dnb.ddb.de>.

CR Subject Classification (1998): G, I.2.1, I.2.6, I.2.8, F.2, E.1

ISSN 0302-9743
ISBN 3-540-20545-4 Springer-Verlag Berlin Heidelberg New York

Springer-Verlag is a part of Springer Science+Business Media

springeronline.com

© Springer-Verlag Berlin Heidelberg 2003
Printed in Germany

Typesetting: Camera-ready by author, data conversion by Olgun Computergrafik
Printed on acid-free paper        SPIN: 10968413        06/3142        5 4 3 2 1 0

# Preface

The Computers and Games (CG) series began in 1998 with the objective of showcasing new developments in artificial intelligence (AI) research that used games as the experimental test-bed. The first two CG conferences were held at Hamamatsu, Japan (1998, 2000). Computers and Games 2002 (CG 2002) was the third event in this biennial series. The conference was held at the University of Alberta (Edmonton, Alberta, Canada), July 25–27, 2002. The program consisted of the main conference featuring refereed papers and keynote speakers, as well as several side events including the Games Informatics Workshop, the Agents in Computer Games Workshop, the Trading Agents Competition, and the North American Computer Go Championship.

CG 2002 attracted 110 participants from over a dozen countries. Part of the success of the conference was that it was co-located with the National Conference of the American Association for Artificial Intelligence (AAAI), which began in Edmonton just as CG 2002 ended.

The CG 2002 program had 27 refereed paper presentations. The papers ranged over a wide variety of AI-related topics including search, knowledge, learning, planning, and combinatorial game theory. Research test-beds included one-player games (blackjack, sliding-tile puzzles, Sokoban), two-player games (Amazons, awari, chess, Chinese chess, clobber, Go, Hex, Lines of Action, Othello, shogi), multi-player games (Chinese checkers, cribbage, Diplomacy, hearts, spades), commercial games (role-playing games, real-time strategy games), and novel applications (Post's Correspondence Problem).

The Computers and Games conference has traditionally appealed to researchers working on artificial intelligence problems that have been motivated by the desire to build high-performance programs for the classic board and card games. However, the commercial games industry has its own set of challenging AI problems that need to be investigated. These problems are difficult, in part because they need solutions that have tight real-time constraints. Further, from the research point of view, many of these research problems have "fuzzy" outcomes. Our community is used to classic performance metrics such as tree size or winning percentage, whereas the commercial games community values the intangible "fun factor" and is not interested in world-beating programs. In an attempt to get more communication between these two communities, we strove to diversify the range of interest for CG by trying to attract more commercial involvement. We were partially successful, enjoying invited talks from Scott Grieg (BioWare, Corp.) and Denis Papp (TimeGate Studios), as well as three refereed papers.

We want to thank our keynote speakers for their excellent presentations:

- Murray Campbell (IBM T.J. Watson Research Center): "Deep Blue: Five Years Later."
- Matt Ginsberg (Computational Intelligence Research Laboratory, University of Oregon): "GIB: Imperfect Information in a Computationally Challenging Game."
- Scott Grieg (BioWare, Corp.): "Tales from the Trenches: Practical AI in Video Games."
- John Romein (Free University, Amsterdam): "Solving Awari Using Large-Scale Parallel Retrograde Analysis."
- Peter Stone (University of Texas at Austin): "The Trading Agent Competition: Two Champion Adaptive Bidding Agents."

All the keynote presentations were conference highlights – a reflection of the quality of the speakers and their talks.

This conference would not have been possible without the tireless efforts of many people. The quality of the papers presented at Computers and Games 2002 is a reflection of the excellent job done by the program committee and the referees. Numerous other people helped make this event a success: Darse Billings, Jim Easton, Amanda Hansen, Akihiro Kishimoto, Tony Marsland, Louise Whyte, Peter Yap, and Ling Zhao. Thank you!

Finally, we thank the Department of Computing Science (University of Alberta), the University of Alberta (Research), and BioWare Corp. for their sponsorship.

August 2003

Yngvi Björnsson,
Martin Müller,
Jonathan Schaeffer

# Organization

## Executive Committee

Co-chairs                    Jonathan Schaeffer (University of Alberta)
                             Martin Müller (University of Alberta)
                             Yngvi Björnsson (University of Alberta)

## Program Committee

Mark Brockington (BioWare)
John Buchanan (Electronic Arts)
Michael Buro (NEC)
Murray Campbell (IBM)
Ian Frank (Future University, Hakodate)
Matt Ginsberg (University of Oregon)
Reijer Grimbergen (Saga University)
Robert Holte (University of Alberta)
Hiroyuki Iida (Shizuoka University)
Andreas Junghanns (DaimlerChrysler)
Graham Kendall (University of Nottingham)
Richard Korf (UCLA)
John Laird (University of Michigan)
Mike Littman (AT&T Research)
Denis Papp (TimeGate Studios)
Duane Szafron (University of Alberta)
Jaap van den Herik (University of Maastricht)

Computers and Games 2002 conference attendees

# Table of Contents

## Part 1: Evaluation and Learning

Distinguishing Gamblers from Investors at the Blackjack Table .......... 1
  David Wolfe

$MOUSE(\mu)$: A Self-teaching Algorithm that Achieved Master-Strength
at Othello ........................................................ 11
  Konstantinos Tournavitis

Investigation of an Adaptive Cribbage Player ........................ 29
  Graham Kendall and Stephen Shaw

Learning a Game Strategy Using Pattern-Weights and Self-play .......... 42
  Ari Shapiro, Gil Fuchs, and Robert Levinson

## Part 2: Search

PDS-PN: A New Proof-Number Search Algorithm...................... 61
  Mark H.M. Winands, Jos W.H.M. Uiterwijk, and Jaap van den Herik

A Generalized Threats Search Algorithm............................. 75
  Tristan Cazenave

Proof-Set Search ................................................. 88
  Martin Müller

A Comparison of Algorithms for Multi-player Games .................. 108
  Nathan Sturtevant

Selective Search in an Amazons Program............................ 123
  Henry Avetisyan and Richard J. Lorentz

Playing Games with Multiple Choice Systems ........................ 142
  Ingo Althöfer and Raymond Georg Snatzke

The Neural MoveMap Heuristic in Chess............................. 154
  Levente Kocsis, Jos W.H.M. Uiterwijk, Eric Postma,
  and Jaap van den Herik

Board Maps and Hill-Climbing for Opening
and Middle Game Play in Shogi .................................... 171
  Reijer Grimbergen and Jeff Rollason

# Part 3: Combinatorial Games/Theory

Solitaire Clobber .................................................. 188
    *Erik D. Demaine, Martin L. Demaine, and Rudolf Fleischer*

Complexity of Error-Correcting Codes Derived
from Combinatorial Games .......................................... 201
    *Aviezri S. Fraenkel and Ofer Rahat*

Analysis of Composite Corridors .................................... 213
    *Teigo Nakamura and Elwyn Berlekamp*

# Part 4: Opening/Endgame Databases

New Winning and Losing Positions for 7×7 Hex ....................... 230
    *Jing Yang, Simon Liao, and Miroslaw Pawlak*

Position-Value Representation in Opening Books ..................... 249
    *Thomas R. Lincke*

Indefinite Sequence of Moves in Chinese Chess Endgames ............. 264
    *Haw-ren Fang, Tsan-sheng Hsu, and Shun-chin Hsu*

# Part 5: Commercial Games

ORTS: A Hack-Free RTS Game Environment ............................ 280
    *Michael Buro*

Causal Normalization: A Methodology for Coherent Story Logic Design
in Computer Role-Playing Games .................................... 292
    *Craig A. Lindley and Mirjam Eladhari*

A Structure for Modern Computer Narratives ........................ 308
    *Clark Verbrugge*

# Part 6: Single-Agent Search/Planning

Tackling Post's Correspondence Problem ............................ 326
    *Ling Zhao*

Perimeter Search Performance ...................................... 345
    *Carlos Linares López and Andreas Junghanns*

Using Abstraction for Planning in Sokoban ......................... 360
    *Adi Botea, Martin Müller, and Jonathan Schaeffer*

# Part 7: Computer Go

A Small Go Board Study of Metric
and Dimensional Evaluation Functions................................ 376
    *Bruno Bouzy*

Local Move Prediction in Go ....................................... 393
    *Erik van der Werf, Jos W.H.M. Uiterwijk, Eric Postma,
    and Jaap van den Herik*

Evaluating Kos in a Neutral Threat Environment: Preliminary Results.... 413
    *William L. Spight*

**Author Index** ................................................. 429
**Game Index** ................................................. 431

# Distinguishing Gamblers from Investors at the Blackjack Table

David Wolfe

Gustavus Adolphus College, Saint Peter, Minnesota
wolfe@gustavus.edu

**Abstract.** We assess a player's long-term expected winnings or losses at the game of blackjack without knowing the strategy employed. We do this by comparing the player's moves to those of a baseline strategy with known expected winnings. This allows an accurate estimate of the player's expectation to be found hundreds of times faster than the naive approach of using the average winnings observed as an estimate.

## 1 Background

A skillful blackjack player, one who counts cards, maintains some information about the distribution of cards remaining in the deck at all times. The player adjusts both wager and play decisions based on this *count* information. Depending on the rules used by a particular casino, the skillful player may have a slight edge over the casino. Without knowing exactly what the player is counting, we would like to assess a blackjack player's playing skill.

There are two potential benefits from this research. First and foremost, this is related to the much harder problem of assessing the quality of decisions people make under uncertainty. For example, a pension fund manager tries to distinguish a good portfolio manager from a lucky one. Second, there are many gamblers who deceive themselves into thinking they are able to play blackjack well enough to beat the casino. In fact, casino blackjack revenues skyrocketed [1] after Thorp published his landmark book, *Beat the Dealer* [2], which explained how to effectively count cards[1]. Players who discover their true skill (usually very poor) will hopefully be deterred from gambling. (As an aside, the author suspects this sort of research is conducted by casinos who, due to their financial interests, are disinclined to publish results in the area.)

### 1.1 Blackjack Rules

Blackjack rules vary from casino to casino. We will summarize one possible set of rules, and mention some common variations.

---

[1] The actual history is a bit more complex. The first response of casinos to Thorp's book was to panic and change the rules to remove any player advantage, at which point angered players stopped playing. The casinos then changed the rules back and players, mostly poor ones thinking they could exploit their *edge*, returned to the tables in greater numbers than before.

J. Schaeffer et al. (Eds.): CG 2002, LNCS 2883, pp. 1–10, 2003.

Blackjack is played with a shuffled *shoe* consisting of one to eight standard 52-card decks. Aces are worth 1 or 11, face cards (and 10s) are worth 10, and all other cards are worth their face value. The player is playing only against the dealer (or house), and the dealer's strategy is fixed by casino rules[2]. Hence, discounting the second-order effects of other players playing at the same table and drawing from the same shoe, we can assume blackjack is a one player game[3]. The goal is to come closer than the dealer to 21 without going over.

First the player places a wager. She is then dealt two cards face up, and the dealer is dealt one card up and one down. If the player (respectively, dealer) is dealt 21 (or *blackjack*) she immediately wins her wager times 1.5 (respectively, loses her wager). If both are dealt *blackjack*, the hand is a tie (or *push*) and the player keeps her wager. If neither has blackjack, play continues; achieving 21 later in play no longer pays 3-to-2.

The player then has three (or sometimes more) options[4]:

- *Stand* to accept no more cards.
- *Hit* to receive an additional card. The option to hit or stand is then repeated until she chooses to stand.
- *Double down* to double her wager and receive exactly one additional card face up.
- *Split* if the two cards are of the same value. The player doubles her wager and the two cards are separated into two hands. Each hand receives one card. After splitting, doubling down is no longer permitted, blackjack no longer pays 3 : 2, but re-splitting to a maximum of four hands is usually allowed. Aces are often treated specially, with only one card dealt to each. Each split hand independently plays one wager against the dealer.

If the player goes over 21 (or *busts*), the dealer immediately wins her wager, giving the dealer his sole advantage.

The dealer then turns up his second card and draws cards until his total is 17 or higher[5]. The player with the higher total then wins the amount of the wager, where a tie is a *push*.

---

[2] In this paper, the player is female and the dealer is male.

[3] In truth, this assumption is somewhat suspect since the dealer can, in fact, affect the game. The dealer could cause irregularities by misdealing or cheating and he has some control over when to shuffle the shoe.

[4] There are many variations in casino rules. It may be that only some hands can be *doubled down* such as totals of 9, 10 or 11. Splitting may only be allowed up to four hands. The player may have an additional option to *surrender* half her bet, ending the hand. She may also be able to *purchase insurance* for half her wager to protect against the dealer's potential blackjack.

[5] In some casinos the dealer also draws a card when the total is a *soft 17*, (i.e., a total of 17 with an ace counting as 11).

## 1.2  Expected Winnings under Perfect Play

There are a number of approaches listed in the literature for estimating a player's perfect strategy (from a given deck situation) and her expected winnings under that perfect strategy. (Here, winnings could be negative if she loses money.)

The first approach was given in Baldwin, Cantey, Maisel and McDermott in 1956 [3] and was honed by others [4, 5, 6]. The deck is assumed to be infinite and each card appears with probability $1/13$. In this case, dynamic programming can be employed to quickly and accurately calculate both perfect play and expected winnings.

A variant of this approach (and the one we use) is to assume the *remaining* deck is infinite, with the probability of drawing a card given by the fraction of its appearance in the current deck. Before the initial deal, this is identical to the first approach. We will call this estimate *semi-perfect,* and we will refer to the player who makes decisions based on this calculation a *semi-perfect player.*

Thirdly, it is nearly possible to calculate expected winnings exactly by exploiting the fact that there are only, for instance, 2527 possible player hands (that haven't busted) when using a four deck shoe [7, 8]. This permits exact calculations when splitting is forbidden, and for all practical purposes is exact.

## 2  Assessing a Player's Skill

We wish to assess a player's skill by observing her play over a short period of time. Initially, we will also assume she plays with unit wagers. For simplicity of discussion, assume she is playing a fixed strategy. By *fixed strategy,* we mean that the player's wagers and decisions are completely determined by the cards she has seen since the last shuffle. While the strategy could easily be extended to be random, we assume it is independent of external factors such as her bankroll and how many complementary drinks she's had.

### 2.1  The Challenge

One approach for assessing a player's skill would be to watch the player play for a while and report her total winnings divided by the number of hands she played. If we assume unit wagers, a single hand's winnings is a random variable, with variance typically exceeding 1.3. Her expected winnings, if she is good, are typically between $-.01$ and $.01$ assuming reasonable favorable rules and decent skill [8, 2]. Therefore, a reasonable goal is to be 95% confident that a measured mean is within, say, $.002$ of the true mean. This goal demands she play at least $1.3 \cdot (1.96/.002)^2 \approx 1,250,000$ hands[6].

In short, the challenge of assessing a player's skill is that her risk is so much higher than her expectation. The key idea to accelerating skill assessment is to consider the difference in expected winnings for a player's actual play (in each situation she is exposed to) and a known baseline strategy. This difference

---

[6] 95% of a normal distribution is between $-1.96\sigma$ and $1.96\sigma$ of the mean.

has much smaller variance than that of a single hand, and we can recover the measured mean by adding this difference to the expectation of the baseline.

We will formalize this plan in the next two subsections.

## 2.2   Estimating Skill with Unit Wagers

We first address the case when a player always wagers one unit. This section will be a bit informal in the hopes of building intuition.

The player's situation before any cards are dealt is determined by the order of the remaining cards in the deck. We will separate this sample space into two components: $C$ is the cards remaining (i.e., number of aces, 2s, 3s, ..., ten-cards), and $O$ is the *order* of the cards remaining.

Again, a player's strategy $S$ is a mapping of player situations to actions. The player's strategy cannot depend upon the order of unseen cards, though it may depend upon the distribution of unseen cards since that is completely determined by cards seen since the last shuffle.

We will construct a reasonably good *baseline strategy*, $B$, to which the player's play will be compared. While the actual details of the baseline strategy are irrelevant to the discussion, the measurement will be most accurate if $B$ is close to player's strategy $S$. In particular, since we expect the player to be pretty good (for very poor play is easy to diagnose), strategy $B$ should implement pretty good—if not perfect—play.

Lastly, $W_S$, a random variable over sample space $C \times O$, is the winnings of the player playing strategy $S$. In summary, we have defined

$$C \stackrel{\text{def}}{=} \text{Sample space of cards remaining,}$$

$$O \stackrel{\text{def}}{=} \text{Sample space specifying ordering of the deck,}$$

$$W_S \stackrel{\text{def}}{=} \text{Winnings playing strategy } S, \text{ and}$$

$$B \stackrel{\text{def}}{=} \text{Any automated baseline strategy.}$$

Then, the expected winnings of the player, $\mathbf{E}_C \{\mathbf{E}_O \{W_S\}\}$, can be rewritten as follows:

$$\mathbf{E}_C \{\mathbf{E}_O \{W_S\}\} = \mathbf{E}_C \{\mathbf{E}_O \{W_B\}\}$$
$$+ \mathbf{E}_C \{\mathbf{E}_O \{W_S\} - \mathbf{E}_O \{W_B\}\}.$$

Again, $W_S$ typically has variance far exceeding its expectation. Therefore a large number of sample hands are required to accurately estimate $\mathbf{E}_C \{\mathbf{E}_O \{W_S\}\}$ if we estimate it by simply dividing actual winnings by number of hands played.

However, for any reasonable choice of baseline strategy $B$, $\mathbf{E}_C \{\mathbf{E}_O \{W_B\}\}$ can be calculated almost exactly or estimated by computer simulation. Since $\mathbf{E}_C \{\mathbf{E}_O \{W_S\} - \mathbf{E}_O \{W_B\}\}$ typically has much smaller variance, the right hand side can be estimated by observing the player for far fewer hands.

## 2.3   Estimating Skill with Varied Wagers

In a casino, a player can vary her wager between some minimum and maximum quantity. For instance, she may be able to bet any integer amount between \$2 and \$500 [7].

To estimate the player's skill, we will watch her play. After she wagers, we will estimate how much she *would* make on average if she were playing a known baseline strategy. When she makes a decision (whether to hit, stand, double, or split), we will determine whether her decision differed from the baseline strategy. If it did, we will credit her the difference between her expected winnings according to the decision she made (assuming she plays baseline for the rest of the hand), and the expected winnings under baseline.

It is plausible that two players are (in the long run) likely to encounter very nearly the same *situations* no matter their respective strategies. We will make the simplifying assumption that the situations a player is exposed to are independent of her strategy. In particular, we are assuming the following:

- The distribution of cards remaining are independent of her hit and stand choices on previous hands. While it's conceivable that a player might choose to make a sub-optimal play on the current hand (for instance, by hitting rather than standing) in order to gain more information on future hands, it's hard to believe it could make much difference.
- The number of times a player has the choice to hit or stand with, say, 16 versus dealer's 10 is independent of her strategy. This is false, for a poor player might never hit with 12 versus dealer's 10, and thereby lose opportunities to get 16 versus dealer's 10. For a good player, this assumption is (hopefully) of little consequence.

Now a typical *situation* consists of the cards remaining, the current wager (if it's been made), the cards in front of the player and dealer, and the order of cards remaining in the deck. Focusing on the cards, let's separate the sample space consisting of the cards remaining in the deck and the cards in front of the players. Let $C$ be the distribution of cards remaining in the deck before a hand is dealt. Let $D$ be the initial deal and $O$ the order of the deck after the deal. A strategy $S$ now includes both how the cards should be played (when to hit, stand, split, or double) and how much to wager as a function of the cards remaining in the deck.

We will define the random variable $W_{S_0 S_1 S_2 \ldots}$ as the winnings of a player when wagering according to strategy $S_0$, making her first decision of a hand (to split, stand, double, or split) by strategy $S_1$, her second decision according to $S_2$, and so forth. Again, $B$ is a baseline strategy. In summary,

---

[7] It should be noted that if one is to maximize expected winnings, the only two reasonable wagers are the minimum or the maximum. In simulations, we considered much narrower ranges because, (1) wide bet variation results in far too high risk for most players, (2) casinos tend to evict good players with such high bet variation, and (3) rare events warranting large wagers can contribute more strongly to long-term expected winnings, making measurement more difficult.

$$C \stackrel{\text{def}}{=} \text{Sample space of cards remaining,}$$

$$D \stackrel{\text{def}}{=} \text{Sample space specifying ordering of initial deals,}$$

$$O \stackrel{\text{def}}{=} \text{Sample space specifying ordering of the deck after deal,}$$

$$W_{S_0 S_1 S_2 \ldots} \stackrel{\text{def}}{=} \text{Winnings using wagering strategy } S_0 \text{ and playing strategy } S_i \text{ for the } i^{\text{th}} \text{ decision after the deal, and}$$

$$B \stackrel{\text{def}}{=} \text{Any automated baseline strategy.}$$

The player's long-term winnings per hand are then given by

$$\mathbf{E}_C \{\mathbf{E}_D \{\mathbf{E}_O \{W_{SSSSS\ldots}\}\}\}.$$

We will estimate this quantity using the equation:

$$\mathbf{E}_C \{\mathbf{E}_D \{\mathbf{E}_O \{W_{SSSSS\ldots}\}\}\}$$

$$= \mathbf{E}_C \left\{ \begin{array}{l} \mathbf{E}_D \{\mathbf{E}_O \{W_{SBBBB\ldots}\}\} \\ + \mathbf{E}_D \{\mathbf{E}_O \{W_{SSBBB\ldots}\} - \mathbf{E}_O \{W_{SBBBB\ldots}\}\} \\ + \mathbf{E}_D \{\mathbf{E}_O \{W_{SSSBB\ldots}\} - \mathbf{E}_O \{W_{SSBBB\ldots}\}\} \\ + \qquad\qquad\qquad \vdots \end{array} \right\}.$$

To estimate a player's average winnings, we will watch her play. Before each deal, she fixes a wager. Once the wager is fixed, $\mathbf{E}_D \{\mathbf{E}_O \{W_{SB}\}\}$ depends only on $B$ and the cards remaining, and can be calculated or estimated through simulation with high precision. Once she decides what to do first (hit, stand, split or double down), we can calculate the difference $\mathbf{E}_D \{\mathbf{E}_O \{W_{SSBBB\ldots}\} - \mathbf{E}_O \{W_{SBBB\ldots}\}\}$, and so forth. The sum terminates when she no longer is taking cards.

## 3   Simulation Study

We simulated a six deck shoe with a five deck *penetration*; if, after a hand, there are fewer than 52 cards (one deck) remaining, the shoe is reshuffled. In assessing play, the baseline strategy is the semi-perfect player of Sect. 1.2.

The player's strategy is the *high-low* count system as described by Stanford Wong [9]. Wong attributes the system to Harvey Dubner from 1963. The player's *count* is the number of twos through sixes minus the number of 10s and Aces which have been seen in the deck. (This is a *balanced count*, meaning it has 0 expected value.) The player's *count per deck* is the ratio of the count and the number of decks not yet seen. This *count per deck* is truncated (i.e., rounded to the integer nearest 0). The player wagers 1 unit at a count per deck which is negative, 2.5 units at 0 or 1, 5 units at 2, 7.5 units at 3, and 10 units at 4 or greater. The reader can refer to [9] to see how the count affects play. Wong predicts the player's long-term advantage is around $0.020 \pm .002$ units per hand.

Table 1 shows the results of the simulations. An individual run (say, Run 1) shows a single player's average winnings over a number of hands (Average Actual Winnings). Note that the average is cumulative over a single run. Three

**Table 1.** Three simulated players play at separate blackjack tables. Each plays enough 6-deck shoes to play 10,000,000 hands. Each player's cumulative average after 100, 200, 500, etc., is shown in the first three columns. The simulated players then play an additional 200,000 hands, and their expected winnings are estimated using the techniques of Sect. 2.3

| Hands Simulated | Average Actual Winnings | | | Assessed Expectation | | |
|---|---|---|---|---|---|---|
| | Run 1 | Run 2 | Run 3 | Run 1 | Run 2 | Run 3 |
| 100 | .23750 | −.29750 | .56750 | −.00161 | .11359 | −.00217 |
| 200 | .45625 | −.52125 | .71375 | .00397 | .04827 | −.00914 |
| 500 | .40450 | .03850 | .25850 | .01828 | .02646 | .01560 |
| 1,000 | .09625 | .01375 | .10825 | .01362 | .01893 | .00679 |
| 2,000 | .01013 | .02250 | −.05938 | .01138 | .01805 | .01604 |
| 5,000 | .04065 | −.04090 | −.07930 | .00777 | .01090 | .01263 |
| 10,000 | .05743 | −.06128 | −.04533 | .01510 | .01258 | .01833 |
| 20,000 | .05600 | −.00184 | −.03680 | .01552 | .01316 | .01511 |
| 50,000 | .02950 | .04133 | .01489 | .01537 | .01403 | .01458 |
| 100,000 | .01845 | .03622 | .01924 | .01539 | .01592 | .01415 |
| 200,000 | .03175 | .02756 | .01296 | .01491 | .01528 | .01456 |
| 500,000 | .02516 | .02317 | .00794 | | | |
| 1,000,000 | .02427 | .02221 | .01500 | | | |
| 2,000,000 | .02262 | .02055 | .01460 | | | |
| 5,000,000 | .02059 | .02197 | .01619 | | | |
| 10,000,000 | .02002 | .01873 | .01783 | | | |
| Variance | 20 | | | 0.007 | | |

runs are shown in order that the reader can get some sense for the variability in the average winnings. The right side of the table (Assessed Expectation) shows the assessed winnings for a separate series of three runs using the techniques discussed in Sect. 2.3. The Variance is the variance of a single hand's actual winnings and of a single hand's assessed expectation, respectively.

Observe that using the techniques in this paper, the predicted long-term winnings converge much faster than the player's actual average winnings. To explain the high variance of actual winnings, recall that while typical winnings are ±wager, the actual winnings could be larger or smaller if the hand is a blackjack, or has been doubled or split.

The assessed expectation differs from the average winnings by about .004. This error is primarily due to the inexact calculations described in Sect. 1.2. The author conducted a separate simulation in which the blackjack rules were augmented to match the calculations by dealing cards from the shoe with replacement during the play of each hand. In this case, the difference between assessed expectation and actual were statistically insignificant.

If the hands were independent, one would expect that the assessed expectation would converge 20/.007 or about 2500 times faster than the average winnings. The author has confirmed this assertion by a separate simulation study in which the player plays only one hand (at random penetration from 0 to 5/6) out of each shoe. Here, the assessed expectation converges only a couple hundred times faster than the average winnings. We attribute this loss of power to the correlation between hands of the same shoe; once a shoe is good (or bad) for the player, it typically remains so for a while.

# 4  Further Research

In addition to dispensing more formally with some of the assumptions mentioned in this paper, a number of other (perhaps more interesting) questions remain.

## 4.1   More Accurate Blackjack Assessment

Although our technique allows the player's expectation to be measured hundreds of times faster than the naive technique, it still requires observing upwards of a thousand hands. An experienced player can play about one hundred hands per hour and has practiced for hundreds or thousands of hours to become proficient, so playing a few thousand hands to benchmark her progress is not out of the question. (It should be quicker to assess lack of skill in a poor player.) Further improvement may be possible if we can limit which situations the player is exposed to. Since most decisions a player makes are rote, it should be possible to test the player on only *challenging* hands and more rapidly assess the player's skill.

In this paper we have not addressed the affect of multiple hands, either played by the same player against one dealer, or by different players sitting at the same table against the dealer.

The most serious blackjack players take into account risk as well as expected winnings. One popular approach is the *Kelly criterion* [10, 11] which seeks to maximize the expected logarithm of the player's bankroll. It's therefore of interest to measure a player's skill who is trying to safely increase her bankroll rather than merely maximize her expected winnings. In addition, it might then be more reasonable to assess a player's skill who is permitted wide bet variation without some of the concerns expressed in the footnote on page 5.

There are a number of assumptions which we are less optimistic of addressing. In particular, the real world just doesn't fit the model. In real casinos, the deck's distribution is not uniformly random and depends greatly on the shuffling methods used. Real players don't use a fixed strategy, but change their strategy both deliberately (for instance, to avoid getting spotted as a card-counter by the casino) and inadvertently (due to distractions, exhaustion, etc.).

## 4.2   Other Games

The same techniques are applicable to assessing player skill in other games such as backgammon. Backgammon differs from blackjack in that it's essentially a two-player game. However, once one fixes a specific computer opponent, backgammon can be thought of as a one-player game, and one can judge a player's expected winnings against that fixed program.

The reduction of backgammon to a one-player game is essential, for in typical tournament situations, strong players play *super-optimal* moves. Against weak opponents, a world-class player makes moves tailored to the weaknesses of an opponent, moves which would be deemed poor against another world-class player. Judging super-optimal play would require more subtle techniques than those found in this paper.

## 4.3    Other Domains

In real-world domains, the benefits of simulation are limited by the fitness of computer models. But there are other challenges as well. For example, it is notoriously hard to accurately assess the skill of a fund manager. One reason is that a fund manager's goals differ from those of the fund investors. If the fund manager is judged to be good, she can make a great deal of money in salary and bonuses, while putting none of her own money at risk. Consequently, the fund manager should exploit any assessment strategy, at times risking a dreadful measure in the hopes of a slightly positive assessment. For similar reasons, extensions to decisions outside of games (such as assessing a manager's or investor's skill) is also complicated by the existence of multiple-players, in addition to the inherent difficulty of assessing a reasonable benchmark strategy.

## Acknowledgments

The author would like to thank Brian Kleinke, whose undergraduate research project confirmed the feasibility of the computer calculations necessary to complete this work, and Max Hailperin, whose helpful advice improved this presentation. The author is indebted to two blackjack investors (who will remain anonymous) for educating the author about card counting issues. In addition, a reviewer recommended the application to backgammon assessment.

## References

1. Thomason, W.: The Ultimate Blackjack Book: Basic Strategies, Money Management, and More. Carol Publishing Group (1996)
2. Thorp, E.: Beat the Dealer. Random House (1962)
3. Baldwin, R., Cantey, W., Maisel, H., McDermott, J.: The optimum strategy in blackjack. Journal of the American Statistical Association **51** (1956) 429–439
4. Heath, D.: Algorithms for computations of blackjack strategies. Presented at the Second Conference on Gambling, sponsored by the University of Nevada. Available at http://www.bjmath.com (1975)
5. Benjamin, A., Huggins, E.: Optimal strategy with "lucky bucks". The Journal of Undergraduate Mathematics and its Applications **14** (1993) 309–318
6. Gordon, E.: Optimal strategy in blackjack. Claremont Economic Papers (1973)

7. Manson, A., Barr, A., Goodnight, J.: Optimum zero-memory strategy and exact probabilities for 4-deck blackjack. The American Statistician **29** (1975) 84–88
8. Griffin, P.: The Theory of Blackjack. 6th edn. Huntington Press (1999)
9. Wong, S.: Professional Blackjack. Pi Yee Press (1994)
10. Kelly Jr., J.: A new interpretation of information rate. IRE Transactions on Information Theory **2** (1956) 185–189
11. Thorp, E.: The Kelly criterion in blackjack, sports betting, and the stock market. Paper presented at The 10th International Conference on Gambling and Risk Taking, June 1997. Revised May, 1998 in `http://www.bjmath.com/bjmath/thorp/paper.html` (1997)

# $MOUSE(\mu)$: A Self-teaching Algorithm that Achieved Master-Strength at Othello

Konstantinos Tournavitis

Institute for Knowledge Based Systems (WBS), Technical University Berlin
tournavitis@fokus.gmd.de

**Abstract.** This paper discusses an experimental comparison of supervised and reinforcement learning algorithms for the game of Othello. Motivated from the results, a new learning algorithm MOUSE($\mu$) (*MOnte-Carlo learning Using heuriStic Error reduction*) has been developed. MOUSE uses a heuristic model of past experience to improve generalization and reduce noisy estimations. The algorithm was able to tune the parameter vector of a huge linear system consisting of about 1.5 million parameters and to end up at the fourth place in a recent GGS Othello tournament[1], a significant result for a self-teaching algorithm. Besides the theoretical aspects of the used learning methods, experimental results and comparisons are presented and discussed. These results demonstrate the advantages and drawbacks of existing learning approaches in strategy games and the potential of the new algorithm.

## 1 Introduction

Many artificial intelligence (AI) systems use evaluation functions for guiding search tasks. In the context of strategy games they usually score positions with a heuristic assessment of the winning chance for the player to move. Research focused in this area has shown how hard the construction of a good evaluation function is. A domain expert has to identify a set of *features*, i.e. a numerical property of a position which is somehow correlated with the game outcome. These features are combined with each other using weights which have to be found in a second step.

The fitting of evaluation weights is relatively well understood, ever since Samuel proposed ways for automatically tuning weights [1]. Later research focusing on this topic produced LOGISTELLO and TD-GAMMON, which are world-class programs in their domains. These programs use different ways to tune their evaluation functions. For LOGISTELLO Buro used a linear evaluation function and tuned the parameter vector using linear regression on training examples that were marked with the actual game outcome (disk differential). The examples were taken from games played by top programs [2]. For TD-GAMMON Tesauro

---

[1] The GGS Internet Server is home to many of the world's strongest Othello players – both human and program. Details on the programs that have participated and the final standings can be found in http://www.btinternet.com/~chris.welty/Tournaments/2002_01_19.htm

J. Schaeffer et al. (Eds.): CG 2002, LNCS 2883, pp. 11–28, 2003.

used a neural network and a reinforcement learning approach, the TD($\lambda$) – algorithm formalized by Sutton, to fit the weights using self-play [3]. The first approach, also known as supervised learning, treats each sequence of observations and its outcome as a sequence of observation-outcome pairs. The second approach uses prediction differences of successive observations in order to change the parameter vector. A more comprehensive review of automatic learning in all areas of computer game programming can be found in [4, 5].

In this paper we discuss the application of supervised and reinforcement learning to Othello. We provide experimental results and comparisons and we conclude that the main problem of reinforcement learning when used with function approximation is generalization. Motivated by the experimental results, the MOUSE($\mu$)-algorithm *"MOnte carlo learning Using heuriStic Error reduction"* has been developed. MOUSE uses ideas from reinforcement learning and a heuristic model of past experience, in order to improve generalization.

We first introduce the evaluation features of the game-playing program we used for our experiments. Next we provide the theoretical aspects of the different learning methods used. After this, we discuss the advantages and drawbacks of these methods and we provide ideas and experimental results on how to overcome them. Finally the MOUSE algorithm is introduced and we provide local and internet tournament results of each learning method.

## 2   Evaluation Features

The evaluation function features used in this work are based on Michael Buro's research on his program LOGISTELLO. The features are patterns of squares comprising combinations of corners, diagonals, and rows. These patterns capture important Othello concepts, such as mobility, stability and parity.

Eleven such patterns are used. Rotations and reflections yield a total of 46. The patterns are a 3x3 and a 5x2 configuration of stones anchored in a corner, all rows, all verticals, and all diagonals of length greater than 3.

The evaluation function is completely table-driven. Given a position, all 46 patterns are matched against the position, with a successful match returning the associated weight. These weights are added to get the overall evaluation, which approximates the final disc differential[2]. In order to increase the evaluation quality, twelve game stages are used, each of which has a different parameter vector. In Othello the number of discs on the board is a reasonable measure for a game stage [6]. In addition to the pattern features, a bias feature is used. This leads to over 1.5 million table entries which must be tuned.

## 3   Weight Fitting

The classic approach for constructing evaluation functions for game-playing programs is to combine win correlated evaluation features of a position linearly:

---

[2] The disc differential is the difference between the number of the player's discs and the number of the opponent's discs at the end of the game (i.e. victory margin).

$$f(p) = \sum_{i=1}^{n} \omega_i f_i(p) \tag{1}$$

This type of evaluation function is often chosen because the combination overhead is relatively small and there exist methods for determining the feature weights. In case of more than one game stage this simple model can be generalized to:

$$f(p) = \sum_{i=1}^{n} \omega_{s,i} f_{s,i}(p), \quad s = stage(p) \tag{2}$$

Significant progress has been made the last years in tuning the weights automatically [7, 3]. In order to make an experimental comparison of existing learning approaches in the game of Othello possible, every evaluation function that we have created uses the same evaluation features, as described in the previous section.

In order to test the learning speed and quality of each learning method, each evaluation function competes against the same benchmark opponent. This benchmark opponent that we created achieved on GGS a rating of 1516 points which corresponds to a rating of an intermediate human player. We have tried to make all figures in this paper comparable to each other in the following way: after some learning time, which is the same for every learning algorithm, the new evaluation function had to play a few hundred games against the benchmark opponent. For each of these games, a random starting position[3] had been generated. Starting from this position every evaluation function had to play the game twice – one as black and one as white – against the benchmark player using same search depths. To avoid advantages for one evaluation function through the random start position, the same starting positions have been used for each program in one tournament cycle. The average endgame result differential is then plotted on the y-axis. When possible, figures are also provided that show the average absolute error between endgame results and current evaluations for each game stage.

## 3.1   Supervised Learning in the Game of Othello

Historically the most important learning paradigm for automatically tuning feature weights has been supervised learning. In this case the learning agent uses a set of training positions that are labeled by a teacher. The training positions can be generated and scored in several ways. The simplest scoring procedure assigns the final game result to all positions occurring in a game [6].

One might think that selecting games between good players as training input would be the most ideal. However, that approach has the limitation that the machine learns from such games only the finer points of play. Good players know the important evaluation features and keep them mostly balanced in their

---

[3] For the generation of a random starting position the system played few random moves at the beginning of the game.

games. However, an evaluation function must also be aware of the most important features. The training should also contain games in which at one point a player makes a serious mistake. Thus, a reasonable strategy for generating training positions from a game database is to select games played by at least one good player and to score game positions according to the final game result. This technique, introduced by Buro, is used by most the top programs and has also been used in this work [2]. Due to the past research and the popularity of Othello, several thousand of games exist in electronic format. About 11 million positions could be found marked with the game result and used to tune the parameters of the evaluation function. The training set consists of 60,000 games played between early versions of Igor Durdanovic's program KITTY and Buro's LOGISTELLO, 20,000 additional games that were generated by LOGISTELLO while extending its opening book and, finally, about 100,000 games that have been played by at least one good player on IOS[4]. All positions were labelled with the disc differential at the end of the game. This procedure is accurate for endgame positions since most of the example games are played perfectly in this stage, whereas labels assigned to opening and middle-game positions are only approximations.

Using these examples, an appropriate parameter vector can be determined by means of linear regression [2]. In order to avoid big evaluation jumps when crossing different game stages, it is helpful to smooth evaluations across adjacent games stages. A second problem is generalization quality. To improve generalization, only weights of pattern configurations that match at least $n$ training positions are changed. In case that there is insufficient data to train a pattern configuration its value is defined to be zero [6].

In Fig. 1 we present the learning curve of the supervised learning approach. The y-axis shows the average absolute prediction error of the evaluation function dependent on the game stage. As it can be seen the prediction quality increases with the number of discs on the board. The resulting program was able to achieve one of the top positions in standard and "synchro-rand"-Othello[5] on GGS. Detailed information on the local and Internet tournament comparisons of this program will be given in the last section.

This learning approach is commonly used for the construction of top Othello programs. Unfortunately, there are some limitations. Rare pattern configurations lead to evaluation errors, especially when a random position has to be evaluated. Secondly, to find or generate good training examples is not an easy task for every game domain. On the other hand, reinforcement learning is an elegant way of tuning the evaluation function, and does not rely on existing training positions.

---

[4] IOS was the Internet Othello Server, the first server allowing players from all over the world to play each other on the Internet.

[5] Each game between two players consists of two simultaneous Othello games with the same randomly chosen starting position; each player is white in one game and black in the other. The winner is the player with the largest disc total in both games.

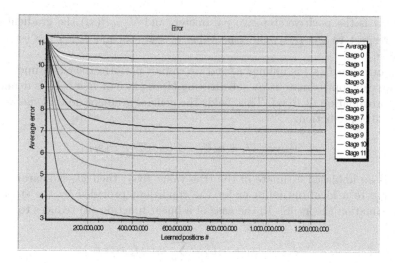

Fig. 1. Supervised learning.

## 3.2 Reinforcement Learning

Reinforcement learning has become one of the most active research areas in machine learning, artificial intelligence, and neural-network research. The defining feature of reinforcement learning algorithms is that they involve learning predictions on the basis of other predictions. One of the challenges that arises in every reinforcement learning approach is the tradeoff between exploration and exploitation. In order to obtain big reward, a reinforcement learning agent must prefer actions that it has tried in the past and found to be effective in producing reward. However to discover such actions it has to try actions that it has not selected before. The agent has to exploit what it already knows to obtain reward but also has to explore to make better action selections in the future. This means that to find an appropriate parameter vector both exploration and exploitation strategies are important [8].

## 3.3 Reinforcement Learning with Function Approximation

Function approximation is an instance of supervised learning, and the primary topic studied in machine learning, artificial networks and pattern recognition. One class of learning methods for function approximation, the gradient descent method, has been used for value prediction. The gradient descent methods are among the most widely used for function approximation. They are adjusting the parameter vector by a small amount in the direction that would most reduce the training error:

$$\overrightarrow{\omega}_{t+1} = \overrightarrow{\omega}_t - \frac{1}{2}\alpha\nabla_{\overrightarrow{\omega}_t}(R(s_t) - P(s_t))^2 = \overrightarrow{\omega}_t + \alpha(R(s_t) - P(s_t))\nabla_{\overrightarrow{\omega}_t}P(s_t) \quad (3)$$

where $\alpha$ is the positive step-size parameter, and $\nabla_{\vec{w}_t} P(s_t)$ the gradient of $P$ with respect to $\vec{w}_t$, $R(s_t)$ the true value of state $s_t$ and $P(s_t)$ the estimated value [8].

In our case $R(s_t)$ is unknown, and we cannot perform the exact update (3). But we can approximate it by substituting $U_t$ in place of $R(s_t)$. $U_t$ is defined as an approximation of $R(s_t)$. This yields the general gradient descent method for state value prediction:

$$\vec{w}_{t+1} = \vec{w}_t + \alpha(U_t - P(s_t))\nabla_{\vec{w}_t} P(s_t) \tag{4}$$

If $U_t$ is an unbiased estimate of the true value $R(s_t)$ then $\vec{w}_t$ is guaranteed to converge to a local optimum under the conditions provided by the stochastic approximation theory [8]. In the case of a linear function (4) reduces to:

$$\vec{w}_{t+1} = \vec{w}_t + \alpha(U_t - \sum_{i=1}^{n} \omega_t(i)f_{s_t}(i))\vec{f}_{s_t} \tag{5}$$

In case of patterns that use exclusively binary features, the above formula is almost trivial to compute. Rather than performing $n$ multiplications and additions, one simply computes the indices of the $m \ll n$ present pattern instances – only for $m$ pattern instances is $\nabla_{\vec{w}_t} P(s_t) = 1$ in one position (otherwise 0) – and then adds up the $m$ corresponding components of the parameter vector.

Different reinforcement learning approaches, such as the TD($\lambda$) method, now have an unified view; they provide a way for computing $U_t$.

**The TD($\lambda$) Method.** Temporal difference learning TD($\lambda$) is perhaps the best known reinforcement learning algorithm. It provides a way of using a scalar reward for approximating $R(S_t)$ so that existing supervised learning technique can be used to tune the function approximator [8].

The gradient descent form of TD($\lambda$) uses the $\lambda$-return, defined as:

$$R_t^\lambda = (1 - \lambda) \sum_{n=1}^{\infty} \lambda^{n-1} R_t^{(n)} \tag{6}$$

with $R_t^{(n)} = r_{t+1} + \gamma r_{t+2} + \ldots + \gamma^{n-1} r_{t+n} + \gamma P(s_{t+n})$, as an approximation of $R(s_t)$, yielding the forward-view update:

$$\vec{w}_{t+1} = \vec{w}_t + \alpha(R_t^\lambda - P(s_t))\nabla_{\vec{w}_t} P(s_t) \tag{7}$$

This quantity is also called the *"corrected n-step truncated return"* or simply the *"n-step"* return, because it is a return truncated after $n$ steps and then approximately corrected for the truncation by adding the estimated value of the $n$-th next state. $\gamma$ is called the *discount rate* and can be in the range $0 \leq \gamma \leq 1$. $r_t$ is the reward at time step $t$ [8]. In our case the reward has been defined to be zero for every position, except for the terminal position where it is defined

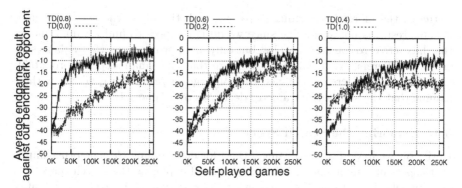

**Fig. 2.** TD($\lambda$) for different values of $\lambda$.

to be the disc differential. The positive parameter $\alpha$ controls the learning rate, and the parameter $\lambda \in [0..1]$ controls temporal difference propagation.

In the special case of $\lambda = 0$, TD(0), the parameter vector is adjusted in such a way as to move $P(s_t)$, the predicted reward at time $t$, closer to $P(s_{t+1})$, the predicted reward at time $t + 1$. This is also known as the *1-step return*. The prediction of $P(s_t)$ depends entirely on the prediction of $P(s_{t+1})$, i.e. bigger *bias* is introduced, since the error in the current value function predictions have a bigger influence.

In contrast, for $\lambda = 1$, TD(1), the parameter vector in adjusted in such a way as to move the predicted reward at time step $t$ closer to the endgame result. This is also known as the Monte-Carlo return. This suffers larger variance in the updates, since more stochastic reward terms appear. In order to reduce the trade-off between bias and variance in these extreme cases, the TD($\lambda$) algorithm uses exponentially weighted sums of n-step updates with decay parameter $\lambda$ [9, 10]. Other values of $\lambda$ interpolate between these behaviors.

Figure 2 shows the learning curves for different values of $\lambda$. All parameters were set to zero in the beginning. The $x$-axis represents the number of self-played games and the $y$-axis the average endgame disc differential, after playing a few hundred games against the benchmark opponent. For all the curves the discount factor was set to $\gamma = 1.0$ and the learning factor to $\alpha = 0.01$. Other values of $\gamma$ lead to the same results but need more time to achieve it. It is interesting that TD($\lambda$) achieves best results for $\lambda = 0.8$. More experiments have shown that for $0.7 < \lambda < 0.95$ there is no difference on the learning performance. We decided to keep $\lambda = 0.8$ for other comparisons as the most suitable value for $\lambda$.

One would expect that to start with some previous knowledge would increase the learning performance. We decided to start with the benchmark program's parameters and to let TD(0.8) improve these parameters further. After 200,000 self-played games the new parameters had shown no significant improvements against the benchmark player.

One could also ask why to perform only one and not more small steps in the direction of the negated gradient. Our experiments have shown that increasing the amount of learning steps for one sample deteriorates the learning quality.

Actually, the more the learning steps are used, the worse the learning quality is. This result is not surprising since we do not expect to find a value function that has zero error on all examples, but only an approximation that balances the errors in different examples. If we do more than one small step the probability of achieving this decreases. In fact, the convergence results for gradient methods assume that the step-size parameter decreases over time [11].

### 3.4   Exploring the Search Space

Until now we have discussed using TD($\lambda$) for determining $U_t$. The other important aspect of reinforcement learning is that of exploring the search space. The experiments showed that the learning quality is very sensitive to the exploration strategy.

The easiest way to explore a big fraction of the search space is to select random next moves. Although this strategy seems to explore the biggest possible search space in a fixed time, it lacks exploitation since it is unlikely that the same state will arise frequently in order to obtain big reward. On the other hand, always selecting the move that seems to be best (*greedy* strategy) obtains more reward but lacks exploration. The $\epsilon$-*greedy* strategy compromises between these two extremes. Here, the agent follows the *greedy* strategy but with a probability $\epsilon$ selects another move.

After experimental tests on these exploration strategies and obtaining poor results, we decided to use the $\epsilon$-*greedy* strategy and to modify it in the following way: *the first n moves in a game are played randomly, and the remaining moves are played using the $\epsilon$-greedy policy.* The value of $n$ had to be manually tuned. The best value was between 4 and 8 moves depending on the learning parameters.

The $\epsilon$-greedy policy is defined as follows: play the move that seems to be the best after searching the game tree at a fixed depth $d$ using the current evaluation function, but with a probability of $\epsilon = 30\%$ choose the second best move. Choosing a large depth $d$ leads to a smaller exploration during a fixed time. An interesting experimental result is that an even $d$ produced better learning results. This may be due to the tree search. Searching at an even depth increases the probability of finding a "quiescent" position, i.e. an immediately obtained "advantage" for one player (odd depth) can be compensated after the opponent's move (even depth). The value of $d = 2$ was found to be best due to the trade-off between generating better games and generating a lot of games in the same amount of time.

Figure 3 shows the learning curves of TD(0.8) using different numbers of random starting moves. It can be seen that using 8 random initial moves increases the learning performance, toward to zero random initial moves, at about an average of 2.5 discs per game, i.e. a significant improvement.

## 4   Discussion of the Results

Unfortunately, none of the reinforcement learning algorithms was able to improve to more than an average of about -2 discs against our benchmark opponent. The

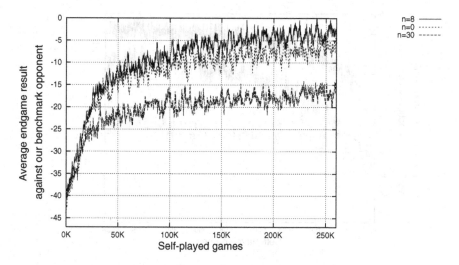

**Fig. 3.** TD(0.8) for different numbers of random starting moves.

best result was obtained by the TD($\lambda$) algorithm, with $\lambda = 0.8$, after a lot of experiments in finding a good exploration strategy. The algorithm was able to start with all parameter set to zero and to improve its playing strength to about somewhere below an average Othello player. This is still not good enough for expert play. We decided to take a deeper look at the learning procedure and to see where the problems are and how to overcome them. There are mainly two problems: *noisy estimations* and *generalization*.

When using reinforcement learning together with function approximation this is one of the key issues. Changing the parameters at one state affects also predictions for other states. One solution for improving generalization in the game of Othello has been proposed by Buro. Instead of changing the parameter vector of every pattern instance that exists in our examples, we change only these that exist at least $MinCount$-times, where $MinCount \gg 1$. This also reduces simultaneously the variance of noisy estimates. For all our experiments until now we have changed the parameter vector after each game, so that most of the pattern instances occur once. In order to increase the probability of finding the pattern instances more than $MinCount$-times, we have increased the number of self-played games before the parameters have to be changed (1 epoch). Even for $MinCount = 1$ this would decrease noise.

In Fig. 4 we show the learning curves for TD(0.8) for different epoch lengths. Three different epoch lengths have been used. The first one consists of 10 games with $MinCount = 1$, the second one of 100 games with $MinCount = 2$ while the third one of 1000 games with $MinCount = 4$. For an epoch of about 10 games we found an optimum. In this case the learning improved further and the resulting agent was able to win against our benchmark player – an important result.

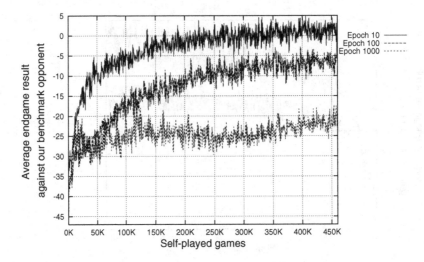

**Fig. 4.** TD(0.8) for different epoch lengths.

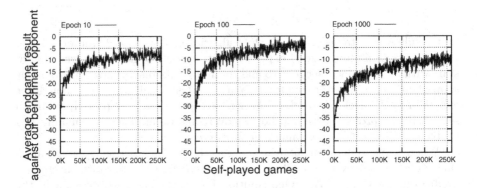

**Fig. 5.** TD(1.0) for different epoch lengths.

Unfortunately, as the epoch length grows the learning speed becomes slower. Our experiments have shown that this depends on $\lambda$. The smaller $\lambda$ and the longer the epoch is, the worse the learning speed. We assume that this has to do with the fact that increasing the epoch length can reduce noise but not bias. In case of $\lambda = 1$, where only noise affects the rewards, the learning quality should increase for increasing the epoch length.

Figure 5 depicts the resulting learning curves of the same experiment for $\lambda = 1$. Although for an epoch length of 10 games the algorithm was not able to achieve the same good results as for $\lambda = 0.8$, the performance was better for longer epochs. The natural explanation why for an epoch length of 1000 games with $MinCount = 4$ the algorithm performs not as well as for an epoch of 100 with $MinCount = 2$ is that $MinCount = 4$ is too big for an epoch length of

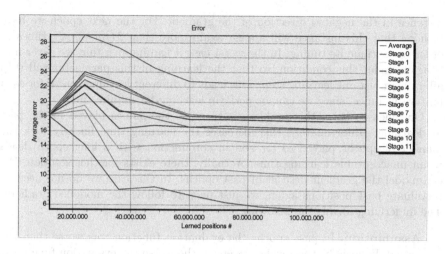

**Fig. 6.** Learning curve of MC-100000. The $y$-axis presents the average absolute prediction error and the $x$-axis the number of positions that occurred while learning.

1000 games. The epoch should be longer in order to increase the probability of finding sufficient number of pattern instances that occur more than 4 times.

We decided to use an epoch consisting of 100,000 self-played games and $MinCount = 50$. These values have not been chosen accidentally but were based on experiments for tuning $MinCount$ and epoch length for the CPU-time that was available. Figure 6 shows the learning process for this program, named MC-100000.

Although the program could achieve a rating of a strong human player on GGS, an important result for learning through self-play, it was not able to come even close on the playing strength of our program which used supervised learning to tune its parameters.

## 5    *MOUSE($\mu$)*

We took again a deeper look at the learning procedure of MC-100000. The problem is that even with an epoch length of 100,000 games it is unlikely to find a sufficient number of pattern instances that occur more than $MinCount = 50$ times in this epoch. One idea is to reduce $MinCount$, but this leads to weaker generalization and noise. Another idea is to increase the length of an epoch. This leads to an increased learning time and we still cannot guarantee that sufficient number of pattern instances will match this condition.

In both cases the main problem still remains. We have different learning speeds for each parameter. In order to understand this, let us take an example of a pattern instance, let's say the first row with an occupied corner. Let us also assume that in the $n$-th epoch we found this instance in our positions more than $MinCount$-times. At the end of this epoch we will change the parameter of this instance in the negated direction of the gradient. For the next epoch we have to hope that this instance will appear again more than $MinCount$ – times in order to continue its tuning. In the worst case it can happen that we never find it again under this condition, so that the learning of this parameter will stop after doing just one learning step. This leads to different learning speeds for each pattern instance, since we cannot perform learning for infinite time.

One idea to solve this problem is to reuse the games played in the past for learning. Unfortunately, this increases the learning time linearly with the number of epochs played so far and in conjunction with a long epoch, leads to an unacceptable learning time. Actually, there exists another solution. We can model the gradient descent of past epochs. In this case we do not need to reevaluate past positions again, we just need to follow the modelled gradient. The underlying idea is the following:

- **Assumption:** In past epochs the evaluation function was worse than the actual. Especially, the earlier the epoch the worse the evaluation function.
- **Heuristic model:** The evaluation errors of the last epoch are at a factor $k = \frac{1}{\mu}$ worse, with $0 < \mu < 1$, where $\mu$ is a heuristic parameter for how much the agent should trust older evaluations. This linear model can be assumed as a good approximation for sufficiently small learning factors $\alpha$.
- **Algorithm:** Multiply the evaluation errors of the last epoch with $\mu$, add them to the evaluation errors of the current epoch and do a small step in the direction of the negated gradient. In case of patterns this is trivial to compute. We just have to multiply the evaluation errors with $\mu$ for every pattern instance.

According to this algorithm the update of the $i$-th parameter at the end of the epoch $t$ is defined as:

$$w_i(t+1) = w_i(t) + \frac{\alpha}{\sum_{j=1}^{t} c_{i,j}} g_t(\overrightarrow{\omega}_t) \tag{8}$$

with

$$g_t(\overrightarrow{\omega}_t) = \mu \cdot g_{t-1}(\overrightarrow{\omega}_t) + \sum_{k=1}^{N} (U_t - P(s_k))\nabla_{\overrightarrow{\omega}_t} P(s_k) \tag{9}$$

where $\sum_{j=1}^{t} c_{i,j}$ counts the number of occurrences of the pattern instance $i$ in all epochs played so far, $N$ the number of positions $s_k$ visited in the $t$-th epoch, $P(s_k)$ the estimated value, $\nabla_{\overrightarrow{\omega}_t} P(s_t)$ the gradient of $P$ with respect to $\overrightarrow{\omega_t}$ and $U_t = R_t^\lambda$ as defined in the $TD(\lambda)$ algorithm.

We decided to use $\lambda = 1$, for the following reasons:

1. In this case the learning rewards suffers a larger variance. Increasing the epoch length is a natural way of reducing variance by the law of large numbers.
2. Unbiased estimations ($\lambda=1$) are important for smoothing parameters between game stages.

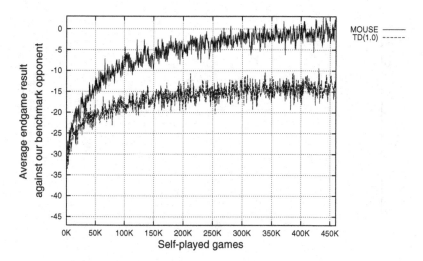

**Fig. 7.** Learning curves for TD(1.0) and MOUSE. The parameter vector is adjusted every 500 games and $MinCount = 20$.

3. Unbiased estimations are important for *Multi-Prob-Cut* [2], a selective extension for tree search, because estimates of different depths have to be compared to each other.
4. Position evaluations for $\lambda < 1.0$ are not always easy to understand.

We call this algorithm MOUSE($\mu$) for **MO**nte-Carlo learning **U**sing heuri**S**tic **E**rror reduction.

Figure 7 depicts an example of how MOUSE can improve learning against standard TD(1.0). The length of the epoch is chosen to be 500 games and $MinCount = 20$, big for this epoch length. $\mu$ is set to 0.95 and the learning factor $\alpha$ to 0.01.

We decided to increase the epoch length to 100,000 self-played games (as in MC-100000) and $MinCount$ set to 100. After five days of learning, the rating of the resulting program on GGS was not only better than the rating of any top human player, but it could also win against our best program we had so far, the one which used supervised learning for tuning its parameters. This program, under the name CHARLY+, ended in the fourth place of the last GGS Othello tournament competing against most of today's top programs, an impressive result for learning only through self-play.

Figure 8 shows the learning process of this program. The $x$-axis shows the number of positions played and the $y$-axis the average absolute errors between current predictions and endgame results for each game stage.

Motivated by the result, we decided to improve MOUSE, so that it can be used in a cluster of computers. This decreases the learning time linearly with the number of computers the cluster has.

One computer acts as both a master and a client whereas the others are only clients. The clients are responsible for self-playing games, while the master

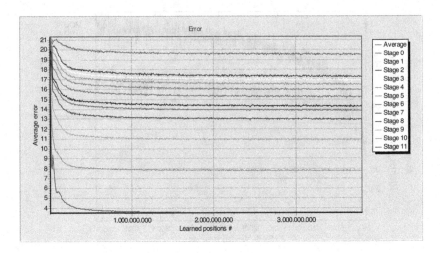

**Fig. 8.** Learning curve of MOUSE.

is responsible for tuning the evaluation function. At the beginning of an epoch the master sends to the clients the new evaluation parameters. According to the new evaluation function each client self-plays a fixed number of games, using the exploration policy as described in the previous section. After generating the games, the client sends the evaluation errors and number of occurrences of each pattern instance back to the master. The master adds the information from every client together and it changes the parameters of each pattern instance according to MOUSE. The new coefficients are sent back to the clients and a new epoch starts.

In order to avoid the bigger network traffic at the beginning and the end of each epoch, Michael Buro suggested a more sophisticated way of the parallelization. In this case for each client the number of self-played games are not longer fixed. The master is now responsible to start and stop a client, depending on the network traffic. More experiments have to be done in this direction [12].

Finally we used MOUSE to create an evaluation function for each Othello type that is played on GGS, i.e every Othello board size from 4x4 up to 12x12, octagonal Othello, as well as their "anti" types[6].

The pseudo-code for MOUSE($\mu$) is shown in the Appendix.

## 6  Tournament Results

### 6.1  Local Tournaments

We provide now the local tournaments results of each program we have generated in standard Othello. Every program used a fixed search depth of 8 plies in the

---

[6] In this type of Othello the winner is the player with the least discs at the end of the game.

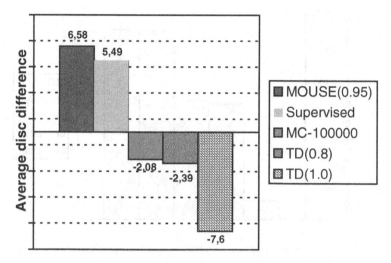

**Fig. 9.** Local tournament results.

middle game and played perfectly the last 16 moves. The first 20 plies have been played randomly by the system. Every combination of two programs had to play two games, one playing as black and one as white. This procedure has been repeated a few hundred times and the average disc differential for every program is used as a predictor for its playing strength.

Figure 9 shows the results of the local tournaments. *TD(1.0)* used TD($\lambda$) with $\lambda = 1.0$ to learn, an epoch of 1 game and *MinCount* = 1. *TD(0.8)* used TD($\lambda$) with $\lambda = 0.8$ to learn, an epoch of 10 game and *MinCount* = 1. *MC-100000* has been described in the previous section. *Supervised* and MOUSE*(0.95)* learned using supervised learning and MOUSE with $\mu = 0.95$ respectively.

## 6.2   GGS Tournaments

In order to test the playing strength of each program in relation to other people's programs, thousands of games have been played on GGS the past 1.5 years. Different names have been used on GGS. *TD(0.8)* is now named DUFFY(1). The MOUSE(0.95) program competes with two versions. Under the name CHARLY it uses 8-ply middle game search without selective extensions and plays 18 empties perfectly. A second time dependent version, CHARLY+, uses Multi-Prob-Cut and selective endgame extensions. Our supervised learning program also comes in two versions: under the name SOFIA, with same search parameters as CHARLY, and as CAESAR with same search parameters as CHARLY+.

Figure 10 shows the rating that each program achieved on GGS (October 2001). With this rating CAESAR achieved the fourth position and CHARLY+ the fifth in this Othello type.

In a recent synchro-rand Othello tournament that took place on January 19, 2002, both programs ended in the fourth place competing against most of

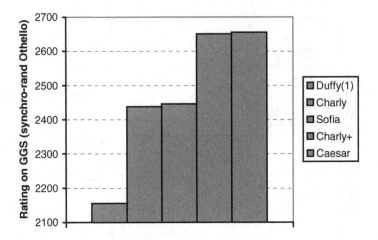

**Fig. 10.** Ratings on GGS in synchro-rand Othello.

today's top programs. This is an impressive result, especially for CHARLY+ that learned using self-play.

## 7    Conclusions

In this paper we have provided an experimental comparison of supervised and reinforcement learning algorithms. We have discussed drawbacks and advantages of each learning method and we provided ideas on how to overcome them.

Additionally, we introduced MOUSE($\mu$), a new learning algorithm that uses ideas of reinforcement learning to tune the parameters of an evaluation function through self-play without any previous knowledge. To improve generalization and reduce variance of noisy estimations, MOUSE uses a heuristic model of past experience.

The program CHARLY+ which used MOUSE to learn, achieved the fourth place in the recent Othello tournament competing against most of today's top programs. This is to our knowledge the first time that a program that learned only through self-play has managed to reach master-level in a strategy game with perfect information.

Further work can show if a more sophisticated heuristic model of past experience would be possible to achieve a better learning performance. More research is also necessary to gauge the learning performance of MOUSE in other game domains.

## Acknowledgments

I would like to thank Michael Buro for his very important support in this work. I would also like to thank the anonymous referees for suggesting improvements to the presentation of this paper.

# References

1. Samuel, A.: Some studies in machine learning using the game of checkers. IBM Journal of Research and Development **3** (1959) 210—229
2. Buro, M.: Experiments with multi-probcut and a new high-quality evaluation function for Othello. In van den Herik, J., Iida, H., eds.: Games in AI Research, Universiteit Maastricht (2000)
3. Tesauro, G.: TD-GAMMON, A self-teaching backgammon program, achieves master-level play. Neural Computation **6** (1994) 215–219
4. Fürnkranz, J.: Machine learning in games: A survey. In Fürnkranz, J., Kubat, M., eds.: Machines that Learn to Play Games. Nova Science Publishers (2001) 11–59
5. Schaeffer, J.: The games computers (and people) play. In Zelkowitz, M., ed.: Advances in Computers 50, Academic Press (2000) 189–266
6. Buro, M.: From simple features to sophisticated evaluation functions. In van den Herik, J., Iida, H., eds.: First International Conference on Computers and Games (CG 1998). Volume 1558 of Lecture Notes in Computer Science. Springer-Verlag (1999) 126–145
7. Buro, M.: Techniken fuer die Bewertung von Spielsituationen anhand von Beispielen. PhD thesis, University of Paderborn (1994) In German.
8. Sutton, R., Barto, A.: Reinforcement Learning: An Introduction. MIT Press (1988)
9. Watkins, C.: Models of Delayed Reinforcement Learning. PhD thesis, Psychology Department, Cambridge University (1989)
10. Kearns, M., Singh, S.: Bias-variance error bounds for temporal difference updates. In: 13th Annual Conference on Computational Learning Theory, Morgan Kaufmann, San Francisco (2000) 142–147
11. Tsitsiklis, J., van Roy, B.: An analysis of temporal-difference learning with function approximation. IEEE Transactions on Automatic Control **42** (1997) 674–690
12. Buro, M.: (2002) Personal communication.

# Appendix: Pseudo-Code of *MOUSE*($\mu$)

```
void MOUSE(real mi)
{
  real Errors = 0,Counts = 0; //Help variables
  int RandomMoves = 8; // Initial random moves
  int Epoch = 100000; // Length of an epoch
  int MinCount = 100;
  TGame Game;
  Init_Parameters;    // set parameters to zero
  for (Iteration=0 ; Iteration < Max_Iterations ; Iteration++) {
    Game = SelfPlay_Game (RandomMoves); // Self-play a game
    SetErrors(Game); // Evaluate evaluation errors
    if (Iteration MOD Epoch = 0) {
      Change_Parameters(MinCount); // change the weights
      ModelError(mi); // the heuristic model
    }
  }
}
```

```
void SetErrors(TGame Game)
{
 for  (int i = 0;I<Game.NumberOfMoves; i++) {
   real Reward = EndGame_Result-Current_Evaluation;
       // Learn from the end game result
   for (every pattern instance occurring in this game) {
     Error[Pattern_Instance] = Error[Pattern_Instance]+Reward;
     Count[Pattern_Instance] ++;
   }
 }
}


void Change_Parameters(int MinCount)
{
 real alpha = 0.01; // Learning factor
 for (int i = 0; i<number of pattern instances; i++) {
   if (count[i]>=MinCount) {
     // Only in this case change the parameters
     Parameter[i] = Parameter[i]+alpha*Error[i]/count[i];
   }
 }
}


void ModelError(real mi)
{
 for (i=0; i<number of pattern instances; i++) {
   Error[i] = mi*Error[i];
   // The counters Count[i] carry their value forward
      from one epoch to the next
 }
}
```

# Investigation of an Adaptive Cribbage Player

Graham Kendall and Stephen Shaw

School of Computer Science & IT, The University of Nottingham, United Kingdom
{gxk,sds98c}@cs.nott.ac.uk

**Abstract.** Cribbage is (normally) a two-player card game where the aim is to score 121 points before your opponent. The game has four stages, one of which involves discarding two cards from the six cards you are dealt. A later stage scores the four cards in your hand together with a card cut randomly from the deck after the discards have been made. The two cards that were discarded are used to form another hand, when combined with the two discards from your opponent. This additional hand is referred to as the crib or box and is scored alternatively by you and your opponent. In this work, we investigate how a strategy can be evolved that decides which cards should be discarded into the crib. Several methods are investigated with the best one being compared against a commercially available program.

## 1 Introduction

Game playing has a long research history. Chess has received particular interest culminating in DEEP BLUE beating Kasparov in 1997, albeit with specialized hardware [1] and brute-force search. Chess is, arguably, a solved game but it is still of interest as researchers turn to adaptive learning techniques which allow computers to 'learn' to play chess, rather than being 'told' how it should play [2]. Adaptive learning was being used for checkers as far back as the 1950's with Samuel's seminal work ([3], re-produced in [4]). Checkers research would lead to Jonathan Schaeffer developing CHINOOK, which claimed the world title in 1994 [5]. Like DEEP BLUE, it is arguable as to whether or not CHINOOK used AI techniques. CHINOOK had an opening and ending database. In certain games it was able to play the entire game from these two databases. If this could not be achieved, a form of mini-max search, with alpha-beta pruning was used. Despite CHINOOK becoming the world champion, the search has continued for an adaptive checkers player. Chellapilla and Fogel's [6] ANACONDA was named due to the stranglehold it placed on its opponent. It is also named BLONDIE24, this being the name it used when competing in Internet games [7]. Anaconda uses an artificial neural network (ANN), with approximately 5000 weights, which are evolved by an evolutionary strategy. The inputs to the ANN are the current board position and it outputs a value which is used in a mini-max search. During the training period, using co-evolution, the program is given no information other than a point score over a series of games. Once ANACONDA is able to play at a suitable level, it often searches to a depth of 10, but depths of 6 and 8 are also common in play.

J. Schaeffer et al. (Eds.): CG 2002, LNCS 2883, pp. 29–41, 2003.
© Springer-Verlag Berlin Heidelberg 2003

Poker also has an equally long research history with von Neumann and Morgenstern [8] experimenting with a simplified, two-player version of poker. Findler [9] studied poker over a 20 year period. He also worked on a simplified game, based on 5-card draw poker with no ante and no consideration of betting position due to the computer always playing last. He concluded that dynamic and adaptive algorithms are required for successful play and static mathematical models were unsuccessful and easily beaten. In more recent times three research groups have been researching poker. Jonathan Schaeffer (of CHINOOK fame) and a number of his students have developed ideas which have led to POKI, which is, arguably, the strongest poker playing program to date [10]. It is still a long way from being able to compete in the World Seies of Poker (WSOP), an annual event held in Las Vegas, but initial results are promising. Schaeffer's work concentrates on two main areas. The first research theme makes betting decisions using probabilistic knowledge [11] to determine which action to take (fold, call, or raise) given the current game state. Billings et al. also uses real time simulation of the remainder of the game that allows the program to determine a statistically significant result in the program's decision making process. Schaeffer's group also uses opponent modeling [12]. This allows POKI to maintain a model of an opponent and use this information to decide what betting decisions to make. See [13] and [14] for good discussions on automated poker.

Koller and Pfeffer [15], using their GALA system, allow games of imperfect information to be specified and solved, using a tree-based approach. However, due to the size of the trees they state "...we are nowhere close to being able to solve huge games such as full-scale poker, and it is unlikely that we will ever be able to do so." Luigi Barone and Lyndon While recognize four main types of poker player; Loose, Tight, Passive, and Aggressive [16, 17]. These characteristics are combined to create the four common types of poker players: Loose Passive, Loose Aggressive, Tight Passive and Tight Aggressive players. A Loose Aggressive player will overestimate their hand, raising frequently, and their aggressive nature will drive the pot higher, increasing their potential winnings. A Loose Passive player will overestimate their hand, but due to their passive nature will rarely raise, preferring to call and allow other players to increase the pot. A Tight Aggressive player will play to close constraints, participating in only a few hands which they have a high probability of winning. The hands they do play, they will raise frequently to increase the size of the pot. A Tight Passive player will participate in few hands, only considering playing those that they have a high probability of winning. The passive nature implies that they allow other players to drive the pot, raising infrequently themselves. In their first paper [18] they suggest evolutionary strategies as a way of modeling an adaptive poker player. They use a simple poker variant where each player has two private cards, there are five community cards and one round of betting. This initial work incorporates three main areas of analysis: hand strength, position and risk management. Two types of tables are used, a loose table and a tight table. The work demonstrates how a player that has evolved using evolutionary strategies can adapt its style to the two types of table. In [16] they develop their work by introducing a hypercube, an $n$-dimensional vector, used to store candidate solutions. The hypercube has one dimension for the betting position (early, middle and late) and another dimension for the risk management (selected from the interval 0..3). At each stage of the game the relevant candidate solutions are selected from the hypercube (e.g., middle betting position and risk management) and the decision is made whether to fold, call or raise. They extend the dimensions of the hypercube to include four betting rounds (pre-flop, post-flop, post-turn and post-river) and an opponent dimension so that the evolved player can choose which type of player it is up against [17]. The authors report that this player out performs a competent static player.

Cribbage is a two player card game where the aim is to get 121 points before your opponent. One of the distinguishing features of the game is the board that is used to "peg" the points. The board has sixty holes for each player, arranged in two rows. A player has to complete two circuits of the board to make 121 points (the final point comes from "pegging off"). Like Bridge, cribbage is a multi-stage game. Unlike Bridge, which has two stages, cribbage has four stages.

The first stage is concerned with discarding some of the cards you are dealt into a "crib" (or box). In the second stage, players alternate playing cards onto the table, trying to earn points which are determined by the rules of the game (for example, playing sequences of cards which total fifteen or 31, playing pairs of cards etc.). The third stage allows the players to use their cards (which they retrieve from the table) to make various card combinations (cards which total fifteen, pairs of cards, runs of cards etc.). In playing these cards a community card, which is cut from the deck after the discards have been made, is also used (see below for an example of the play and scoring). The final stage allows one of the players to play the cards in the crib, creating card combinations as they did in the previous stage for their own hand. Playing of the crib alternates between the players. For a two player game (by far the most common) the players are dealt six cards and have to discard two. This leaves each player with four cards and also gives the crib the same number of cards. As an example of the play and scoring, consider the hand {9♣, 9♢, 9♠, 6♡, 5♣, 5♠}. The player holding this hand might decide to discard the 5♣ and 5♠ into the crib. If the cut card came as 6♢ the player would score two points for each total of fifteen they could achieve; in this case twelve points as each nine can be paired with each six. They would also score two for the pair of sixes and six for the three nines (i.e. three ways to make a pair of nines). Therefore this hand would score $(12+2+6) = 20$, a very good score. However, if it were not your turn to play the crib, you may not want to discard the two fives into it as you immediately give your opponent two points and if they also discard cards valued at ten or five (or one of these cards is cut) then their points start to accumulate rapidly. It can be appreciated that there is a certain amount of luck involved in the game, a certain amount of skill, and some strategy as you need to be aware who will play the crib and play accordingly. For a more complete description of the rules see any book on cards games (e.g., [19]). The rules are readily available on the Internet (e.g., http://www.pagat.com/adders/crib6.html).

The academic literature for cribbage is limited. O'Connor applied temporal difference reinforcement learning to teach a multi-layer perceptron to play cribbage through self-play [20]. A technique that has also been successfully applied to backgammon is temporal difference learning [21]. Martin considers the optimal hand values [22] and states that there has been little previous work carried out for cribbage (his thesis contains no references!). This paper aims to create a cribbage player that evolves a good discard strategy (stage one from above) and to maximize the score in the players hand (stage three). In conducting these experiments we will ignore the suit of the cards. That is, the deck is made up from cards of ace through king but the cards have no suit. This decision was made as the number of possible hand combinations is 18,395 by not including suits but prohibitively large if we include suits. In fact, for cribbage, the suits play a minor part in the game so not including them does not prevent the proposed technique from being investigated. Assuming the technique is successful, the method could be extended to include suits for a future implementation.

## 2      Evolutionary Strategies

Evolutionary strategies (ES) are closely related to genetic algorithms and evolutionary programming. Originally they used only mutation, only used a population of size one, and were used to optimize real-valued variables. More recently, ES's have used a population size greater then one, they have used crossover and have also been applied to discrete variables [23, 24]. However, their main use is still in finding values for real variables by a process of mutation, rather than crossover.

An individual in an ES is represented as a pair of real vectors, $v = (x, s)$. The first vector, $x$, represents a point in the search space and consists of a number of real valued variables. The second vector, $s$, represents a vector of standard deviations.

Mutation is performed by replacing $x$ by

$$x^{t+1} = x^t + N(0, \sigma)$$

where $N(0, \sigma)$ is a random Gaussian number with a mean of zero and a standard deviation of $s$. This mimics the evolutionary process that small changes occur more often than larger ones. In evolutionary computation there are two variations with regard to how the new generation is formed. The first, termed $(\mu + \lambda)$, uses $\mu$ parents and creates $\lambda$ offspring. Therefore, after mutation, there will be $\mu + \lambda$ members in the population. All these solutions compete for survival, with the $\mu$ best selected as parents for the next generation. An alternative scheme, termed $(\mu, \lambda)$, works by the $\mu$ parents producing $\lambda$ offspring (where $\lambda > \mu$). Only the $\lambda$ compete for survival. Thus, the parents are completely replaced at each new generation. Or, to put it another way, a single solution only has a life span of a single generation. In this initial investigation, we use a 1+1 ($\mu = 1$ and $\lambda = 1$) strategy but plan to investigate other strategies in the future. Good introductions to evolutionary strategies can be found in [25, 26, 27, 28, 29].

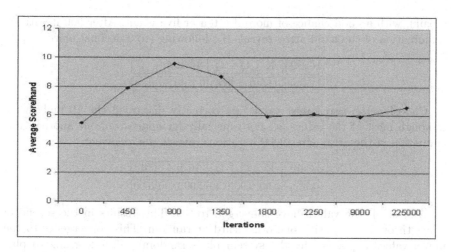

**Fig. 1.** Learning which cards to discard and constraining variables to 0.1..0.9.

## 3    Experiments

Each of the 18,395 possible six card hands was represented by the cards that
make up that hand along with a real-valued variable for each card in that hand.
These variables are used in the decision as to which cards to discard. The aim of
an initial experiment was to ascertain if any learning could take place. A small
set, $n$, of training hands was used. One iteration consisted of playing a single
hand. It therefore takes $n$ iterations to play the entire training sample. In this
case $n = 9$ (although $n$ is relatively small the hands were carefully chosen to
represent both potentially low and high scoring hands). The values associated
with each card for each hand were randomly set. The values were constrained
between 0.1 and 0.9. In order to decide on the discard, the two lowest values
from the hand were chosen. If the discard choice was deemed unsuccessful (see
below), the values relating to the two cards that were discarded are updated
by adding a random Gaussian number (standard deviation = 0.1). The discard
was deemed unsuccessful if the point count was less than 75% of the optimal
point count by considering all possible discards. The results arising from this
experiment are revealing (see Fig. 1).

At the start of learning the average score per hand was approximately 5.5.
After each hand has been played 50 times (450 iterations), there was an improvement with the average score being about 7. This increases further until
the average score peaks at 9.56. After this the average decreases until it almost
returns to the initial value.

Examination of one hand reveals why the rise and subsequent drop occurs.
Hand seven, for example, consisted of a king, two queens, a ten, a five and a
four. With no consideration for crib scoring, the optimum cards to discard are
the ten or king and the four. This leaves a guaranteed 8 points (three 15s and

a pair), with a good chance of more if a ten or five is cut. After 900 iterations, examination of the hand array reveals the following (to 2 decimal places):

| Card | K | Q | Q | 10 | 5 | 4 |
|------|------|------|------|------|------|------|
| Value | 0.90 | 0.90 | 0.88 | 0.79 | 0.80 | 0.11 |

Choosing the two lowest values we correctly discarded the 10 and the 4, although most of the values are not especially far apart from one another. Examination of the array after 9000 iterations reveals the following:

| Card | K | Q | Q | 10 | 5 | 4 |
|------|------|------|------|------|------|------|
| Value | 0.90 | 0.90 | 0.49 | 0.90 | 0.90 | 0.90 |

Almost all the values have converged to 0.9. This results in the second Q being thrown, along with one other card at random. This convergence to the upper values explains why the system plays randomly after learning to play relatively well.

In an attempt to overcome the problem of the values converging to the upper bound the same experiment was conducted but without placing bounds on the variables. The results are shown in Fig. 2. They appear to support the argument that constraining the values held in the statistical array had a detrimental effect on the performance of the player after a certain period of learning time.

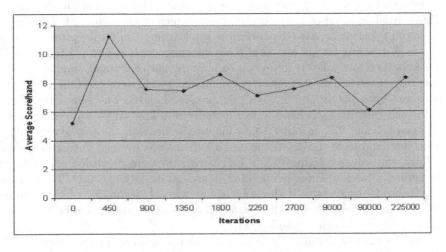

**Fig. 2.** Learning which cards to discard with unconstrained variable.

Discards are still at a reasonable level even after 225,000 iterations, with an average score of between 6 and 8. The peaks and troughs in the graph are explainable by the fact that cribbage relies on a certain degree of luck. The cut card plays a large role in determining the final score for each hand, with good discards sometimes not rewarded, and bad ones "getting a lucky cut." The figure

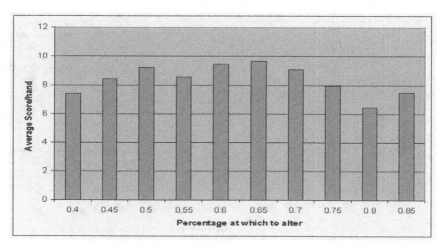

**Fig. 3.** Testing for the best cut off point $p$.

of 75% as being the cut off point, $p$, between a good discard and a bad discard was arbitrarily chosen. Intuitively, the value of p is important, as it determines how often the value relating to the discarded cards are altered, and therefore how often the discard pattern for that hand are changed. It seems likely that if $p$ is set too low, then non-optimal discards will allow low scores to be seen as acceptable. If $p$ is too high, then even good discards may be altered, if the cut card is an adverse one. Therefore, if $p$ is set too high, the statistical array is too volatile. This means that a "final" discard pattern is never reached, as the player constantly evolves, trying to find a non-existent discard pattern that satisfies the over-stringent evaluation function.

In order to try and find a good quality value for $p$, a series of experiments were run that varied $p$ and recorded the average score. The same set of training hands were used for each value of $p$, 250,000 iterations were executed. The results obtained are shown in Fig. 3. The best value for $p$ appears to be 0.65. As suspected, if the value is too high or too low the scoring, in general, tails off. $p = 0.65$ can be further verified as a reasonable value by comparing the number of hands over which it appears to make good quality discards. With $p = 0.65$, 6 of the 9 test hands are played perfectly, resulting in an average score of 9.63 points per hand. Although this figure is a lot higher than the average score when the player has not been evolved, it is not clear why some hands have been learned and others have not.

Further analysis, over a greater number of test hands, showed that 0.65 is not optimal for every hand and it would appear that each hand has its own optimal value. It would appear that 0.65 is the value at which the largest number of hands can be learned optimally. Some hands will need values lower than this and some higher. One suggestion is to evolve the percentage value, as well as the card values, for each hand. However, this was discounted, preferring instead to investigate an approach, which did not rely on percentage values.

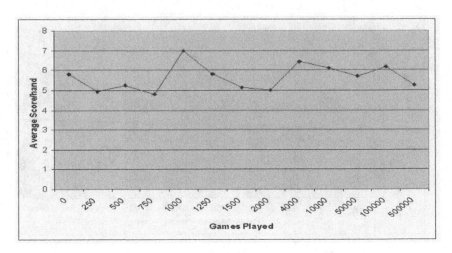

**Fig. 4.** Discard strategy using co-evolution.

A co-evolutionary approach was introduced by dealing one of the test hands to two players, so that they receive (normally) different hands from the training set. The players were then allowed to discard and the cut was made to reveal the community card before calculating the scores. The loser then altered their statistical array (i.e. added a Gaussian random number with a standard deviation of 0.1 to the variable associated with the cards that were discarded). The results can be seen in Fig. 4. This data was collected by dealing, at random, one of the test hands to two players and executing 500,000 iterations.

These results are disappointing as the player never gets close to the average previously recorded (about 9.63 using the strategy based around Fig. 2 with $p = 0.65$) and there exists no general upward trend. This would appear to indicate that learning is not taking place, although this is not entirely true.

When explaining why the mechanism for learning employed here is not successful, we need to examine in more depth the game of cribbage. As explained by Dan Barlow on his website (http://zone.msn.com/cribbage/tips.asp), cribbage involves a high degree of luck. This luck is mostly present in the form of the cards you are dealt, and the cut received. In other words, given better cards, the authors are capable of beating the best cribbage player in the world, although, given the same cards as an expert, the authors would be defeated. This leads to the conclusion that to be successful at cribbage depends not only on playing the hand you are dealt well, but also on getting good hands in the first place.

The nine test hands are all capable of achieving different scores. Consequently, the hands with the least potential are forced to constantly evolve, as they inevitably lose to better hands, no matter how well they are played. Throughout the learning process, we can see that hand four (which consists of 9,9,9,6,Ace,2) quickly converges to optimal values (that is discarding the Ace and 2). All the other hands, when dealt against this hand, will be forced to evolve, no matter how well they are played. Therefore, it seems sensible to deal both players the same hand (even though this is sometimes not normally possible if playing with a physical deck of cards, for example when a player is dealt three nines) and let the player that plays it worse evolve. However, to stop the players converging to the same values and then playing the same way, but maybe sub-optimally, one of the players decides which discards to make based on the values stored in its array and the other player plays randomly. Only the player playing from the array will update its values should it lose (which is certainly possible especially given that the cut card adds a high degree of luck), on the basis that if it plays well it will beat the random player but should the random player win, then the player should evolve.

As the array will be altered less frequently, this means that the training period for each hand will be longer. These tests were conducted on the same set of training hands as previous experiments, and the results are summarized in Fig. 5.

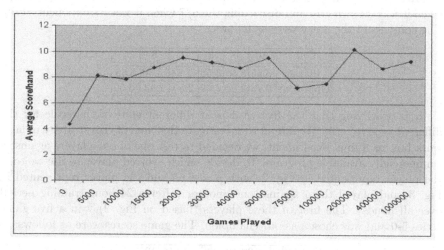

**Fig. 5.** Discard strategy using co-evolution (same hands dealt to both players).

The results for this experiment are comparable to the strategy based around Fig. 2 and with $p = 0.65$. There is an upward trend with peaks and troughs corresponding to the element of luck introduced by the cut card. Undeveloped, the player scored an average of 4.33 points per hand which increased to 9.33 after the training period.

In cribbage, the cards you discard are not just based on maximizing your own hand score but some attention must be paid to the crib. If it is your crib you also want to maximize that score but if it is your opponent's crib you will want to minimize the crib score which may involve reducing the score in your own hand. In order to model this, each hand now requires two sets of values, one representing your turn to play the crib, the other set of values being used when it is your opponent's crib.

To test this idea, two random cards were placed into the crib to simulate the cards discarded by the opponent. If it is your crib, then the crib score is added to the hand score, but the crib score is deducted from your score if it is the opponents crib. As before, the best overall score is calculated, and if the score is not within 65% of the best score, then the array is altered. One test hand was created, which has different discard requirements depending on whose box it is. This hand consisted of 10, 9, 8, 7, 6, 5. If it is your crib then the best discard is likely to be 10/5. If it is the opponent's crib the best discard is probably 6/10 or 6/7, which preserves some points in your hand: and avoids giving your opponent a 5 (notionally the best card in cribbage as it allows scores of 15 to be easily made). In this initial investigation we only tested with one hand. However, if the adaptive player can be shown to adapt to different discards depending on whose crib it is for this single hand, then we believe it will be possible to adapt across a wide range of hands.

After playing the hand for 5000 iterations, the player was discarding the 10/5 when it was its own crib and discarding the 6/7 when it was its opponent's crib. The contents of the array were as follows:

| Crib | 10 | 9 | 8 | 7 | 6 | 5 |
|---|---|---|---|---|---|---|
| Player | 3.55 | 3.58 | 3.58 | -1.83 | -3.50 | 5.56 |
| Opponent | 0.73 | 2.54 | 2.60 | 2.50 | 2.52 | -3.50 |

It is gratifying that the five card is at different ends of the scale for the different types of crib. We intend to investigate this area of crib strategy in later work but, as a final experiment, we decided to test one of our players against a commercially available program. To decide which evolved player to use we took the player represented in Fig. 2, with $p = 0.65$ and the player represented in Fig. 5, and played them against one another (after a 24 hour training period over all hands). The first of these players (based on Fig. 2) won a five game series 5-0 and was chosen as our champion. The game scores were as follows:

| 121-98 | 121-100 | 121-89 | 121-109 | 121-92 |
|---|---|---|---|---|

There are a number of versions of cribbage available, all of which are capable of playing at quite a high level. The version we decided to use was ULTIMATE CRIBBAGE (UC) by Keith Westley. This decision was taken as Keith was the only person who replied to our help for assistance [30]. UC uses statistical methods and heuristics in order to calculate which cards to discard into the crib. The 'easy' level discards the first two cards from the hand, while the 'medium' level applies a sorting heuristic to find the best discards. The 'hard' level introduces additional rules to make the card selection even better. The 'harder' level reviews all possible card combinations and applies probabilities and observed card discard frequencies to calculate the best discards. In order to play against UC, a pegging routine was implemented for the evolved player. This algorithm was of poor quality with regards to playing cribbage, as we are really only interested in the discard strategy and just needed a pegging algorithm to allow us to play the game against an opponent. It plays the first available card supplemented with simple heuristics (make 15 if possible, and lead from a pair).

## 3.1  Evolved Player versus UC (Easy)

Playing ULTIMATE CRIBBAGE at its 'easy' level resulted in an easy win for the evolved player. This is not surprising as the commercial program simply takes the first two cards from its hand, therefore playing randomly. The results were as follows (evolved player first).

| 121-78 | 121-50 | 121-91 | 121-84 | 121-78 |
|---|---|---|---|---|

## 3.2  Evolved Player versus UC (Medium)

During this test, it was noticeable that the discards made by the evolved player were usually better than those made by UC, but the weakness of the pegging algorithm was laid bare, resulting in a tight game. However, the evolved player still won by three games to two. The scores for this match were as follows (evolved player first).

| 121-110 | 112-121 | 121-119 | 90-121 | 121-115 |
|---|---|---|---|---|

## 3.3  Evolved Player versus UC (Hard)

During this match it became apparent how important pegging is. The discards were of a comparable quality each time, but UC would usually score around 4-6 points more than the evolved player in each hand. This inevitably led to the evolved player being defeated; 5-0 (evolved player first).

| 100-121 | 98-121 | 110-121 | 96-121 | 115-121 |
|---|---|---|---|---|

It was suspected that this defeat was due to the evolved player's poor pegging so a rematch was held, this time discounting the pegging points. This resulted in a much tighter game with the evolved player winning 3-2 (evolved player first).

| 121-120 | 121-118 | 111-121 | 121-116 | 109-121 |
|---|---|---|---|---|

## 3.4  Evolved Player versus UC (Harder)

This match was one step too far for the evolved player and it lost the match, even based solely on discards, 5-0.

## 4   Discussion

It is gratifying to see that a player that evolves its discard strategy is able to compete with a commercial application and it demonstrates that players that have no strategy programmed into them are able to evolve strong playing styles that are able to compete with players that have been explicitly programmed with game strategy.

Now that this technique has shown promise in this game of incomplete information we plan to apply the same techniques (and developments of this technique) to other games of incomplete information (such as poker). In addition, we would hope (in the longer term) to use the same techniques in real-world "games" such as stock market prediction. However, we recognize this is a long term aim and one which is far from guaranteed to work. With regards to cribbage we also plan to develop this work further so that we are able to compete with the highest levels of play with regards to the discards and we are also planning to develop these techniques so that we can play the other phases of cribbage and thus compete against other programs and humans in all aspects of the game.

## Acknowledgments

The authors would like to thank the referees for the helpful comments and also to the editors and their staff for their help in improving this paper.

## References

1. Hamilton, S., Garber, L.: Deep Blue's hardware-software synergy. IEEE Computer **30** (1997) 29–35
2. Kendall, G., Whitwell, G.: An evolutionary approach for the tuning of a chess evaluation function using population dynamics. In: Congress of Evolutionary Computation. (2001) 995–1002
3. Samuel, A.: Some studies in machine learning using the game of checkers. IBM Journal of Research and Development **3** (1959) 210–229
4. Samuel, A.: Some studies in machine learning using the game of checkers. IBM Journal of Research and Development **44** (2000) 207–226
5. Schaeffer, J.: One Jump Ahead: Challenging Human Supremacy in Checkers. Springer-Verlag (1997)
6. Chellapilla, K., Fogel, D.: Anaconda defeats Hoyle 6-0: A case study competing an evolved checkers program against commercially available software. In: Congress on Evolutionary Computation. (2000) 857–863
7. Fogel, D.: Blondie24: Playing at the Edge of AI. Morgan Kaufmann (2001)
8. von Neumann, J., Morgenstern, O.: Theory of Games and Economic Behavior. Princeton University Press (1944)
9. Findler, N.: Studies in machine cognition using the game of poker. Communications of the ACM **20** (1977) 230–245
10. Billings, D., Davidson, A., Schaeffer, J., Szafron, D.: The challenge of poker. Artificial Intelligence **134** (2002) 201–240

11. Billings, D., Peña, L., Schaeffer, J., Szafron, D.: Using probabilistic knowledge and simulation to play poker. In: Sixteenth National Conference of the American Association for Artificial Intelligence (AAAI-99), AAAI Press (1999) 697–703

12. Billings, D., Papp, D., Schaeffer, J., Szafron, D.: Opponent modelling in poker. In: Fifteenth National Conference of the American Association for Artificial Intelligence (AAAI-98), AAAI Press (1998) 493–499

13. Billings, D., Papp, D., Schaeffer, J., Szafron, D.: Poker as a testbed for AI research. In Mercer, R., Neufeld, E., eds.: Advances in Artificial Intelligence, Springer-Verlag (1998) 228–238

14. Schaeffer, J., Billings, D., Papp, D., , Szafron, D.: Learning to play strong poker. In Fürnkranz, J., Kubat, M., eds.: Machines That Learn To Play Games, Nova Science Publishers (2001) 225–242

15. Koller, D., Pfeffer, A.: Representations and solutions for game-theoretic problems. Artificial Intelligence **94** (1997) 167–215

16. Barone, L., While, L.: An adaptive learning model for simplified poker using evolutionary algorithms. In: Congress of Evolutionary Computation. (1999) 153–160

17. Barone, L., While, L.: Adaptive learning for poker. In: Genetic and Evolutionary Computation Conference. (2000) 560–573

18. Barone, L., While, L.: Evolving adaptive play for simplified poker. In: IEE International Conference on Computational Intelligence. (1998) 108–113

19. Buttler, F., Buttler, S.: Cribbage: How to Play and Win. Cassell Illustrated (2000)

20. O'Connor, R.: Temporal difference reinforcement learning applied to cribbage (2000) http://www.math.berkeley.edu/~roconnor/cs486.

21. Tesauro, G.: Temporal difference learning and TD-GAMMON. Communications of the ACM **38** (1995) 58–68

22. Martin, P.: Optimal expected hand values for cribbage. Technical report, Department of Mathematics, Harvey Mudd College (2000) (http://www.math.hmc.edu/seniortheses/00/philip-martin-00.pdf).

23. Bäck, T., Hoffmeister, F., Schwefel, H.: A survey of evolution strategies. In: International Conference on Genetic Algorithms, Morgan Kaufmann Publishers (1991) 2–9

24. Herdy, M.: Application of the evolution strategy to discrete optimization problems. In Schwefel, H., Männer, R., eds.: International Conference on Parallel Problem Solving from Nature, Springer-Verlag (1991) 188–192

25. Bäck, T., Fogel, D., Michalewicz, Z.: Handbook of Evolutionary Computation. Oxford University Press (1997)

26. Fogel, D.: Evolutionary Computation: The Fossil Record. IEEE Press (1998)

27. Fogel, D.: Evolutionary Computation: Toward a New Philosophy of Machine Intelligence. IEEE Press (2000)

28. Michalewicz, Z.: Genetic Algorithms + Data Structures = Evolution Programs. Springer-Verlag (1996)

29. Michalewicz, Z., Fogel, D.: How to Solve It. Springer-Verlag (2000)

30. Westley, K.: (2000) Personal communication.

# Learning a Game Strategy
# Using Pattern-Weights and Self-play

Ari Shapiro[1], Gil Fuchs[2], and Robert Levinson[2]

[1] Computer Science Department, University of California, Los Angeles
ashapiro@cs.ucla.edu
[2] Computer and Information Sciences, University of California, Santa Cruz
{gil,levinson}@cse.ucsc.edu

**Abstract.** This paper demonstrates the use of pattern-weights in order to develop a strategy for an automated player of a non-cooperative
version of the game of Diplomacy. Diplomacy is a multi-player, zero-
sum and simultaneous move game with imperfect information. Pattern-
weights represent stored knowledge of various aspects of a game that are
learned through experience. An automated computer player is developed
without any initial strategy and is able to learn important strategic aspects of the game through self-play by storing pattern-weights and using
temporal difference learning.

## 1   Introduction

Automated computer players have been designed for many games including
chess, backgammon, Go and card games such as bridge. For games such as
chess and backgammon, the best computer programs can play at the level of the
best humans. However, programs for games such as Go can be beaten handily
by human players [1]. Typically, computer players can beat human players in
games where the branching factor of the search space is small. The approaches
for creating computerized players vary and can be categorized as follows: brute
force searching approaches, learning approaches, knowledge approaches.

The game of Diplomacy was chosen due to its unique combination of features and difficulty of computerized play. Diplomacy is a multi-player, zero-sum
and simultaneous move game with imperfect information. Past attempts at developing an automated player have applied the searching approach and the expert knowledge approach in order to develop a game strategy. In this paper, a
strategy for game play is developed automatically through self-play of a non-
cooperative version of Diplomacy. A learning-based approach was taken rather
than a search-based approach because the tremendous search space makes a
purely search-based approach intractable. Rules of the game and descriptions of
the state of the game were added as knowledge without assigning any value to
these features. The automated player then extracted the important features of
the game and generated a strategy for playing. This strategy was then used to
generate starting moves in a game, which were then compared to a human-expert
created opening book based on tournament play.

J. Schaeffer et al. (Eds.): CG 2002, LNCS 2883, pp. 42–60, 2003.
© Springer-Verlag Berlin Heidelberg 2003

Games were played utilizing a database of knowledge that was consulted for decision-making on every turn. Partial state information was stored in a structure called a pattern-weight pair [2]. For each move, the automated player searched the database for partial state information that best matched the current state, then calculated the best move from that state. Once a game had been completed, temporal difference learning was applied to the pattern-weights and the process was repeated.

We chose to implement a non-cooperative version of the game where players do not form alliances with each other. Real games of Diplomacy typically involve cooperative strategies and negotiation tactics between players. Our simplified version of the game focusses on the strategic aspects of position and movement. Ultimately, a negotiating agent would need to be coupled with our automated player in order for our automated player to compete in real tournament play. However, a negotiating agent would utilize the information from this automated player in order to determine the benefits of a proposed cooperative strategy.

## 1.1   Rules of Diplomacy

Diplomacy is a seven-player board game that is based on the struggle of the major European powers during World War I. The countries played are: England, France, Germany, Russia, Italy, Austria-Hungary and Turkey [3].

**Game Board.** The board consists of seventy-three adjacent provinces and each player starts with pieces representing military units in their respective home countries. Thirty-five of the seventy provinces on the board are called supply centers (see Appendix A, Fig. 2). The object of the game is to control eighteen of the thirty-five supply centers. A player will have as many pieces on the board as he owns supply centers. A supply center is acquired by moving a piece from an adjacent province to the supply center. The game advances by years, starting in 1901. There are three different turns every year: spring, fall and winter. Supply centers are owned if they are occupied after the fall turn. During the winter turn, additional units are built in the players' home country and then the turns advance to the subsequent year. No moves are made during the winter turn.

**Pieces.** Pieces are represented as one of two types of military units: armies and fleets. Fleets are allowed to move across bodies of water and coastal provinces, while armies can move onto any adjacent land province. Both types of units have equal strength in the game.

**Actions and Orders.** On each spring or fall turn, a piece can be ordered to perform one of the following actions: *move, hold* or *support*. A *move* order will transport a piece from one province to an adjacent province. A *hold* order will keep the piece in its current location. A *support* order will assist another piece that is moving from one province to a province adjacent to the piece executing the support order. This assistance allows the moving piece to take control of a province during conflict situations as described below.

Fleets are also allowed to order a convoy action, which transports the army piece across the body of water that is occupied by the fleet to any coastal province adjacent to that body of water. This is the only exception to the rule that states that pieces may only move one province at a time.

**Game Play.** A face-to-face game of Diplomacy involves a negotiation period during each turn of approximately 30 minutes where players attempt to gain the trust and cooperation of the other players. Players meet with each other in private rooms to discuss collaborative strategy and to establish alliances. Alliances can be made so as to coordinate the moves of each player's countries, as well as to define 'demilitarized zones' (DMZs) which are provinces deliberately kept unoccupied by both players. Once the 30 minutes has expired, each player secretly writes their actions on a piece of paper and the actions are resolved simultaneously. The following is an example of a set of orders for the French player for the initial turn:

```
Fleet in Brest to Mid-Atlantic Ocean
Army in Marseilles to Piedmont
Army in Paris to Burgundy
```

**Simultaneous Moves and Standoffs.** Since the moves are recorded in secret, no player knows the moves of the other players until after they have committed to their moves for the turn. This differs from turn-based games where the current state of the game and subsequent state after applying the operators is known. In addition, the simultaneous resolution of moves creates potential conflicts, as two pieces may move into the same province. In order to resolve conflicts, pieces that move into the same province create a standoff, where both pieces remain in their original provinces and do not move. This, in turn, could create additional standoffs, as pieces moving into the provinces of the pieces that were involved in a standoff would also generate a standoff, thus a single stand-off could create a chain of multiple standoffs.

**Supports and Cooperation.** A standoff may be overcome with the aid of a third piece that is not involved in the standoff. This piece issues a *support* action which effectively aids the movement of one of the pieces into the contested province. This *support* action gives the moving piece the additional strength necessary to usurp an occupied province, or to win a contest between another piece moving into the same province. However, if each piece moving into a province is subsequently supported by another piece, then another standoff is created. Whichever piece retains a greater amount of support will be allowed to move into the province. This aspect leads to cooperation among players, since the initial positions of the players leave them very little room to move without coming

Table 1. Average number of moves.

| Type of Units | Avg. No. of Moves | Avg. No. of Supports | Avg. No. of Holds | Avg. No. of Convoys |
|:---:|:---:|:---:|:---:|:---:|
| Army | 3.22 | 1.11 | 1 | .56 |
| Fleet | 4.15 | 1.58 | 1 | .79 |

into conflict with another player. Often, two players will cooperate by supporting each other's pieces against a third player. In the example below, the Russian and Turkish players are coordinating their moves against an enemy piece in Budapest:

*Russia*: Army in Ukraine to Galicia
*Turkey*: Army in Rumania Support Army Ukraine to Galicia

The effect of supporting a piece will be nullified if another piece moves into the province occupied by the piece giving the support. In effect, the piece performing the support must 'defend itself' against its attacker and will be unable to aid another piece.

**Popularity.** Diplomacy is played by tens of thousands of people worldwide in both face-to-face, as well as email and web-based versions. Email and web-based versions typically take one week between turns.

## 2   Analysis of the Search Space

Search-based approaches to game playing utilize a search tree where each node represents the state of the game and each edge represents an operator available from that node [1]. An evaluation function is applied to the leaves on the tree to determine the value of choosing the operators. The state of the game can be represented by the locations of each player's pieces on the board and the supply centers controlled by each country.

The Diplomacy game board can be viewed as two undirected graphs, one that indicates army adjacencies and one that indicates fleet adjacencies. The average degree of each node in the fleet adjacency graph is 4.15 and for the army adjacency graph is 3.22.

The board consists of 73 provinces, so assuming an even distribution of pieces on the board, each piece will have the average moves per turn shown in Table 1.

Since the game is a seven-player simultaneous move game, an operator is the combination of all moves by all players during a turn. On average, each piece has approximately six moves each turn. During most of the game, all thirty-five supply centers are occupied and there are thirty-five pieces on the board. Since each piece can have six different orders, there are approximately $6^{35}$ move combinations, which is approximately $2^{91}$. In the beginning of the game, there are fewer pieces on the board, since most countries start with three pieces (Russia starts with four). It has been shown that there are over four trillion initial move combinations [4]. This huge search space makes a brute-force search intractable without using heuristics. Clearly, this search space must be reduced in order to perform a search in a reasonable amount of time.

Simultaneous move games, such as the Prisoner's Dilemma or Chicken can be modeled with a payoff matrix. Small matrices can be solved by finding dominant strategies in order to generate Nash Equilibra. However, a payoff matrix in Diplomacy would be a seven-dimensional matrix with $6^{35}$ entries, which would require an extremely long time to search fully. In addition, the payoff matrix would only represent the next move in the game and not the entire game.

A simple way to reduce the search space is to only consider other pieces that could affect the results of an order. Since pieces can only move one province at a time, each piece needs only to evaluate the moves of the other pieces within a distance of two. However, the nature of the heuristic evaluation function will have a direct effect on the size of the search space. Heuristics that include all pieces on the board will require a search of all combinations. Heuristic evaluation functions that examine only one player's pieces will require a smaller subset of the board.

### 2.1   Use of Knowledge in a Heuristic Evaluation

In order to determine a good heuristic evaluation function, knowledge of the game is needed [5]. A simple evaluation function would be to generate the number of supply centers occupied by the player. Loeb [6] utilizes a heuristic evaluation function based on rating the provinces according to their 'height'. The 'height' of a province is based on a mountain metaphor where the peaks of the mountain reside in the player's home country and the mountain descends into the provinces of the other players. The height of a province is reduced when other player's pieces are adjacent to the province and increased when the player's own pieces are adjacent. The orders are then made based on the provinces with the greatest height.

This heuristic clearly utilizes knowledge about the game. By rating the home provinces higher than neighboring provinces, the heuristic values protection of the homeland. Also, by reducing the value of the provinces due to other player's pieces, the heuristic encodes the probability of success of a move due to the presence of other pieces that could stop the move. However, the effectiveness of Loeb's heuristic is not clear.

Experts often spend a great deal of time working in conjunction with the software designer in order to obtain the proper heuristic evaluation. A self-learning approach removes the need for an expert in order to develop a game-playing strategy.

## 3   Pattern-Weights

Levinson describes patterns in [7] as representations of previous experience during game play. These patterns represent partial positions on the game board.

**Table 2.** Features used.

| Feature | Values |
|---|---|
| Province Occupation | *supply center, non-supply center* |
| Supply Center Control | *you, enemy, no one* |
| Piece Type | *army, fleet* |
| Order Type | *hold, move, support, convoy* |
| Adjacent Province Occupation | *none, you, enemy* |

The pattern-weights were used for a chess program called MORPH as a way to remember game experiences. Instead of saving the entire chess board, a reduced board was saved to a database. This reduced board showed only the attack patterns in chess, for example. In this representation, a white queen that is capable of attacking a black knight is represented as a directed graph from a node representing the white queen to a node representing the black knight. Thus, this pattern can represent any number of game states that reflect that attack situation.

Each pattern is associated with a weight that is used as a heuristic evaluation of the actual state. The advantage of using pattern-weights is that through self-play and learning mechanisms, the weights are tuned so as to develop a game strategy without any initial value of the various board positions. The learning mechanism will evaluate all the knowledge in the game and choose only the most important ones. Similarly, strategic aspects of the game can be individually analyzed by examining the pattern-weight database.

Pattern-weights can only be ordered in a generalization-specialization hierarchy where the least specific patterns lead to more specific ones [8]. Thus general patterns are applicable to many states but are less accurate than more specific ones that apply to fewer states. Pattern-weights that consist of $n$ features will be called $n$-pattern-weights.

The pattern-weights used in Levinson's chess program were of two types: material patterns and attack patterns. Material patterns represented the relative assessment of black and white pieces remaining in the game. Thus, "down one bishop" received a pattern assessment. Attack patterns are the subsets of the game graph as described above.

Diplomacy utilizes a board that can be viewed as an undirected graph. Pieces can only move one space at a time, so while a pattern-weight for chess represents attack formations, the pattern-weight for Diplomacy uses the adjacent nodes in the graph and examines both the adjacent pieces and the aspects of the provinces themselves.

Distinct features of the state were chosen that describe the state of the board and the influences on the specific orders given to each piece. Each of the features that are represented in the pattern-weight database are shown in Table 2.

These features can be combined and are independent of each other. Thus, by examining the pattern-weight database we can see the relative values of each of these features. For example, the Province Occupation, Order Type and Piece

**Fig. 1.** Feature example.

Type pattern can be combined to determine the value placed on moving into a supply center that is occupied by an enemy army. Additional features of the game can be added. Each feature will produce a pattern that will be stored as a pattern-weight whose value can be learned.

For example, in Fig. 1, the Russian player in Silesia, represented by the white tank, could move to Berlin, a German supply center. The order would be written as follows:

```
Army in Silesia to Berlin
```

Since the German player (represented in black) is considered to be an enemy by the Russian player, the following features are utilized from the Russian pattern-weight database: *supply center, owned by an enemy, unoccupied, army, move, adjacent to two enemies*.

Similarly, the German player could move from Munich to Silesia, yielding the following patterns from its pattern-weight database: *not a supply center, occupied by enemy, army, move, adjacent to no one*.

## 3.1   Feature-Based vs. Featureless Patterns

While a pattern is a proper subset of a state, a feature can be viewed as a directed abstraction of the state. A state indicates a board position, a turn, ownership of supply centers and so on, while the features were chosen as representations of perceived useful aspects of the state. The inclusion of the features listed above to define a pattern-weight has both benefits and drawbacks. Features will help guide the automated player to make decisions about game play. No value was initially attached to any feature, so it is up to the learning process to decide which ones to utilize and which ones to avoid. We attempted to choose the features of the game neutrally, to avoid biasing the computer player towards preselecting one feature over another.

The initial attempts to create patterns utilized a featureless approach, where the pattern-weights did not include any features, but rather stored specific moves for each piece. Since the moves were stored, information about the state of the game was included, such as the locations of the units on the board.

A computer player utilizing a feature-based approach learns a strategy more quickly than one that uses a featureless approach. However, by restricting the search domain to that of specified features, we leave open the possibility that the features chosen do not describe the semantics of the game with the proper amount of detail. As a result, an automated player might miss some important aspect of strategy. For example, by indicating the presence of an enemy by a single value, the computer player is unable to distinguish between enemies of two different countries. Another example is a province that is surrounded by three enemies (as noted by a feature) but the actual positions of those enemies are not known as it is not recorded in the pattern.

Given long enough game play, a featureless approach has the potential of producing more surprising strategy, since the preconceived notions of its creator do not confine it.

## 3.2   Temporal Difference Learning

Temporal difference learning has been shown to be an effective technique for learning game play when feedback occurs only after an extended period of time [9]. All pattern-weights start with an initial weight of .5, indicating that the pattern is neither good nor bad, and range from zero to one. The pattern-weights also utilize an age field that indicates the frequency of occurrence of that pattern. Since patterns can occur frequently in one game, the age of the pattern for some patterns will be greater than the number of games played. The age is not used during the learning process, but rather when evaluating a pattern-weight during game play. Older pattern-weights are given slightly higher consideration than newer patterns.

Temporal difference learning is applied using a method similar to MORPH [7] as follows:

$$Weight_n := \frac{Weight_{n-1} * (n-1) + k * v}{n + k - 1} \tag{1}$$

$Weight_i$ is the weight after the $i$th update, $v$ is final value of the game and $k$ is the learning rate. $v$ is 1 when the winning conditions were met, and 0 otherwise. When $k = 0$, only past experiences are considered. When $k = 1$, a new experience is averaged with all previous ones. With a larger $k$, the system values the latest experience more than previous ones. An external counter varied $k$ from 1 to 5, thus simulated annealing was utilized by keeping a global temperature that varied the intensity of the learned experience. Levinson [7] showed that the effect of simulated annealing on the development of pattern-weights was to hone in on specific values. Without simulated annealing, the weights were shown to fluctuate between certain ranges, while the effect in MORPH with simulated annealing was a smooth transition to the final value of a material pattern-weight.

Credit assignment within the same game was handled by assigning credit equally among all patterns found by their frequency. The pattern-weights were updated at the end of each game, which typically lasted 20 moves. The pattern-weights that were updated corresponded to patterns that occurred in the game.

## 4 Tests

A separate pattern-weight database was generated for each player through self-play. Since the game graph is asymmetric, a separate pattern-weight database for each player was necessary. However, all the pattern-weight databases could be generated simultaneously.

The first test was done using featureless pattern-weights, where moves were explicitly stored as 1-pattern-weights and combinations of two moves were stored as 2-pattern-weights. The goal of the game was to possess the greatest amount of supply centers. The game ended when one of the players reached seven supply centers.

The second test utilized feature-based pattern-weights. No cooperation was assumed in the game; each automated player was playing against all the other automated players. The goal of the game was the same as above.

The game was run utilizing the FREEDIP software, a web-based version of Diplomacy written in Java [10]. The software was modified by adding the logic for an automated player, pattern databases and self-play.

### 4.1 Game Play

The following sequence of actions was taken for each automated player of the game. Each step is examined in more detail below:

1. Generate all individual piece legal orders.
2. Generate all order combinations.
3. Match the moves against a pattern-weight database.
4. Weigh the moves according to the pattern-weights.
5. Choose among the top-rated moves.
6. Submit the moves and advance the turn.

**1. Generate All Individual Piece Legal Orders.** All possible moves, holds, supports and convoys were entered for each individual piece. At this stage, the move combinations are not generated, but only the potential orders for each piece. No cooperation was performed; so orders that involved other player's pieces were not generated. For example, it is possible to support another player's piece movement into an adjacent province, but this option was not considered.

**2. Generate All Move Combinations.** The cross product of all the orders for all the pieces was applied to generate a move set. Since each piece has approximately six orders per turn, a move set consisted of approximate $6^N$ move combinations where $N$ is the number of pieces. This number becomes very large as N increases. Once N reached six, the number of move combinations was approximately 50,000, which took approximately 5 minutes to analyze per move. In order to reduce this number and maximize the number of games played, fewer orders for each piece were generated in (1) above, which yielded fewer move combinations. Randomness was used to decide which orders to generate.

**3. Match the Moves against a Pattern-Weight Database.** Each move was compared against the pattern-weight database in search of pattern-matches. Features of the individual orders were independently compared to the 1-pattern-weights. Next, pairs of features were combined together and compared to the 2-pattern-weights and so forth.

**4. Weigh the Moves According to the Pattern-Weights.** Each pattern-weight that matched a move is associated with a value. The value of each pattern-weight found $w_1$ was then compared to the current total value for that move $w_2$ by averaging them and weighing them by their size. Thus, a 1-pattern-weight contributed half of its value as did a 2-pattern-weight. This value was then averaged with the weights of all the individual moves. The final value was between 0 and 1. Any move that did not match a pattern in the pattern-weight database was assigned the value of .5.

The pattern-weights can be seen as advisors. Each advisor suggests a result for the pattern. The smaller pattern-weights, representing less specific information, were valued less than the more specific patterns. Also, the averaging of pattern-moves was done to avoid the dominance of any one order for a single piece to the pattern.

**5. Choose among the Top-Rated Moves.** The weighted moves were sorted by the highest value. The top five moves were selected and one of them was chosen randomly based on their weighted value. Thus, a move with twice as much weight as another was chosen twice as often.

## 4.2  Self-learning and Pattern-Weight Updates

Temporal difference learning was applied to the pattern-weight database by examining the relative success of the sequence of moves and then updating the weights of all of those moves. Success was determined by retaining all the home supply centers and by capturing the greatest, second greatest or third greatest number of supply centers.

If the player's move sequence was considered a success, then the sequences of moves were decomposed into 1-pattern-weights, 2-patterns-weights and n-pattern-weights containing all the features of the game state. Pattern-weights

**Table 3.** 2-pattern-weight featureless test (1042 games).

| Player | Number of 1-Pattern-Weights | Number of 2-Pattern-Weights |
|---|---|---|
| England | 434 | 6,681 |
| France | 502 | 8,744 |
| Germany | 588 | 9,761 |
| Italy | 530 | 8,073 |
| Austria-Hungary | 580 | 6,898 |
| Russia | 589 | 12,397 |
| Turkey | 359 | 5,031 |

higher than 2-pattern-weights and lower than n-pattern-weights were not considered. The weights of all the pattern-weights that matched the move sequences were reinforced according to temporal difference learning as indicated above.

### 4.3   Test 1 Results – Featureless Pattern-Weights

Game strategies were learned via self-play as follows:

Table 3 shows the number of 1- and 2-pattern-weights generated for each country in the featureless test of 1042 games of self-play. The featureless players were also tested against random players and beat them in nearly all of the games tested.

The number of 1-pattern-weights differed between the countries due to the asymmetry of the game board. Initially, Turkey has very few moves and is confined to a corner of the board, so it tended to produce fewer patterns than other countries. Russia, on the other hand, starts with four pieces instead of three, thus generated a greater number of patterns. The size of the 1-pattern-weights is expected to grow only as large as the number of possible moves for the pieces in the game. The size of the 2-pattern-weights are limited by the number of pairs of individual moves.

Initially, moves were chosen at random, so *support*, *hold* and *move* orders were frequently seen. As more games were played, fewer support moves were chosen, perhaps due to the fact that success involved attaining supply centers quickly. The players learned to move as quickly as possible to the other provinces in order to claim the supply centers.

Competition between the automated featureless pattern-weight player showed that it learned basic strategies. Moves tended to alternate with hold orders to prevent other units from entering key locations.

There were many shortcomings to this approach. Units tended to move in line behind each other to the same locations as their predecessors. This is logical since their movements are stored and valued and units are indistinguishable from each other. Also, the units had no spatial concept of the game board and essentially played a blind game against each other and ignored the positions of other units. This would be easily fixed by storing larger than $n = 2$ n-patterns.

The games played with the pattern-weight databases revealed interesting biases in the game. Different countries tended to choose different directions to move and thus different players to move against. For example, Italy tended to ignore its neighbor, Austria-Hungary, and tended to move its fleets to Greece and Turkey in order to capitalize on those open supply centers. Similarly, France tended to invade Germany, while England found its benefits by invading France. Russia and Germany were constantly at battle with each other, particularly over Silesia, which borders both German and Russian homes. The moves chosen by the automated player seemed to reveal the relative benefit of attacking another country due to the asymmetry of the players and the board.

## 4.4   Test 2 Results – Feature-Based Pattern-Weights

Feature-based pattern-weight tests developed a strategy much more quickly than the featureless tests. The strategy of moving quickly towards their opponents and then retaining their positions was adopted in the same way, but in a shorter amount of time. In addition, the feature-based test players did not tend to repeat the same actions, since their knowledge was based on relative positioning of the units on the board. Like the automated player in the first test, initial versions of this player suffered from a lack of a spatial concept of the game board. Since only units in immediately surrounding provinces were analyzed, the automated player was at a particular disadvantage when another unit was two spaces away,

In early tests of featureless pattern-weights, the feature that indicated the presence of units in adjacent provinces was left out. As a result, some moves were misguided. Since the pattern-weights represented only the locations of pieces, not the positions of the other pieces. In the absence of this information, a piece could not distinguish between a state where it was surrounded by enemy pieces and one where it was far away from any enemies. Like the earlier featureless tests, supports were used only rarely. Once the feature of adjacent units was added, the automated players began to implement more complicated strategies, as described below.

In general, the feature-based patterns were able to conceptually abstract large states into relatively small data sets. This had a distinct advantage over the featureless patterns that required a great amount of storage space.

## 4.5   1-Pattern-Weights

Table 4 shows the highest valued 1-pattern-weights in the pattern-weight database for the Turkish player after 252 games.

The highest valued moves are related to moves and supports. This is perhaps because supported moves will have a tendency to displace other units by breaking standoffs. In fact, the highest rated move for four of the seven players was *1.1 -* Supporting a piece into a supply center owned by an enemy. It appeared as one of the top ten pattern-weights for the other three players.

The 1-pattern-weight *1.2* in Table 4 can be seen as a wasted move. The piece performing the support is helping its own piece move into a province that cannot possibly be occupied by an enemy on the next turn. Perhaps this is a by-product of valuing all support moves very highly, and the player has not

**Table 4.** Highest weighted 1-pattern-weights.

| Num | 1-Pattern-Weights | Weight |
|---|---|---|
| 1.1 | Supporting a piece into a supply center owned by an enemy | .9872 |
| 1.2 | Supporting a piece into province that has no adjacent units | .9377 |
| 1.3 | Supporting a piece into province that is not a supply center | .9028 |
| 1.4 | Convoy from a supply center that is owned by an enemy | .8807 |
| 1.5 | Supporting a piece into a province that is only adjacent to your own units | .8665 |
| 1.6 | Supply center currently occupied is controlled by an enemy | .8618 |
| 1.7 | Province moving to is occupied by an enemy | .8371 |
| 1.8 | Province occupied is not adjacent to any units | .8347 |
| 1.9 | Province moving to is occupied by a another one of your own pieces | .8264 |
| 1.10 | Province moving to is unoccupied | .8206 |

experienced enough situations to understand that this is unnecessary, since the same move without support is guaranteed to succeed. In fact, this is a classical example of how features may produce a biased strategy. A featureless pattern would not 'over assume' the need for support. The philosophical goal of our method is to emphasize knowledge storing as opposed to the power of search. Of course a real system would benefit from doing both. However we want to see how far we can get without doing any search and relying solely on stored knowledge. To this end, we avoid the *explore* all together and concentrate on the *exploit*. So, like any young system, which has not yet had enough exposure to experience naïve mistakes, our agent would not yet know many useful facts, although more experience would eventually teach it.

The 1-pattern-weight *1.4* in Table 4 is clearly wrong. This shows a willingness to leave valuable spaces before taking control of them. Convoys are much rarer in the game, since they require a fleet in an ocean province and an army on a coastal province.

## 4.6    2-Pattern-Weights

In the 2-pattern-weights of Table 5, there is a dominance of *Piece has no neighbors* in conjunction with a support move. This demonstrates the strategy of using support move by pieces whose support cannot be disabled by the enemy. Since the supporting pieces have no adjacent neighbors, their support cannot be cut by another piece, thus potentially leading to a better success rate with the supported move.

The top weighted n-pattern-weights represented the values of pattern-weights that matched features of the entire order. Although the analysis of these features would be complicated, the results of the opening moves can be generated and compared to an opening book of moves that has been generated by human experts through tournament play [11].

Table 5. Highest weighted 2-pattern-weights.

| Num | 2-Pattern-Weights | Weight |
|---|---|---|
| 2.1 | Piece has no neighbors AND Piece it is supporting will move into province adjacent to an enemy | 1.000 |
| 2.2 | Piece has no neighbors AND Piece it is supporting will move into province adjacent to no one | .9546 |
| 2.3 | Piece has no neighbors AND Piece it is supporting is a fleet | .9500 |
| 2.4 | Piece has no neighbors AND Piece supported will move into a location occupied by your own piece | .9207 |
| 2.5 | Piece has no neighbors AND Piece supported will move into a location that is not a supply center | .9115 |
| 2.6 | Piece supported will move into a supply center controlled by an enemy AND Piece supported is a fleet | .9009 |
| 2.7 | Piece has no neighbors AND Piece it is supporting will move into supply center owned by an enemy | .8895 |
| 2.8 | Piece has no neighbors AND Piece it is supporting will move into supply center | .8864 |
| 2.9 | Piece is a fleet AND Piece supported is adjacent to no one | .8551 |
| 2.10 | Piece is an army AND Piece it is supporting will move into a supply center owned by an enemy | .8478 |

## 4.7   N-Pattern-Weights

Since all moves have approximately equal weight, they will be chosen approximately equally. Table 6 shows a comparison of the top ten pattern-weight moves to a human-developed opening book and to random play. The leftmost column indicates the Turkish pieces in their original positions. The right columns show the frequency of moving to a specific province or performing a different action, such as a *hold*.

The random player does not choose equally among the adjacent provinces since the orders are selected from all legal moves which include *holds* and *supports*.

The support strategy seems to dominate the opening moves for the Turkish player. However, if we ignore the supports, we can see that the proportions of moves do correspond reasonably well with the opening book. For example, the army in Constantinople moves to Bulgaria only 40% of the time if supports are included, but 80% of the time if we ignore the support moves. In addition, although the piece in Constantinople has three choices of moves (Smyrna, Ankara, Bulgaria) it never chooses to move deeper into its own province. Again, this is an independently discovered strategy.

Clearly, the automated player has not yet learned that supporting a unit when it cannot be attacked is useless. This simple description requires at least a 3-patten-weight to be cognizant of the three elements: the self, a friendly

**Table 6.** Comparison of human expert opening book to pattern-weight moves for Turkey.

| Piece | Human Expert Opening Book | Pattern-Weight Player | Random Player |
|---|---|---|---|
| Army Constantinople | Bulgaria 98.1% <br> Holds 1.9% | Bulgaria 40% <br> Holds 10% <br> Supports 50% | Bulgaria 24.3% <br> Holds 25.6% <br> Supports 27.0% <br> Smyrna 13.7% <br> Ankara 9.6% |
| Fleet Ankara | Constan 36.5% <br> Black Sea 57.7% <br> Holds 5.8% | Constan 10% <br> Black Sea 20% <br> Holds 20% <br> Supports 50% | Constan 9.6% <br> Black Sea 25.6% <br> Holds 27.0% <br> Supports 17.6% <br> Armenia 20.2% |
| Army Smyrna | Ankara 26.9% <br> Holds 9.6% <br> Armenia 32.7% <br> Constan 26.9% <br> Syria 1.9% <br> Supports 1.9% | Ankara 0% <br> Holds 20% <br> Armenia 20% <br> Constan 10% <br> Syria 10% <br> Supports 40% | Ankara 10.8% <br> Holds 21.6% <br> Armenia 16.2% <br> Constan 6.8% <br> Syria 21.6% <br> Supports 23.0% |

support and an enemy. For this reason, any 2-pattern-weight is unable to ever learn that one need not defend against an absent attacker. This can be shown by demonstrating four possible scenarios after a foreign supply center has been conquered:

*Case 1*- Enemy is present to dislodge us AND we have defenses.

*Case 2*- Enemy is present to dislodge us AND we do not have defenses.

*Case 3*- Enemy is not present to dislodge us AND we have defenses.

*Case 4*- Enemy is not present do dislodge us AND we do not have defenses.

*Case 1* would get a good positive score for the defense of our new supply center, while a bad score would be given to *Case 2* since we would be unable to defend against the attack. Thus, we have the ability to learn the need for defense in the presence of attack. However, the situation is different for learning whether a defense was necessary in the absence of an attack. Since an enemy is not present in either *Case 3* or *Case 4* regardless of our choice of defending the province or not, we would obtain a good resulting score for either pattern. Thus either action would be sanctioned as a good one. It would appear that our temporal differencing approach to learning would only learn what is a punishment. In the absence of punishment, we would not learn what is not useful, merely what is not harmful. However, this would lead us to a wrong conclusion, since the fact that an action yields a good score whether it is chosen or not would mean that this action is unnecessary. Furthermore, if a defending unit could be deployed more effectively elsewhere, then some other 3-pattern-weights (or even higher order ones) would warrant that a defending piece could be used with a higher affinity elsewhere. So in other words, the sum of all the patterns is greater the individual values. Knowledge is not left alone in individual patterns (of course kernels of knowledge are stored in individual patterns) but it is the global mutual maximization of all patterns in concert that would yield the highest score. So, with sufficient experience, the total knowledge of the database would cause the right event to occur, even if any one of the patterns alone might the point in the wrong direction.

# 5    Conclusions and Future Work

The discovery of strategic use of supports by the automated player seems to indicate the potential for using pattern-weights for the game of Diplomacy. In addition, the strategy was refined upon inclusion of 2-pattern-weights, specifically the use of supports only when the supporting piece is not accessible to enemy players. This seems to indicate that more specific patterns lead to better strategy. In addition, feature-based patterns seemed to acquire this higher level strategy, whereas featureless patterns did not.

Comparisons to a human-expert generated opening book of moves show that the pattern-weights recognize certain strategies in opening moves, such as moving to unoccupied supply centers. However, the strategy of using supports seemed to dominate the opening moves, indicating the shallow understanding of the use of *support* moves.

Since the inclusion of features added to the discovery of strategic knowledge, it would seem that adding additional features would allow the system to discover even more complex strategies. One such experiment is to include recognition of both allies in addition to enemies. Thus the automated player could react differently, depending upon whether or not they are cooperating with another country.

As previously mentioned, this method serves as a partial strategy since it only models the non-cooperative aspects of the game. For this reason, it cannot be accurately compared with other complete systems or compete in tournament play. Past attempts to create an automated computer player have separated the task into two aspects: the negotiating agent and the strategy finder. The negotiating agent communicates with other players and develops alliances. The negotiating agent determines the value of a cooperative strategy by consulting the strategy finder, which indicates the value of various moves in the game. Our system can be coupled with a negotiating agent, thus acting as a strategy finder, by adding an *ally* feature (in addition to the *enemy* feature) and then receiving a list of enemies and allies from the negotiating agent. These features in turn would be added to the pattern-weight database and utilized in the same way as the other features.

One important consideration in these experiments was the choice to model pattern-weights after the moves, rather than the state of the board. Levinson's original work on MORPH utilized the position of the chess pieces on the board after choosing one particular move. This is more difficult to implement in Diplomacy since the game is not turn-based, so the final positioning of the pieces on the board is not known due to the uncertainty in predicting the opposing player's moves. Modeling the pattern-weights after moves instead of board states also reduced the size of the pattern-weight database significantly. All the pattern-weight databases were less than 100k in size and no pruning or removal of pattern-weights was necessary.

Many temporal difference learning experiments with games, such as TD-GAMMON [9] have required a large number of games to be played before developing complex strategies. For example, Tesauro observed that TD-GAMMON learned basic strategies early on after only a few hundred games and complicated strategies over the course of 300,000 self-played games. Clearly, more games need to be played in order to evaluate the effectiveness of pattern-weights and features on game play.

### 5.1   Future Directions

The following list potential future directions that can be leverages from the work described in this paper:

*Future 1.* Provide a performance comparison of a random player against another learning agent, an agent using Loeb's heuristic and a human player.

*Future 2.* Perform temporal difference learning of games played by humans and store that knowledge in a pattern-weight database. Compare the pattern-weight database learned from human players against the pattern-weight database obtained from playing a random player.

*Future 3.* Analyze how two different 3-pattern-weights differ greatly in weight, even though the common 2-pattern-weights to both of them would not be able to capture that nuance. Generate an English description of a pattern which would explain why it is a good one or not a good one, based on this $n + 1$ child pattern-weight analysis.

*Future 4.* The current patterns do not take into account specific locations of other pieces. These should be incorporated as part of the pattern-weight (larger patterns-weights would be captured without size limit) and the resulting strategy would be compared to the results of this experiment.

## 6   Related Work

Kraus [4] creates an automated negotiating agent that collaborates with other players, makes alliances, assesses the values of the various alliances and determines which ones to keep and which ones to break. The computer is able to negotiate with the other players through the use of a negotiation language that contains semantics for starting an alliance, creating DMZs and proposing collaborative moves. Once the negotiations are finished, the agent passes this information to the strategy finder module that finds the moves that best satisfy the various alliances made during the last negotiation period. The strategy finder was written by E. Ephrati and published in Hebrew.

Loeb [6] created a negotiating agent and a strategy finder that searched for the best moves for a country during a turn. The strategy finder would execute seven searches in parallel, one for each player. Each search process would periodically poll the other six search processes to determine the current best move for the other players. All seven moves would then be combined and evaluated using a heuristic evaluation.

Lorber [12] has created a negotiating agent that uses an opening book, expert knowledge and statistical analysis in order to determine the best move.

Microprose has developed a computerized version of Diplomacy. It utilizes both a negotiating agent and an opening book. It is widely regarded as a poor strategist and easy to fool through negotiations.

## Acknowledgements

Special thanks to Guy Tsafnat who toiled many long hours in the co-development of the FREEDIP software which provided the foundation and inspiration for developing an automated computer player.

## References

1. Korf, R.: Problem solving and search (2002) CS261A Course Reader, UCLA.
2. Gould, J., Levinson, R.: Experience-based adaptive search. In Michalski, R., Tecuci, G., eds.: Machine Learning 4: A Multi-Strategy Approach, Morgan Kaufman Publishers (1994) 579–604
3. Hasbro: Rules of Diplomacy, 4th edition. http://www.hasbro.com/instruct/Diplomacy.PDF (2000)
4. Kraus, S., Lehmann, D.: Designing and building a negotiating automated agent. Computational Intelligence 11 (1995) 132–171
5. Kraus, S., Lehmann, D., Ephrati, E.: An automated Diplomacy player. In Levy, D., Beal, D., eds.: Heuristic Programming in Artificial Intelligence, Ellis Horwood Limited (1989) 136–153
6. Hall, M., Loeb, D.: Thoughts on programming a diplomat. In van den Herik, J., Allis, V., eds.: Heuristic Programming in Artificial Intelligence 3, Ellis Horwood Limited (1992) 123–145
7. Levinson, R.: Experience-based creativity. In Dartnall, T., ed.: Artificial Intelligence and Creativity: An Interdisciplinary Approach, Kluwer Academic Press (1994) 161–179
8. Levinson, R., Fuchs., G.: A pattern-weight formulation of search knowledge. Computational Intelligence 17 (2001) 783–811
9. Tesauro, G.: Temporal difference learning and TD-GAMMON. Communications of the ACM 38 (1995)
10. Shapiro, A., Tsafnat, G.: FREEDIP, open source software. http://freedip.sourceforge.net (2002)
11. Nelson, M.: Opening's custodian report for 1995. http://devel.diplom.org/DipPouch/Zine/F1996M/Nelson/Part3.html (1996)
12. Lorber, S.: Diplomacy AI software. http://www.stud.uni-bayreuth.de/~a0011/dip/ai/ (2002)

# Appendix: Diplomacy Game Board

The following is the board and initial positions of the pieces of the standard Diplomacy game. Lines delineate the provinces and the supply centers are indicated by the presence of a colored dot in the interior of the province.

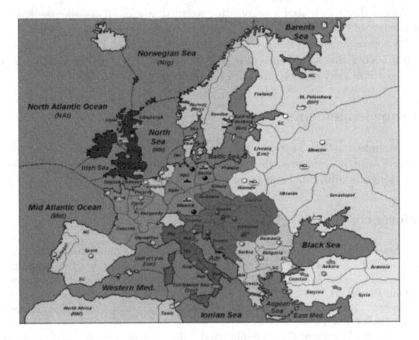

**Fig. 2.** Diplomacy game board.

# PDS-PN:
# A New Proof-Number Search Algorithm
## Application to Lines of Action

Mark H.M. Winands, Jos W.H.M. Uiterwijk, and Jaap van den Herik

Department of Computer Science, Universiteit Maastricht, The Netherlands
{m.winands,uiterwijk,herik}@cs.unimaas.nl

**Abstract.** The paper introduces a new proof-number (PN) search algorithm, called PDS-PN. It is a two-level search, which performs at the first level a depth-first Proof-number and Disproof-number Search (PDS), and at the second level a best-first PN search. First, we thoroughly investigate four established algorithms in the domain of Lines of Action endgame positions: PN, PN$^2$, PDS and $\alpha\beta$ search. It turns out that PN$^2$ and PDS are best in solving hard problems when measured by the number of solutions and the solution time. However, each of those two has a practical disadvantage: PN$^2$ is restricted by the working memory, and PDS is relatively slow in searching. Then we formulate our new algorithm by selectively using the power of each one: the two-level nature and the depth-first traversal, respectively. Experiments reveal that PDS-PN is competitive with PDS in terms of speed and with PN$^2$ since it is not restricted in working memory.

## 1 Introduction

Most modern game-playing computer programs successfully use $\alpha\beta$ search with enhancements for online game-playing [1]. However, the enriched $\alpha\beta$ search is sometimes not sufficient to play well in the endgame. In some games, such as chess, this problem is solved by the use of endgame databases [2]. Due to memory constraints this is only feasible for endgames with a relatively small state-space complexity although nowadays the size may be considerable. An alternative approach is the use of a specialized binary (win or non-win) search method, such as proof-number (PN) search [3]. In some domains PN search outperforms $\alpha\beta$ search in proving the game-theoretic value of endgame positions. PN search or a variant thereof has been applied successfully to the endgame of Awari [3], chess [4], and shogi [5]. In this paper we investigate several PN algorithms in the domain of Lines of Action (LOA). It turns out that the algorithms are restricted by working memory *or* by searching speed. To remove both restrictions we introduce a new PN algorithm, called PDS-PN.

The remainder of this paper is organized as follows. Section 2 explains the rules of LOA. Section 3 describes PN, PN$^2$, PDS; and examines their solution power and solution time, in relation to that of $\alpha\beta$. In Sect. 4 we explain the

J. Schaeffer et al. (Eds.): CG 2002, LNCS 2883, pp. 61–74, 2003.

working of PDS-PN. Subsequently, the results of the experiments with PDS-PN are given in Sect. 5. Finally, in Sect. 6 we present our conclusions and propose topics for further research.

## 2   Lines of Action

Lines of Action (LOA) [6] is a two-person zero-sum chess-like connection game with perfect information. It is played on an $8 \times 8$ board by two sides, Black and White. Each side has twelve pieces at its disposal. The black pieces are placed in two rows along the top and bottom of the board (see Fig. 1a), while the white pieces are placed in two files at the left and right edge of the board. The players alternately move a piece, starting with Black. A move takes place in a straight line, exactly as many squares as there are pieces of either color anywhere along the line of movement (see Fig. 1b). A player may jump over its own pieces. A player may not jump over the opponent's pieces, but can capture them by landing on them. The goal of a player is to be the first to create a configuration on the board in which all own pieces are connected in one unit (see Fig. 1c). In the case of simultaneous connection, the game is drawn. The connections within the unit may be either orthogonal or diagonal. If a player cannot move, this player has to pass. If a position with the same player to move occurs for the third time, the game is drawn.

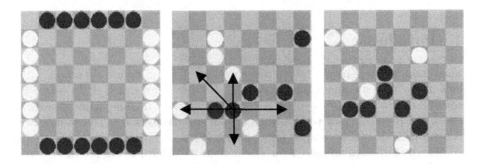

**Fig. 1.** (a) The initial position of LOA. (b) An example of possible moves in a LOA game. (c) A terminal LOA position

An interesting property of the game is that most terminal positions still have more than ten pieces remaining on the board [7], which makes the game not suitable for endgame databases. Although reasonable effort has been undertaken to construct adequate evaluation functions for LOA [8], experiments suggest that these are not very good predictions in the case of forced wins. Therefore, LOA seems an appropriate test domain for PN search algorithms.

# 3 Three Proof-Number Search Algorithms

In this section we give a short description of PN search, $PN^2$ search and PDS. We end with a comparison between PN, $PN^2$, PDS and $\alpha\beta$.

## 3.1 Proof-Number Search

Proof-number (PN) search is a best-first search algorithm especially suited for finding the game-theoretical value in game trees [9]. Its aim is to prove the true value of the root of a tree. A tree can have three values: *true, false* or *unknown*. In the case of a forced win, the tree is *proved* and its value is true. In the case of a forced loss or draw, the tree is *disproved* and its value is false. Otherwise the value of the tree is unknown. In contrast to other best-first algorithms PN search does not need a domain-dependent heuristic evaluation function to determine the most-promising node to be expanded next [3]. In PN search this node is usually called the most-proving node. PN search selects the most-proving node using two criteria: (1) the shape of the search tree (the number of children of every internal node) and (2) the values of the leaves. These two criteria enable PN search to treat game trees with a non-uniform branching factor efficiently.

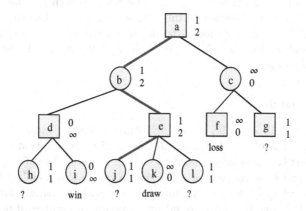

**Fig. 2.** An AND/OR tree with proof and disproof numbers

Below we explain PN search on the basis of the AND/OR tree depicted in Fig. 2, in which a square denotes an OR node, and a circle denotes an AND node. The numbers to the right of a node denote the proof number (upper) and disproof number (lower). A *proof number* represents the minimum number of leaf nodes which have to be proved in order to prove the node. Analogously, a *disproof number* represents the minimum number of leaves which have to be disproved in order to disprove the node. Because the goal of the tree is to prove a forced win, winning nodes are regarded as proved. Therefore, they have proof number 0 and disproof number $\infty$ (e.g., node $i$). Lost or drawn nodes are regarded as disproved (e.g., nodes $f$ and $k$). They have proof number $\infty$ and disproof number 0. Unknown leaf nodes have a proof and disproof number of unity (e.g., nodes $g$, $h$, $j$ and $l$). The proof number of an internal AND node is equal to the sum of its childrens' proof numbers, since to prove an AND node all the children have to be proved. The disproof number of an AND node is equal to the minimum of its childrens' disproof numbers. The disproof number of an internal OR node is equal to the sum of its childrens' disproof numbers, since to disprove an OR node all the children have to be disproved. Its proof number is equal to the minimum of its childrens' proof numbers. The procedure of selecting the most-proving node to expand is the following. We start at the root. Then, at each OR node the child with the lowest proof number is selected as successor, and at each AND node the child with the lowest disproof number is selected as successor. Finally, when a leaf node is reached, it is expanded and its children are evaluated. This is called *immediate evaluation*. The selection of the most-proving node ($j$) in Fig. 2 is given by the bold path.

In the naive implementation, proof and disproof numbers are each initialized to unity in the unknown leaves. In other implementations, the proof number and disproof number are set to 1 and $n$ for an OR node (and the reverse for an AND node), where $n$ is the number of legal moves. In LOA this initialization leads to a speed-up by a factor of 6 in time [10].

A disadvantage of PN search is that the whole search tree has to be stored in memory. When the memory is full, the search process has to be terminated prematurely. A partial solution is to delete proved or disproved sub-trees [9]. In the next subsections we discuss two variants of PN search that handle the memory problem more adequately.

## 3.2   PN² Search

PN² is first described in [9], as an algorithm to reduce memory requirements in PN search. It is elaborated upon in [11]. Its implementation and testing for chess positions is extensively described in [12]. PN² consists of two levels of PN search. The first level consists of a PN search ($pn_1$), which calls a PN search at the second level ($pn_2$) for an evaluation of the most-proving node of the $pn_1$-search tree. This $pn_2$ search is bound by a maximum number of nodes that can be stored in memory. The number is a fraction of the size of the $pn_1$-search tree. The fraction $f(x)$ is given by the logistic growth function [13], $x$ being the size of the first-level search:

$$f(x) = \frac{1}{1 + e^{\frac{a-x}{b}}} \tag{1}$$

with parameters $a$ and $b$, both strictly positive. The number of nodes $y$ in a $pn_2$-search tree is restricted to the minimum of this fraction function or the number of nodes which can still be stored. The formula to compute $y$ is:

$$y = min(x \times f(x), N - x) \tag{2}$$

with $N$ the maximum number of nodes to be stored in memory.

The $pn_2$ search is stopped when the number of nodes stored in memory exceeds $y$ or the sub-tree is (dis)proved. After completion of the $pn_2$ search, the children of the root of the $pn_2$-search tree are preserved, but sub-trees are removed from memory. The children of the most-proving node (the root of the $pn_2$-search tree) are not immediately evaluated by a second-level search, only when they are selected as most-proving node. This is called *delayed evaluation*. Note that immediate evaluation is used for $pn_2$-search trees.

As we have seen in Sect. 3.1, proved or disproved sub-trees can be deleted. If we do not delete proved or disproved sub-trees in the $pn_2$ search the number of nodes searched is the same as $y$, otherwise we can continue the search longer. Preliminary results have shown that deleting proved or disproved sub-trees in the $pn_2$ search causes a significant reduction in the number of nodes investigated [10].

## 3.3   Proof-Number and Disproof-Number Search

In 1995, Seo formulated a depth-first iterative-deepening version of PN search, later called PN* [5]. Nagai [14, 15] proposed a depth-first search algorithm, called Proof-number and Disproof-number Search (PDS), which is a straight extension of PN*. Instead of using only proof numbers such as in PN*, PDS uses disproof numbers too. PDS uses a method called *multiple-iterative deepening*. Instead of iterating only at the root node such as in the ordinary iterative deepening, it iterates at *all* nodes. To keep iterative deepening effective, the method is enhanced by storing the expanded nodes in a TwoBig transposition table [16]. PDS uses two thresholds in a search, one for the proof numbers and one for the disproof numbers. Once the thresholds are assigned to a node, the sub-tree rooted at that node is continued to be searched as long as either the proof or disproof number is below the assigned thresholds. Each OR (AND) node assigns the thresholds to its children with minimum proof (disproof) number. If the threshold of the (dis)proof number is incremented in the next iteration, the search continues mainly using the (dis)proof number to find a (dis)proof. If the proof number is smaller than the disproof number, it means that it seems to have a proof solution and the threshold of the proof number is incremented. Otherwise, it seems to be a disproof solution and the threshold of the disproof number is incremented. When PDS does not (dis)prove the root given the thresholds, it increases one of the threshold values and continues searching. Finally, note that we check whether nodes are terminal when they are expanded. This is called delayed evaluation. The expanded nodes are stored in a transposition table. The proof and disproof number of a node are set to unity when not found in the transposition table.

PDS is a depth-first search algorithm but behaves like a best-first search algorithm. It is asymptotically equivalent to PN search regarding the selection of the most-proving node. In passing, note that PDS using transposition tables suffers from the graph-history-interaction (GHI) problem (cf. [17]). The GHI evaluation problem can occur in LOA too. For instance, draws can be agreed upon due to the three-fold-repetition rule. Thus, dependent on its history a node

**Table 1.** Comparing the search algorithms on 488 test positions

| Algorithm | # of positions solved (out of 488) | 314 positions | |
|---|---|---|---|
| | | Total nodes | Total time (ms.) |
| $\alpha\beta$ | 383 | 1,711,578,143 | 22,172,320 |
| PN | 356 | 89,863,783 | 830,367 |
| PDS | 473 | 118,316,534 | 6,937,581 |
| $PN^2$ | 470 | 139,254,823 | 1,117,707 |

can be a draw or can have a different value. In the current PDS algorithm this problem is ignored.

## 3.4 Comparison

In this subsection we compare PN, $PN^2$, PDS and $\alpha\beta$ search with each other. All experiments have been performed in the framework of the tournament program MIA (Maastricht In Action)[1]. The program has been written in Java and can easily be ported to all platforms supporting Java. MIA performs an $\alpha\beta$ depth-first iterative-deepening search, and uses a TwoDeep transposition table [16], neural-network move ordering [18] and killer moves [19].

For the $\alpha\beta$ depth-first iterative-deepening searches nodes at depth $i$ are counted only during the first iteration that the level is reached. This is how the comparison is done in [9]. For PN, $PN^2$ and PDS search all nodes evaluated for the termination condition during the search are counted. For PDS this node count is equal to the number of expanded nodes (function calls of the recursive PDS algorithm). For PN and $PN^2$ this node count is equal to the number of nodes generated. The maximum number of nodes searched is 50,000,000. The limit corresponds roughly to tournament conditions. The maximum number of nodes stored in memory is 1,000,000. The parameters $(a,b)$ of the growth function used in $PN^2$ are set at (1800K, 240K) according to the suggestions in [12].

PN, $PN^2$, PDS and $\alpha\beta$ are tested on a set of 488 forced-win LOA positions[2]. In the second column of table 1 we see that 470 positions were solved by the $PN^2$ search, 473 positions by PDS, only 356 positions by PN, and 383 positions by $\alpha\beta$. In the third and fourth column the number of nodes and the time consumed are given for the subset of 314 positions, which all four algorithms could solve. If we have a look at the third column, we see that PN search builds the smallest search trees and $\alpha\beta$ by far the largest. PDS and $PN^2$ build larger trees than PN but can solve significantly more positions. This suggests that both algorithms are better suited for harder problems. $PN^2$ investigates 1.2 times more nodes than PDS, but $PN^2$ is six times faster than PDS for this subset.

For a better insight into the relation between $PN^2$ and PDS we did another comparison. In Table 2 we compare $PN^2$ and PDS on the subset of 463 positions,

---

[1] MIA can be played at the website: http://www.cs.unimaas.nl/m.winands/loa/
[2] The test set can be found at
http://www.cs.unimaas.nl/m.winands/loa/tscg2002a.zip

**Table 2.** Comparing PDS and PN$^2$ on 463 test positions

| Algorithm | Total nodes | Total time (ms.) |
|-----------|-------------|------------------|
| PDS | 562,436,874 | 34,379,131 |
| PN$^2$ | 1,462,026,073 | 11,387,661 |

which both algorithms could solve. Now, PN$^2$ searches 2.6 times more nodes than PDS. The reason for the decrease of performance is that for hard problems the $pn_2$-search tree becomes as large as the $pn_1$-search tree. Therefore, the $pn_2$-search tree is causing more overhead. However, if we have a look at the CPU time we see that PN$^2$ is still three times faster than PDS. The reason is that PDS has a relatively large time overhead because of the delayed evaluation (see Sect. 3.3). Consequently, the number of nodes generated is higher than the number of nodes expanded. In our experiments, we observed that PDS generated nodes 7 to 8 times slower than PN. Such a figure for the overhead is in agreement with experiments performed in Othello and Tsume-shogi [20]. The difference between our LOA results and Nagai's [15] Othello results are mainly caused by domain-dependent heuristics used for the initialization of the proof and disproof numbers.

From the experiments we draw three conclusions. First, PN-search algorithms clearly outperform $\alpha\beta$ in solving endgame positions in LOA. Second, the memory problems make the plain PN search a weaker solver for the harder problems. Third, PDS and PN$^2$ are able to solve significantly more problems than PN and $\alpha\beta$. Finally, we note that PN$^2$ is restricted by its working memory, and that PDS is considerably slower than PN$^2$.

## 4   PDS-PN

In the previous section we have seen that an advantage of PN$^2$ over PDS is that it is faster. The advantage of PDS over PN$^2$ is that its tree is constructed as a depth-first tree, which is not restricted by the available working memory. To combine the advantages of both algorithms we propose an algorithm, called PDS-PN, which does not suffer from memory problems and has potentially the speed of PN$^2$. PDS-PN is a two-level search as is PN$^2$. At the first level a PDS search is performed. When a node has to be expanded, which is not stored in the transposition table, a PN search is started instead of the recursive call of the PDS algorithm. The $pn_2$ search is stopped as soon as (1) the sub-tree is (dis)proved or (2) the number of the stored nodes exceeds the number obtained by formula 2, where $x$ equals the number of non-empty positions in the transposition table. After completion of the $pn_2$-search tree, only the root of the $pn_2$-search tree is stored in the transposition table. The PDS-PN algorithm has two advantages. First, the $pn_1$-search is a depth-first search, which implies that PDS-PN is not restricted by memory. Second, in PDS-PN the $pn_1$-search tree is growing slower in size than in PN$^2$. This implies that the focus is on fast PN. Hence, PDS-PN

**Table 3.** Number of solved positions for different $a$ and $b$

| $a$ | $b$ | # of solved positions | $a$ | $b$ | # of solved positions |
|---|---|---|---|---|---|
| 150,000 | 60,000 | 460 | 750,000 | 240,000 | 463 |
| 150,000 | 120,000 | 458 | 750,000 | 300,000 | 460 |
| 150,000 | 180,000 | 466 | 750,000 | 360,000 | 461 |
| 150,000 | 240,000 | 466 | 1,050,000 | 60,000 | 421 |
| 150,000 | 300,000 | 465 | 1,050,000 | 120,000 | 448 |
| 150,000 | 360,000 | 466 | 1,050,000 | 180,000 | 451 |
| 450,000 | 60,000 | 445 | 1,050,000 | 240,000 | 459 |
| 450,000 | 120,000 | 463 | 1,050,000 | 300,000 | 459 |
| 450,000 | 180,000 | 460 | 1,050,000 | 360,000 | 460 |
| 450,000 | 240,000 | 461 | 1,350,000 | 60,000 | 421 |
| 450,000 | 300,000 | 467 | 1,350,000 | 120,000 | 433 |
| 450,000 | 360,000 | 464 | 1,350,000 | 180,000 | 447 |
| 750,000 | 60,000 | 432 | 1,350,000 | 240,000 | 454 |
| 750,000 | 120,000 | 449 | 1,350,000 | 300,000 | 465 |
| 750,000 | 180,000 | 461 | 1,350,000 | 360,000 | 459 |

should in principle be faster than PDS. The pseudo-code of PDS-PN is given in the appendix.

## 5   Experiments

In this section we test PDS-PN with different parameters $a$ and $b$ for the growth function. Next, we evaluate the algorithms PDS-PN and PN$^2$ in solving problems under restricted memory conditions. Finally, we compare PDS-PN with optimized parameters against PN$^2$. Note that in PDS-PN at the first-level, the nodes are counted as in PDS and at the second-level as in PN.

### 5.1   Parameter Tuning

In the following series of experiments we measured the solving ability with different parameters $a$ and $b$. Parameter $a$ takes values of 150K, 450K, 750K, 1050K and 1350K, and for each value of $a$ parameter $b$ takes values of 60K, 120K, 180K, 240K, 300K and 360K. The results are given in Table 3. For each $a$ holds that the number of solved positions grows with increasing $b$, when the parameter $b$ is still small. If $b$ is sufficiently large, increasing it will not enlarge the number of solved positions. In the process of parameter tuning we found that PDS-PN solves the most positions with (450K, 300K). However, the difference with parameters configurations (150K, 180K), (150K, 240K), (150K, 300K), (150K, 360K), (450K, 360K) and (1350K, 300K) is not significant. On the basis of these results we deemed that it is not necessary to perform experiments with a larger $a$.

## 5.2    Memory Results

From the experiments in Sect. 3.4 it is clear that $PN^2$ will not be able to solve really hard problems since it will run out of working memory. To support this statement experimentally, we tested the solving ability of $PN^2$ and PDS with restricted working memory. In these experiments we started with a memory capacity sufficient to store 1,000,000 nodes, subsequently we divided the memory capacity by two at each next step. The parameters $a$ and $b$ were also divided by two. The relation between memory and number of solved positions for both algorithms is given in Fig. 3. We see that the solving performance rapidly decreases for $PN^2$. The performance of PDS-PN remains stable for a long time. Only when PDS-PN is restricted to fewer than 10,000 nodes, does it solve fewer positions. This experiment suggests that PDS-PN is preferable to $PN^2$ for the really hard problems, because it does not suffer from memory constraints.

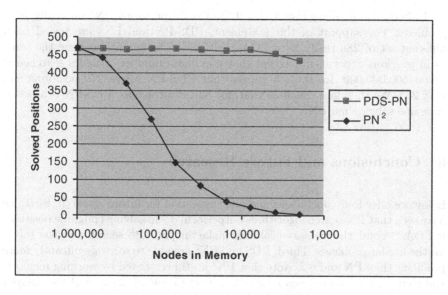

**Fig. 3.** Results with restricted memory

## 5.3    Comparison with $PN^2$

In this subsection we compare PDS-PN (450K, 300K) with $PN^2$. Table 4 shows that PDS-PN was able to solve 467 positions and $PN^2$ 470. The overlap of both sets yields a subset of 461 positions. In the third and fourth column we see that for this subset PDS-PN searches 1.4 times more nodes than $PN^2$. Simple calculation shows that PDS-PN is generating nodes with the same speed as $PN^2$. Because PDS is three times slower than $PN^2$, we may conclude that PDS-PN outperforms PDS in speed.

**Table 4.** Comparing PDS-PN and $PN^2$ on 488 test positions

| Algorithm | # of positions solved (out of 488) | 461 positions | |
|---|---|---|---|
| | | Total nodes | Total time (ms.) |
| PDS-PN | 467 | 1,879,690,850 | 15,887,380 |
| $PN^2$ | 470 | 1,302,157,677 | 11,339,920 |

**Table 5.** Comparing PDS-PN and $PN^2$ on 286 really hard test positions

| Algorithm | # of positions solved (out of 286) | 255 positions | |
|---|---|---|---|
| | | Total nodes | Total time (ms.) |
| PDS-PN | 276 | 16,685,733,992 | 84,303,478 |
| $PN^2$ | 265 | 10,061,461,685 | 57,343,198 |

In Sect. 5.2 it is suggested that PDS-PN outperforms $PN^2$ on really hard problems. For support of this statement, PDS-PN and $PN^2$ are tested on a different set of 286 really hard LOA positions[3]. The conditions were the same as in previous experiments except that maximum number of nodes searched is set at 500,000,000. In Table 5 we see that PDS-PN solves 276 positions and $PN^2$ 265. We therefore conclude that, for harder problems, PDS-PN is a better endgame solver than $PN^2$.

# 6   Conclusions and Future Research

Below we offer four conclusions and one suggestion for future research. First, we have seen that PN-search algorithms outperform $\alpha\beta$ in solving endgame positions in LOA. Second, the memory problems make the plain PN search a weaker solver for the harder problems. Third, PDS and $PN^2$ are able to solve significantly more problems than PN and $\alpha\beta$. Note that $PN^2$ is still restricted by working memory, and that PDS is three times slower than $PN^2$ (Table 2) because of the delayed evaluation. Fourth, the PDS-PN algorithm is almost as fast as $PN^2$ when the parameters for its growth function are chosen properly. PDS-PN performs quite well under harsh memory conditions. Hence, we conclude that PDS-PN is an appropriate endgame solver, especially for hard problems and for environments with very limited memory such as hand-held computer platforms.

We believe that an adequate challenge is testing the PDS-PN in other games, e.g., the game of Tsume-shogi since that game is notoriously known for its difficult endgames. Recently, some of the hard problems including solutions over a few hundred ply are solved by PN* [5] and PDS [21]. It would be interesting to test PDS-PN on these problems.

---

[3] The test set can be found at
http://www.cs.unimaas.nl/m.winands/loa/tscg2002b.zip.

# Acknowledgments

The authors would like to thank the members of the Maastricht Search & Games Group for their useful remarks.

# References

1. Campbell, M., Hoane, J., Hsu, F.: DEEP BLUE. Artificial Intelligence **134** (2002) 57–83
2. Nalimov, E., Haworth, G., Heinz, E.: Space-efficient indexing of chess endgame tables. International Computer Games Association Journal **23** (2000) 148–162
3. Allis, V., van der Meulen, M., van den Herik, J.: Proof-number search. Artificial Intelligence **66** (1994) 91–123
4. Breuker, D., Allis, V., van den Herik, J.: How to mate: Applying proof-number search. In van den Herik, J., Herschberg, I., Uiterwijk, J., eds.: Advances in Computer Chess 7. University of Limburg (1994) 251–272
5. Seo, M., Iida, H., Uiterwijk, J.: The PN*-search algorithm: Application to Tsume-shogi. Artificial Intelligence **129** (2001) 253–277
6. Sackson, S.: A Gamut of Games. Random House (1969)
7. Winands, M.: Analysis and implementation of Lines of Action. Master's thesis, Department of Computer Science, Universiteit Maastricht (2000)
8. Winands, M., Uiterwijk, J., van den Herik, J.: The quad heuristic in Lines of Action. International Computer Games Association Journal **24** (2001) 3–15
9. Allis, V.: Searching for Solutions in Games and Artificial Intelligence. PhD thesis, Department of Computer Science, University of Limburg (1994)
10. Winands, M., Uiterwijk, J.: PN, PN$^2$ and PN* in Lines of Action. Technical report, Department of Computer Science, Universiteit Maastricht (2001)
11. Breuker, D.: Memory versus Search in Games. PhD thesis, Department of Computer Science, Universiteit Maastricht (1998)
12. Breuker, D., Uiterwijk, J., van den Herik, J.: The PN$^2$-search algorithm. In van den Herik, J., Monien, B., eds.: Advances in Computer Games 9, Universiteit Maastricht (2001) 115–132
13. Berkey, D.: Calculus. Saunders College Publishing (1988)
14. Nagai, A.: A new AND/OR tree search algorithm using proof number and disproof number. In: Proceedings of Complex Games Lab Workshop, ETL, Tsukuba, Japan (1998) 40–45
15. Nagai, A.: A new depth-first-search algorithm for AND/OR trees. Master's thesis, The University of Tokyo (1999)
16. Breuker, D., Uiterwijk, J., van den Herik, J.: Replacement schemes and two-level tables. International Computer Chess Association Journal **19** (1996) 175–180
17. Breuker, D., van den Herik, J., Uiterwijk, J., Allis, V.: A solution to the GHI problem for best-first search. Theoretical Computer Science **252** (2001) 121–149
18. Kocsis, L., Uiterwijk, J., van den Herik, J.: Move ordering using neural networks. In Montosori, L., Váncza, J., Ali, M., eds.: Engineering of Intelligent Systems. Springer-Verlag (2001) 45–50 Lecture Notes in Artificial Intelligence, Vol. 2070.
19. Akl, S., Newborn, M.: The principal continuation and the killer heuristic. In: 1977 ACM Annual Conference Proceedings, ACM (1977) 466–473
20. Sakuta, M., Iida, H.: The performance of PN*, PDS and PN search on 6×6 Othello and Tsume-shogi. In van den Herik, J., Monien, B., eds.: Advances in Computer Games 9. Universiteit Maastricht (2001) 203–222

21. Nagai, A.: DF-PN Algorithm for Searching AND/OR Trees and Its Applications. PhD thesis, The University of Tokyo (2002)

# Appendix

The pseudo-code of PDS-PN is given below. For ease of comparison we use similar pseudo-code as given in [14] for the PDS algorithm. The proof number at an OR node and the disproof number at an AND node are equivalent. Analogously, the disproof number at an OR node and the proof number at an AND node are similar. As they are dual to each other, an algorithm similar to Negamax in the context of minimax searching can be constructed. This algorithm is called NegaPDS. In the following, proofSum(n) is a function that computes the sum of the proof numbers of all the children. The function disproofMin(n) computes the minimum of all the children. The procedures putInTT() and lookUpTT() store and retrieve information of the transposition table. isTerminal(n) checks whether a node is a win, a loss or a draw. The function generateChildren(n) generates the children of the node. By default, the proof number and disproof number of a node are set to unity. The procedure findChildrenInTT(n) checks whether the children are already stored in the transposition table. If a hit occurs for a child, its proof number and disproof number are set to the values found in the transposition table. The procedure PN() is just the plain PN search. The algorithm is described in [9] and [11]. The function computeMaxNodes() computes the number of nodes which may be stored for the PN search, according to Equation 2.

```
//iterative deepening at root r
procedure NegaPDS(r){

  r.proof = 1;
  r.disproof = 1;

  while(true){
    MID(r);
    // terminate when the root is proved or disproved
    if(r.proof = 0 || r.disproof = 0)
      break;

    if(r.proof <= r.disproof)
      r.proof++;
    else
      r.disproof++;
  }
}
```

```
//explore node n
procedure MID(n){

  //Look up in the transposition table
  lookUpTT(n,&proof,&disproof)
  if(proof = 0 || disproof = 0 ||
      (proof >= n.proof && disproof >= n.disproof)){
    n.proof = proof; n.disproof = disproof;
    return;
  }

  //Terminal node
  if(isTerminal(n)){
    if((n.value = true  && n.type = AND_NODE) ||
       (n.value = false && n.type =  OR_NODE)) {
      n.proof = INFINITY; n.disproof = 0;
    }
    else{
      n.proof = 0; n.disproof = INFINITY;
    }
    putInTT(n);
    return;
  }

  generateChildren();
  //avoid cycles
  putInTT(n);

  //Multiple iterative deepening
  while(true){
    //Check whether the children are already stored in the TT.
    //If a hit occurs for a child, its proof number
    //and disproof number are set to the values found in the TT.
    findChildrenInTT(n);

    //Terminate searching when both proof and disproof number
    //exceed their thresholds
    if(proofSum(n) = 0 || disproofMin(n) = 0 ||
        (n.proof <= disproofMin(n) && n.disproof <= proofSum(n))){
      n.proof = disproofMin(n);
      n.disproof = proofSum(n);
      putInTT(n);
      return;
    }
    proof = max(proof,disproofMin(n));
```

```
    n_child = selectChild(n,proof);

    if(n.disproof > proofSum(n) &&
        (proof_child <= disproof_child
        || n.proof <= disproofMin(n)))
      n_child.proof++;
    else
      n_child.disproof++;

    //This is the PDS-PN part
    /////////////////////////////////
    if(!lookUpTT(n_child)){
      PN(n_child,computeMaxNodes());
      putInTT(n_child);
    }
    else
    /////////////////////////////////
      MID(n_child);
  }
}

//Select among children
selectChild(n,proof){

  min_proof = INFINITY;
  min_disproof  = INFINITY;
  for(each child n_child){
    disproof_child = n_child.disproof;
    if(disproof_child != 0)
      disproof_child = max(disproof_child,proof);

    //Select the child with the lowest disproof_child (if there
    //are plural children among them select the child with the
    //lowest n_child.proof)
    if(disproof_child < min_disproof ||
        (disproof_child = min_disproof &&
          n_child.proof < min_proof)){
      n_best = n_child;
      min_proof = n_child.proof;
      min_disproof = disproof_child;
    }
  }
  return n_best;
}
```

# A Generalized Threats Search Algorithm

Tristan Cazenave

Labo IA, Université Paris 8, France
cazenave@ai.univ-paris8.fr

**Abstract.** A new algorithm based on threat analysis is proposed. It can model existing related algorithms such as Lambda Search and Abstract Proof Search. It solves 6x6 AtariGo much faster than previous algorithms. It can be used in other games. Theoretical and experimental comparisons with other related algorithms are given.

## 1 Introduction

A new search algorithm based on threats is presented. It is related to other threat based search algorithms. Threat based search algorithms work well in games such as the capture game of Go [1, 2, 3] or Go-Moku [4]. Generalized Threats Search (GTS) is based on the notion of generalized threats. GTS is a generalization of previously published algorithms. It is able to model the other existing threat based algorithms. It also solves problems and games faster than the other threat based algorithms when used with appropriate parameters. An analysis of the well formed generalized threats that give good results is given. Experimental comparisons of the different algorithms on the game of 6x6 AtariGo are detailed. The algorithm can also be used to solve problems related to many other games.

The second section explains the game of AtariGo which has been used for the experiments. The third section discusses some related threat search algorithms. The fourth section defines generalized threats. The fifth section describes the search algorithm based on generalised threats as well as how it can model Abstract Proof Search and Lambda Search. The sixth section details experimental results. The seventh section outlines future work and the eighth section concludes.

## 2 AtariGo

AtariGo is used to teach beginners to play the game of Go. The goal is to be the first player to capture a string. It can be played on any board size. It is usually played on a small board so that games do not take too much time. Teachers also often choose to start with a crosscut in the centre of the board to have an unstable position. We have tested different algorithms on the version with a crosscut in the centre of a 6x6 board.

The rules are similar to Go: Black begins, Black and White alternate playing stones on the intersections of the board, strings of stones are stones of the same

J. Schaeffer et al. (Eds.): CG 2002, LNCS 2883, pp. 75–87, 2003.

color that are linked by a line on the board. The number of empty intersections adjacent to the string is the number of liberties of the string. A string is captured if it has no liberty. For example in Fig. 6, the two black strings each have four liberties. A string that has only one liberty left is said to be in atari and can be captured by the other color in one move.

# 3    Related Work

Search algorithms that develop trees based on threats work well in games such as Go-Moku [4], or the capture search in the game of Go [1, 2, 3].

In this section we give an overview of search algorithms based on threats. The first subsection deals with Threat Space Search that has been used by V. Allis to solve Go-Moku. The second subsection briefly summarizes Abstract Proof Search (APS) which has been used to solve capturing problems in the game of Go. The third subsection is about Lambda Search (LS) which deals with similar problems. The fourth subsection explains Iterative Widening (IW), an improvement over APS. The fifth subsection describes Gradual Abstract Proof Search (GAPS) which introduces some graduality in IW and APS.

## 3.1    Threat Space Search

Go-Moku has been solved by V. Allis and coworkers using a selective proof search algorithm based on threats, and proof number search for the main search when no threats are available [4]. The threats are given names that correspond to patterns: Four, Straight Four, Three, Five. APS, GAPS and lambda search are a generalization of Threat Space Search. They are based on tree search to find threats of increasing orders instead of fixed patterns.

## 3.2    Abstract Proof Search

Abstract Proof Search [3] is a very selective search algorithm that ensures that winning moves are correct. It is much faster than brute force Alpha-Beta. It consists of developing small search trees at the Min nodes of the main search in order to select the interesting moves or to decide to stop search. Given that Left plays at Max nodes, and Right at Min nodes, an Abstract Proof Search of order one consists of verifying at each Min node if the Left player can win in one move. If it is not the case, the search is stopped and the Min node is labeled as lost for Left. Otherwise, if Left can win in one move, only the Right moves that can prevent the win in one move are considered and tried at this node. The search of order one consists of developing small 'trees' consisting of only one Left move at each Min node. A search of order $n$ consists of developing trees with $n$ Left moves (search trees of depth $2n - 1$ ply) at each Min node.

### 3.3   Lambda Search

Lambda Search [1] is a search algorithm that has strong links with Abstract Proof Search. It can be defined using lambda trees and lambda moves. A lambda tree of order $n$ is a search tree that contains lambda moves of order $n$. A lambda move of order $n$ for the attacker is a move that implies that there exist at least one subsequent winning lambda tree of order strictly inferior to $n$. A lambda move of order $n$ for the defender is a move that implies that there is no winning tree of order strictly inferior to $n$.

Abstract Proof Search imposes limits on the depth and the order of the trees developed at each node, whereas Lambda Search imposes limits on the order of these trees. Abstract Proof Search relies more on abstract properties of the game to select a few interesting moves and reduce the number of moves to look at for each order. Apart from these distinctions, they are based on similar ideas.

### 3.4   Iterative Widening

The Iterative Widening algorithm [2] consists of performing a full Abstract Proof Search at a given order, before increasing the order of the search. It has been successfully tested on the capture game in the game of Go. Practically, it means trying an order one Abstract Proof Search, and if it fails trying an order two search, and if it fails trying an order three search, and so on until the time allotted for the search is elapsed. It gives a speed-up of two for the capture game in Go.

### 3.5   Gradual Abstract Proof Search

Gradual Abstract Proof Search (GAPS) [5] is based on gradual games. A gradual game is defined as the shape of a search tree. Gradual games can be deeper for a lower order than the games used in APS. They can also be more shallow for a higher order than the lambda trees used in LS. GAPS iteratively widens the scope of the gradual games, instead of widening the games based on the depth of the games as in [2]. It is a generalization of Iterative Widening.

GAPS mixes the good properties of APS and LS. Instead of selecting moves mainly based on the depth as in APS, or based on the order as in LS, it combines these two criteria. Therefore it enables more control on the search behavior than either APS or LS. It can also easily use abstract properties of a game in order to be very selective on the moves to try as in APS.

## 4   Generalized Threats

GTS is based on the idea of generalized threats. In the first subsection, generalized threats are defined. The second subsection is about the comparison of generalized threats. The third subsection gives the composition operator that can be used to build all the relevant generalized threats. The last subsection gives information on the verification of generalized threats.

## 4.1   Definition of Generalized Threats

A move of order $n$ is a move that wins if it is followed by $n-1$ moves in a row by the same player. Each move in a game can be associated with an order. The order of a move $M$ is denoted $\omega(M)$. The last move of a game directly wins the game, it is always a move of order one.

A generalized tree represents search trees where the players have the possibility to play multiple moves in a row. The two players are named Left and Right. A branch that goes to the left represents some Left moves, and a branch that goes to the right represents some Right moves. A generalized tree is a binary tree. The usual Minimax search algorithm can be represented by a simple generalized tree such as the first tree of Fig. 1, which represents a depth 7 Minimax search tree. A Null-move imax search with a reduction factor of 2 stops searching when the result of a search tree of a depth equal to the depth of the current node minus 2, starting with another move of the same color, does not have an evaluation greater than beta at a max node, and less than alpha at a min node. The second generalized tree in Fig. 1 gives the tree developed by a depth 5 Null-Move Minimax with a reduction factor of 2. Null-move search speeds up Alpha Beta search but does not preserve the correctness of the results of a search. Generalized Threats Search is even faster, and moreover preserves correctness.

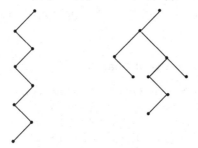

**Fig. 1.** Generalized trees for depth 7 Minimax and depth 5 Null-Move Minimax.

A node of order $n$ in a generalized tree is a node where the number of left branches in a row after the node is $n$. For example the root node of the $(1,0)$ generalized threat in Fig. 2 is of order one. The root node of the $(6,3,2,0)$ generalized threat is of order 3.

A generalized threat is a set of generalized trees that have some special properties. It is represented by a vector of integers. The first element of the vector is the number of order one nodes that are allowed in the verification of the threat. The second element is the number of allowed order two nodes, and the nth element gives the maximum number of order n nodes that can be used to verify the threat.

A generalized threat is defined as $g = (o_1, o_2, ..., o_n, 0)$ where $o_i$ is the maximum number of order i nodes that can be visited during the verification of the threat. It always ends with a zero. For example, in order to verify that a winning move is available, the threat (1,0) has to be verified.

In order for a generalized tree to be a generalized threat, it has to fulfill a special property: at each node of the tree that has a left and a right branch, the left subtree has to be included in the left subtree following the right branch.

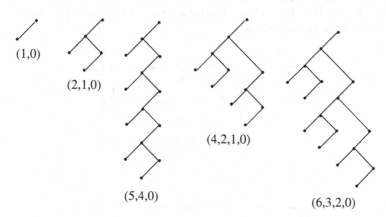

**Fig. 2.** Some trees representing generalized threats.

Figure 2 gives some examples of generalized trees representing different generalized threats. Left tries to win the game and Right tries to prevent Left from winning. Left branches are associated with winning moves for Left, and right branches are associated to the complete set of Right moves that can possibly prevent the win of the corresponding left branch (the left branch directly at the left of the right one with the same parent). All the leaves of the trees are positions won for Left. In order for the threat to be verified, all Left moves have to be winning moves, and all Right moves have to be refuted by Left.

In these trees, the number of leaves is the number of order one threats, as each leaf is a won position for Left. The number of order two nodes is the number of Left branches that are followed by an order one node for Left. More generally, the number of order $n$ nodes is the number of Left branches that are followed by an order $n - 1$ node for Left.

Generalized threats are a generalization of the gradual games used in GAPS [5]. In GAPS, the representation of gradual games is less general. A generalized threat can represent multiple GAPS trees. Moreover, programming generalized threats is easier than programming gradual games as they have a simple definition and nice properties as we will see in the following subsections.

Each order $n + 1$ node is followed by at least one order $n$ node. It is easy to see that $\forall i : o_i > o_{i+1}$. All the values after a zero in a vector representing a threat are also zero. Only the first zero of the vector is written.

## 4.2   Comparison of Generalized Threats

Let $g$ be a generalized game. $\omega(g)$ is the value of the first null element of the vector representing $g$. For example, $\omega(1,0) = 2$, $\omega(2,1,0) = 3$ and $\omega(6,3,2,0) = 4$. $\omega(g)$ is the maximum order of the threat plus one.

Let $g_k = (o_{k,1}, o_{k,2}, ..., o_{k,n}, 0)$. We have $g_a \leq g_b$ if $\omega(g_a) \leq \omega(g_b)$ and $\forall i < \omega(g_b) : o_{a,i} \leq o_{b,i}$.

Figure 3 gives different possible generalized threats. An arrow between two threats indicates that the pointed-to threat is less than the other threat. Some of the threats are incomparable.

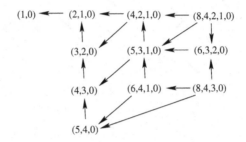

**Fig. 3.** Order between some generalized threats.

When a generalized threat is greater than another one, it means that all the generalized trees that can be built with the smaller one can also be built with the greater one. This partial order between threats is particularly useful for building and verifying threats, because the generalized threat following a Right move is always greater than the generalized threat the Right move tries to prevent.

## 4.3   Composition of Generalized Threats

The basic threat is (1,0). All other threats can be built from this (1,0) threat using a composition operator. Let T be the operator used to compose two generalized threats.

Let $g_l$ and $g_r$ be two games with $g_l \leq g_r$. We can define the T operator by: $g_t = g_l T g_r$, $\forall k \neq \omega(g_l) : o_{t,k} = o_{l,k} + o_{r,k}$ and for $k = \omega(g_l) : o_{t,k} = o_{r,k} + 1$.

For example, we have (1,0) T (1,0) = (2,1,0) as illustrated in Fig. 4. Another example is given in Fig. 5 with the operation (2,1,0) T (4,2,1,0) = (6,3,2,0).

The T operator ensures by construction that for all the nodes in a tree representing a generalized threat, if the node has a left and a right branch, then the left subtree is always smaller than the subtree at the left of its right branch. This property is very important for finding forced moves and for finding won states even if Right is to play. The right branch represents the set of Right moves that can possibly prevent a Left threat. Because we expect Left to have a harder job winning after a Right move than before, the right branch has to be followed by a generalized tree greater than the Left threat tree in order to ensure that all the Right moves fail.

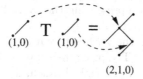

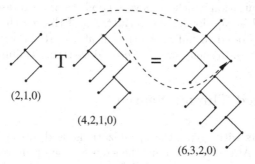

**Fig. 4.** Composition of the two most simple generalized threats gives (2,1,0).

**Fig. 5.** Composition of (2,1,0) and (4,2,1,0) gives (6,3,2,0).

## 4.4   Verification of Generalized Threats

Verifying a generalized threat consists of verifying that for each left branch, there is a winning Left move, and that for each right branch, there are no Right moves that prevent Left from winning. It is possible to verify threats without optimizations. However, we use several optimizations that are described in this subsection.

An optimization used to verify generalized threats is iterative widening on the maximum order of the threat at nodes that have only one left branch. At these nodes, the program starts with trying an order one move. If it does not work, it tries an order 2 move, decrements the number of order 2 nodes in the threat at hand, updates the threat at hand so that every value in the threat vector is less or equal to the following value, and tries to verify the updated threat. If the threat is not verified, it continues to increase the order of the threat until a threat is verified or the maximal order of the threat at hand is reached.

At nodes that contain both a left and a right branch, we can make another nice optimization. We know that the left threat is lower than the threat at the left of the right branch. Therefore, as we know the threat that has to be verified at the current node, we can define a new threat which is the current threat divided by two (dividing by two all the integers in the vector representing the current threat). This new threat is the maximum threat that has to be verified for the left subtree. For example, if the program has to verify a (4,2,1,0) threat, it will only try the (2,1,0) threat for the left subtree. Because the right subtree is greater than the left subtree, the left subtree is at most half of the overall tree (this is why all the integers representing the overall tree are divided by two in order to find the maximum left subtree).

At every node of the tree, the verified threat can be smaller than the maximal threat that was to be verified. The program always memorizes the verified threat. At nodes that contain a left and a right branch, it computes the maximal right threat that has to be tried by subtraction: the vector of the threat to verify, minus the vector of the verified left threat.

Another optimization which gives very good results is to use sets of abstract moves, as defined in [3], during the verification of the search. For example, an order 2 search is bound to fail in AtariGo if all the Right strings have strictly more than 2 liberties. In such cases, the procedure returns failure without search.

# 5    Generalized Threats Search

This section starts with giving the optimizations used in the Alpha-Beta algorithm for solving AtariGo, which are the same as the optimizations used in the main Alpha-Beta used to perform a GTS. Then the second subsection describes how to select the moves at the Min nodes of Alpha-Beta so as to perform a GTS. The third subsection is about the selection of moves at the Max nodes of Alpha-Beta in order to perform a GTS. The fourth subsection shows that it is possible to model the Abstract Proof Search and the Iterative Widening algorithms with the Generalized Threats Search algorithm. The fifth subsection shows how to model Lambda Search with Generalized Threats.

## 5.1    Alpha-Beta

An optimized Alpha-Beta search is used as the core algorithm of GTS. Generalized threats are used in different ways at Max and at Min nodes. At Max nodes, generalized threats are used to find Left moves that prevent Left from losing if Right plays first, and if no threat is verified, all relevant moves are tried. At Min nodes, generalized threats are used to find Right moves that prevent Left from winning if Left plays first, and if no threat is verified, the node is cut.

The optimizations used are:

- The use of transposition tables, containing the score and the best move.
- The memorization and use of two killer moves after the transposition move.
- The history heuristic with a weight of $2^{Depth}$.
- An incremental evaluation function which computes the difference between the number of liberties of the black string that has the least liberties and the number of liberties of the white string that has the least liberties.
- The number of liberties of strings are updated incrementally.

These optimizations are similar to the optimizations used in [6] to solve AtariGo with Alpha-Beta.

## 5.2   Forced Moves for Right

Right is the player that tries to prevent Left from winning. Right moves take place at the Min nodes of the Alpha-Beta search.

When the generalized threat is not verified for Left at a node of the Alpha-Beta search where Right is to move, a cut is performed, Right has prevented Left from winning with this threat.

If the generalized threat is verified, all the right moves that may prevent the threat are tried.

For example, the White move number 2 at E4 in Fig. 6 is found by a (4,3,0) generalized threat (the principal variation of the threat is B(E4), W(D5), B(E5), W(D6), B(E6), W(C6), B(B6) captures the white string). Once this threat is verified, all the relevant White moves are tried, and after each White move, the same threat is checked (in this case the (4,3,0) threat). The only White moves that are kept are the moves that prevent the threat to be verified.

## 5.3   Forced Moves for Left

Left moves take place at the Max nodes of the Alpha-Beta search. In some positions, Left has a limited number of moves if he does not want to lose the game. The generalized threat is tried for Right at each node of the Alpha-Beta search where Left is to play. If the generalized threat is verified for Right, the only moves to be tried for Left are the forced moves of the threat.

In positions where there are no forced moves for Left, all the possible moves for Left are tried. For example, the move number 5 at D2 in Fig. 6 is a forced move for Left (=Black). If Left does not play at move 5, Right can win with a (3,2,0) generalized threat for White (the principal variation for this threat is W(D2), B(F3), W(F4), B(F2), W(F1) capturing a black string).

## 5.4   Modeling Abstract Proof Search and Iterative Widening

It is possible to model Abstract Proof Search (APS) with GTS. We have ip1 = (1,0), ip2 = (2,1,0), ip3 = (4,2,1,0), ip4 = (8,4,2,1,0), and so on. An APS of order one is a GTS with the (1,0) generalized threat. An APS of order three as described in [3] is a GTS with the (4,2,1,0) generalized threat.

The Iterative Widening algorithm [2] performs a GTS (1,0), and if it fails performs a GTS (2,1,0), and if it fails a GTS (4,2,1,0) and so on.

## 5.5   Modeling Lambda Search

$\lambda$-search can be modeled with generalized threats. For example, developing a $\lambda$1-tree is equivalent to verifying a $(\infty,\infty,0)$ generalized threat. The different $\lambda$-trees can be modeled as follow: $\lambda$1-tree = $(\infty,\infty,0)$. $\lambda$2-tree = $(\infty,\infty,\infty,0)$. $\lambda$3-tree = $(\infty,\infty,\infty,\infty,0)$, $\lambda$4-tree = $(\infty,\infty,\infty,\infty,\infty,0)$.

**Table 1.** Solving 6x6 Atari-Go with Alpha-Beta.

| Depth | Value | Move | Time | Nodes |
|---|---|---|---|---|
| 1 | 1 | D5 | 0.00 | 33 |
| 2 | 0 | D5 | 0.00 | 143 |
| 3 | 1 | D5 | 0.01 | 1234 |
| 4 | 0 | C5 | 0.01 | 3177 |
| 5 | 1 | C5 | 0.09 | 25662 |
| 6 | 0 | C5 | 0.29 | 71265 |
| 7 | 1 | C5 | 2.04 | 563k |
| 8 | 0 | C5 | 2.49 | 604k |
| 9 | 1 | C5 | 27.91 | 7442k |
| 10 | 0 | C5 | 44.04 | 10375k |
| 11 | 1 | C5 | 168.06 | 43034k |
| 12 | 0 | C5 | 303.21 | 69300k |
| 13 | 1 | C5 | 2094.46 | 518016k |
| 14 | 500 | C5 | 150.39 | 34178k |
| Total | | | 2793.00 | |

**Table 2.** Solving 6x6 Atari-Go with GAPS.

| Depth | Value | Move | Time | Nodes |
|---|---|---|---|---|
| 1 | 1 | D5 | 0.00 | 33 |
| 2 | 0 | D5 | 1.84 | 47 |
| 3 | 1 | D5 | 2.33 | 235 |
| 4 | 0 | E3 | 4.72 | 325 |
| 5 | 1 | E3 | 5.76 | 594 |
| 6 | 0 | E3 | 11.52 | 861 |
| 7 | 1 | E3 | 11.98 | 2557 |
| 8 | 1 | E3 | 20.58 | 1982 |
| 9 | 2 | E3 | 2.78 | 1838 |
| 10 | 500 | E3 | 0.46 | 53 |
| Total | | | 61.97 | |

## 6   Experimental Results

The computer used for these experiments is a 600 MHz Pentium III with 256 MB of RAM running Linux. We have tested four different algorithms.

Alpha-Beta solves 6x6 Atari-Go at depth 14 in 2793s. Results are given in Table 1. All the optimizations described in the Alpha-Beta subsection are used.

Gradual Abstract Proof Search solves 6x6 AtariGo in 62s at depth 10 with the ip4221 gradual game, results are in Table 2. The Alpha-Beta search used in the GAPS is the same as in the experiments in Table 1. At each node of the Alpha-Beta search, all the gradual ip games are tested and the set of forced moves is the intersection of all the sets of moves sent back by the gradual games.

**Table 3.** Solving 6x6 Atari-Go with Lambda Search.

| Depth | Res | Order1 | Res | Order2 | Res | Order3 |
|---|---|---|---|---|---|---|
| 3 | 0 | 0.01 | 0 | 0.00 | 0 | 0.01 |
| 5 | 0 | 0.00 | 0 | 0.03 | 0 | 0.13 |
| 7 | 0 | 0.00 | 0 | 0.10 | 0 | 1.29 |
| 9 | 0 | 0.00 | 0 | 0.96 | 0 | 12.34 |
| 11 | 0 | 0.00 | 0 | 6.13 | 0 | 106.38 |
| 13 | 0 | 0.00 | 0 | 41.29 | 0 | 1521.08 |
| 15 | 0 | 0.00 | 0 | 108.10 | 500 | 1914.07 |
| Total | | 0.01 | | 156.61 | | 3555.30 |

**Table 4.** Solving 6x6 Atari-Go with Generalized Threats Search.

| Depth | R | (1,0) | R | (2,1,0) | R | (5,4,0) | R | (4,2,1,0) | R | (6,3,2,0) |
|---|---|---|---|---|---|---|---|---|---|---|
| 1 | 1 | 0.00 | 1 | 0.00 | 1 | 0.00 | 1 | 0.00 | 1 | 0.00 |
| 2 | 0 | 0.00 | 0 | 0.00 | 0 | 0.02 | 0 | 0.01 | 0 | 0.14 |
| 3 | 1 | 0.00 | 1 | 0.00 | 1 | 0.02 | 1 | 0.12 | 2 | 0.23 |
| 4 | -1 | 0.00 | -1 | 0.01 | 0 | 0.05 | 0 | 0.07 | 0 | 0.30 |
| 5 | 0 | 0.00 | 0 | 0.00 | 1 | 0.02 | 1 | 0.17 | 1 | 0.56 |
| 6 | -500 | 0.00 | -500 | 0.00 | -1 | 0.09 | -1 | 0.17 | 0 | 1.00 |
| 7 | | | | | 0 | 0.11 | 0 | 0.35 | 1 | 1.56 |
| 8 | | | | | -1 | 0.20 | -1 | 0.69 | 1 | 4.69 |
| 9 | | | | | 0 | 0.06 | 0 | 0.51 | 2 | 1.27 |
| 10 | | | | | -1 | 0.08 | -1 | 0.14 | 500 | 0.08 |
| 11 | | | | | 0 | 0.04 | 0 | 0.13 | | |
| 12 | | | | | -1 | 0.08 | -1 | 1.29 | | |
| 13 | | | | | 0 | 0.04 | -500 | 0.01 | | |
| 14 | | | | | -500 | 0.09 | | | | |
| Total | | 0.00 | | 0.01 | | 0.90 | | 3.66 | | 9.83 |

Lambda Search solves 6x6 AtariGo in 3555s at depth 15 and order 3, results are in Table 3. We did not use the optimization of Alpha-Beta in LS, we have simply reused the code given by Thomas Thomsen on his web page associated to his paper. In order to have a better basis for comparison between the relative merits of LS and GTS, we turned off the Alpha-Beta optimizations in GTS. GTS only takes 115s to solve 6x6 AtariGo when all the Alpha-Beta optimizations are turned off, using the (6,3,2,0) generalized threat. GTS also solves 6x6 AtariGo in 57s without Alpha-Beta optimizations, using the (12,6,4,0) generalized threat.

Another experiment was run with all Alpha-Beta optimizations turned off, and with all the abstract knowledge removed in order to have an algorithm that is even less optimized than Lambda Search (all the forced moves are computed in the threats in GTS even if they do not need to be, whereas Lambda Search stops after the first working forced move without looking for the others). To compare it to LS, we have summed the times used at even depth for this algorithm less optimized than LS, and it solves 6x6 AtariGo in 1731s with the (6,3,2,0)

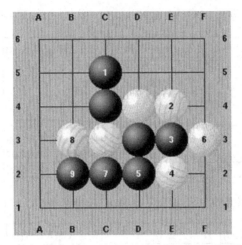

**Fig. 6.** The solution to 6x6 Atari-Go found by GTS(6,3,2,0).

generalized threat. Therefore even with less optimizations than LS, generalized threats still solve 6x6 AtariGo twice as fast.

Another thing that can be noted about LS, is that in Thomsen's code, the iterative deepening LS is called with orders ranging from 1 to (depth-1)/2. We did not use these settings because for AtariGo it would spend a very long time trying to solve order 4 and higher order Lambda Search unsuccessfully. We have voluntarily restricted LS to the order 3 which is the order needed to solve AtariGo.

From a more general point of view, we think it is better to be cautious about increasing the order in LS. A heuristic such as Iterative Widening [2] is more appropriate for LS: start with fully searching at order 1, and if it does not work, search at order 2 and so on.

Using the Alpha-Beta optimizations, Generalized Threats Search solves 6x6 AtariGo in 10s at depth 10 with the (6,3,2,0) generalized threat. The results are in Table 4. The columns with an R are the result of the GTS. The following columns give the time used to search each depth. The solution found by GTS(6,3,2,0) is given in Fig. 6.

## 7   Future Work

GTS works for AtariGo. It has good chances to work in other games. We plan to test it for capture, connection and life and death in the game of Go, Lines of Action, Phutball, Hex, shogi and chess.

Special attention has to be given to the order in which the generalized threats have to be tried. We only have a partial order between generalized threats, so there is room for choice in the order in which generalized threats can be tried. This might be game dependent. However some heuristics on the order of the threats are also probably game independent. It might be possible to find a good game independent order between generalized threats.

An optimization used in LS and not yet used in GTS is to incrementally find the forced moves. GTS searches for all the forced moves before trying them in Alpha-Beta. It would improve the response time to incrementally search for forced moves and to stop as soon as one of them prevents the win.

Using transposition tables and killer moves for the verification of the generalized threats could certainly speed up GTS. These optimizations are currently only used in the main Alpha-Beta search. There are opportunities to integrate Alpha-Beta and GTS more closely.

## 8    Conclusion

We have described GTS, a search algorithm based on the notion of generalized threats. We have given a constructive definition of generalized threats, and we have unveiled some properties of generalized threats that enable to optimize their verification. We have also defined a partial order between generalized threats. Generalized threats can easily be inserted in an existing Alpha-Beta search to speed it up, as described in the GTS section. We have also shown that GTS is a generalization of previous related algorithms such as Abstract Proof Search and Lambda Search. Experimental results for solving the game of 6x6 AtariGo show that it solves the game faster than other related search algorithms. Some further optimizations are still possible, and the algorithm can be used in other games.

## References

1. Thomsen, T.: Lambda-search in game trees - with application to Go. International Computer Games Association Journal **23** (2000) 203–217
2. Cazenave, T.: Iterative widening. In: Seventeenth International Joint Conference on Artificial Intelligence (IJCAI-01), Morgan Kaufmann Publishers (2001) 523–528
3. Cazenave, T.: Abstract proof search. In Marsland, T., Frank, I., eds.: Second International Conference on Computers and Games (CG 2000). Volume 2063 of Lecture Notes in Computer Science. Springer-Verlag (2001) 39–54
4. Allis, V., van den Herik, J., Huntjens, M.: Go-moku solved by new search techniques. Computational Intelligence **12** (1996) 7–23
5. Cazenave, T.: La recherche abstraite graduelle de preuves. In: Proceedings of RFIA 02. Volume 2. (2002) 615–623
6. van der Werf, E.: Message to the computer Go mailing list. (2002)

# Proof-Set Search

Martin Müller

Department of Computing Science, University of Alberta, Canada
mmueller@cs.ualberta.ca

**Abstract.** Victor Allis' proof-number search is a powerful best-first tree search method which can solve games by repeatedly expanding a most-proving node in the game tree. A well-known problem of proof-number search is that it does not account for the effect of transpositions. If the search builds a directed acyclic graph instead of a tree, the same node can be counted more than once, leading to incorrect proof and disproof numbers. While there are exact methods for computing proof numbers in DAGs, they are too slow to be practical.

*Proof-set search (PSS)* is a new search method which uses a similar value propagation scheme as proof-number search, but backs up proof and disproof *sets* instead of numbers. While the sets computed by proof-set search are not guaranteed to be of minimal size, they do provide provably tighter bounds than is possible with proof numbers.

The generalization *proof-set search with (P,D)-truncated node sets* or $PSS_{P,D}$ provides a well-controlled tradeoff between memory requirements and solution quality. Both proof-number search and proof-set search are shown to be special cases of $PSS_{P,D}$. Both PSS and $PSS_{P,D}$ can utilize heuristic initialization of leaf node costs, as has been proposed in the case of proof-number search by Allis.

## 1   Proof Sets and Proof Numbers

Victor Allis' *proof-number search (PNS)* [1] is a well-known game tree search algorithm, which has been successfully applied to games such as connect-four, qubic and gomoku. In contrast to many other methods, PNS does not compute a minimax value based on heuristic position evaluations; rather its aim is to find a proof or disproof of a partial boolean predicate $P$ defined on a subset of game positions. The usual predicate is $CanWin(p)$, but predicates representing other goals such as the tactical capture of some playing piece can also be used with proof-number search.

Proof-number search is a best-first method for expanding a game tree. It computes proof and disproof numbers in order to find a *most-proving node*, which will be expanded next in the tree search. Search continues until the root is either proven or disproven.

There is a simple bottom-up backup scheme for computing proof numbers, which is correct for trees. However, many game-playing programs use a transposition table to detect identical positions reached by different move sequences. Such a table changes the search graph from a tree to a directed acyclic graph

J. Schaeffer et al. (Eds.): CG 2002, LNCS 2883, pp. 88–107, 2003.

(DAG) or even a directed cyclic graph (DCG). If the same backup method for proof numbers is used on a DAG, it fails to compute the correct proof and disproof numbers, since the same node may be counted more that once along different paths. The new algorithm of *proof-set search (PSS)* is designed to reduce this problem and thereby improve the search performance on game graphs containing many transpositions.

The outline of the paper is as follows: the introduction continues with a short description of proof-number search on game trees and on directed acyclic graphs, and with an example that illustrates the problems of proof-number computation in DAGs. Section 2 describes the new method of proof-set search, and characterizes it by a theorem establishing its dominance over PNS on the same DAG. On the other hand, counterexamples show that even PSS cannot always select a smallest proof set. Section 3 describes the algorithmic aspects of PSS in those areas where it differs from PNS. Section 4 introduces the data structure of a *K-truncated node set*, defines the generalization of PSS to *PSS with (P,D)-truncated node sets* or $PSS_{P,D}$, characterizes both PNS and PSS as special cases of $PSS_{P,D}$, and proves a generalized dominance theorem of $PSS_{P,D}$ over PNS. Section 5 describes how to use a heuristic initialization of leaf node costs in PSS and $PSS_{P,D}$. Section 6 closes with a discussion of future work, including the extension of PSS to cyclic game graphs and potential applications of PSS.

## 1.1   Proof-Number Search in a Tree

This introductory section describes the basic procedure of proof-number search. For detailed explanations and algorithms, see [1]. Proof-number search (PNS) grows a game tree by incrementally expanding a *most-proving node* at the frontier. Nodes in a proof tree can have three possible states: *proven*, *disproven*, and *unproven*. Search continues as long as the status of the root is *unproven*. After each expansion, a leaf evaluation predicate $P$ is applied to each new node, to see whether it is defined in the corresponding game position. If yes, the new node can be evaluated as *proven (P = true)* or *disproven (P = false)*, while if $P$ does not apply, the node status becomes *unproven*. Proofs and disproofs are propagated to interior nodes by using proof numbers. Interior nodes are proven by finding a *proof tree* or disproven by finding a *disproof tree*. A proof tree $s$ for a node $r$ is a subtree of the game tree with following properties:

1. $r$ is the root of $s$.
2. In all leaf nodes of $s$, the predicate $P$ is well-defined and evaluates to true.
3. If $n$ is an AND node in $s$, then *all* of its successor nodes in the game tree are also contained in $s$.

Analogous properties hold for a disproof tree $s$ of $r$:

1. $r$ is the root of $s$.
2. In all leaf nodes of $s$, the predicate $P$ is well-defined and evaluates to false.
3. If $n$ is an OR node in $s$, then *all* of its successor nodes in the game tree are also contained in $s$.

PNS maintains proof and disproof numbers for each node in a game tree, and updates them after each expansion. These numbers can be interpreted as the size of a minimal proof or disproof set: a smallest set of currently unproven terminal nodes of the tree with the property that (dis)proving all nodes in that set would create a (dis)proof tree for the root node. An impossible (dis)proof is represented by an infinite (dis)proof number.

Proof and disproof numbers are used to select a *most-proving node* to expand next. They are computed by a simple backup scheme. Each search node $n$ stores a proof number $pn(n)$ and a disproof number $dn(n)$. For an *unproven* frontier (or leaf) node $n$, set $pn(n) = dn(n) = 1$. A *proven* frontier node is assigned $pn(n) = 0, dn(n) = \infty$, while a *disproven* node obtains $pn(n) = \infty, dn(n) = 0$. For non-frontier or interior nodes, let the children of a node $n$ be $n_1, \ldots, n_k$. The backup rules for proof and disproof numbers are as follows: In an AND node, the proof number is computed as the sum of the proof numbers of the children. In an OR node, the proof number becomes the minimum among the proof numbers of all children.

$$\text{AND node: } pn(n) = pn(n_1) + pn(n_2) + \ldots + pn(n_k)$$

$$\text{OR node: } pn(n) = \min(pn(n_1), pn(n_2), \ldots, pn(n_k))$$

Disproof numbers are computed by taking sums at OR nodes and minima at AND nodes.

$$\text{OR node: } dn(n) = dn(n_1) + dn(n_2) + \ldots + dn(n_k)$$

$$\text{AND node: } dn(n) = \min(dn(n_1), dn(n_2), \ldots, dn(n_k))$$

The most-proving node to expand next can be found by traversing the graph from the root to a frontier node, selecting a child with identical proof number at OR nodes and a child with identical disproof number at AND nodes. After expanding a node, proof and disproof numbers of that node must be recomputed, and changes propagate upwards in the tree.

**Heuristic Initialization of Proof Numbers.** Proof numbers can be viewed as a lower bound on the work required to prove a node. The standard algorithm assigns the same estimate of 1 to each unproven leaf node. However, game-specific knowledge can be used to provide different initial estimates. Allis [1] proposes to use a heuristic initialization $h(n)$ for the proof and disproof numbers of new frontier nodes $n$. Such a heuristic initialization can be viewed as a heuristic lower bound estimate of the work required to prove $n$. Using the constant function $h(n) = 1$ yields the standard algorithm.

## 1.2   Proof-Number Search in a Directed Acyclic Graph

The basic algorithm for proof-number search on a tree can also be used for directed acyclic game graphs (DAG) with some small modifications. However,

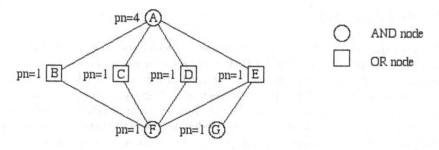

**Fig. 1.** Overestimating proof numbers in a DAG.

proof numbers can overestimate the size of proof sets. Therefore the algorithm cannot always find a minimal proof set, and it can fail to identify a most-proving node, even though such a node always exists even in a DAG [2, 1]. The following example shows how tree backup in a DAG overestimates proof numbers.

*Example 1.* Consider the DAG in Fig. 1, with an AND node $A$ at the top and 4 OR children $B, C, D, E$. Nodes $B, C$ and $D$ have the same single child $F$, and $E$'s two children are $F$ and $G$.

The proof numbers of the leaves $F$ and $G$ are initialized to 1. The proof numbers of $B \ldots E$ are also all equal to 1 since these are OR nodes, which obtain the minimum proof number of their children, and they all have a child with proof number 1 but no child with proof number 0. The proof number of $A$ is greatly overestimated by the tree-backup scheme as $pn(A) = pn(B) + pn(C) + pn(D) + pn(E) = 4$. The true value of $pn(A)$ is 1, since proving the single frontier node $F$ proves $A$, but $F$ is counted four times by the algorithm. The example can easily be extended, to make the difference between true and computed values arbitrarily large, by adding further nodes at the OR level which are only connected to $A$ above and to $F$ below. If such a DAG occurs as part of a larger problem, the overestimate of $A$'s proof number can be very costly. It can greatly delay the expansion of the sub-DAG below $A$, and lead the search into different parts of the DAG for a long time, even though $F$ is a very good candidate node. Expanding $F$ could lead to a quick proof of $F$ and thereby $A$. This may be by far the fastest - or even the only - way to solve the overall problem.

To overcome the problem of overestimation, Schijf [2, 3] has developed exact methods for computing proof numbers in DAGs and identifying a most-proving node. Unfortunately, these methods seem to have a huge computational overhead and have turned out to be impractical even in tests on small Tic-Tac-Toe game DAGs. The new method of *proof-set search* reported here lies in between proof-number search and Schijf's exact method, both in terms of complexity and solution quality. Furthermore, by using *truncated node sets*, the tradeoff between informedness and memory overhead of proof-set search can be controlled precisely.

# 2    Proof-Set Search

*Proof-set search*, or *PSS*, is a new search method which uses the simple children-to-parents propagation scheme for DAGs. Instead of using proof numbers, which are an upper bound on the size of the minimal proof sets in a DAG, PSS backs up *proof sets* directly. While the sets selected by PSS cannot be guaranteed to be minimal, they provide provably tighter bounds than is possible with proof numbers only. This can lead to a better node selection and thereby to a smaller (dis)proof DAG being generated. The prize for better approximation is that more memory and time per expansion step is needed to store and propagate sets of nodes instead of numbers.

## 2.1    Backup Algorithm for Proof Sets

The algorithms for proof-set search are similar to the ones for proof-number search [1]. Each search node $n$ stores both a proof set $pset(n)$ and a disproof set $dset(n)$. For unproven frontier nodes, set $pset(n) = dset(n) = \{n\}$. An impossible (dis)proof is indicated by a set of infinite size, represented by the symbol $\{\}_\infty$. A proved frontier node is assigned $pset(n) = \emptyset, dset(n) = \{\}_\infty$, while a disproved node obtains $pset(n) = \{\}_\infty, dset(n) = \emptyset$. For interior nodes $n$, with children $n_1, \ldots, n_k$, the backup rules for proof and disproof sets are as follows: In an AND node, the proof set is defined to be the *union* of the proof sets of all children. In an OR node, the proof set is the minimal set among the proof sets of all children, computed by a function set-min according to some total ordering of node sets.

$$\text{AND node: } pset(n) = pset(n_1) \cup pset(n_2) \cup \ldots \cup pset(n_k)$$

$$\text{OR node: } pset(n) = \text{set-min}(pset(n_1), pset(n_2), \ldots, pset(n_k))$$

Disproof sets are backed up analogously, taking minima in AND nodes and unions in OR nodes.

$$\text{OR node: } dset(n) = dset(n_1) \cup dset(n_2) \cup \ldots \cup dset(n_k)$$

$$\text{AND node: } dset(n) = \text{set-min}(dset(n_1), dset(n_2), \ldots, dset(n_k))$$

The following rules define a simple total order on node sets, which is close to the spirit of the original PNS:

1. A smaller set is always preferred to a larger one.
2. To break ties between sets of the same size, first define a total ordering of single nodes. For example, the ordering given by a depth first traversal of the DAG, the order of node expansion, or even the memory address of a node can be chosen.

As notation, let $n_1 < n_2$ if $n_1$ precedes $n_2$ in the chosen total order on single nodes. Sort each set $s = \{n_1, \ldots, n_k\}$ such that $n_1 < n_2 < \ldots < n_k$. Then a total ordering of sets of nodes can be defined as follows:

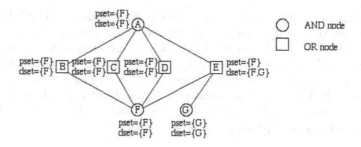

**Fig. 2.** Computing proof and disproof sets in a DAG.

1. $s_1 < s_2$ if $|s_1| < |s_2|$.
2. $s_1 < s_2$ if $|s_1| = |s_2|$ and $s_1$ precedes $s_2$ in lexicographical ordering. So for two sorted sets of equal size $s_1 = \{n_1, \ldots, n_k\}$ and $s_2 = \{m_1, \ldots, m_k\}$, $s_1 < s_2$ iff there is an $i, 1 \leq i \leq k$, such that $n_j = m_j$ for all $j, 1 \leq j < i$, and $n_i < m_i$.

The function set-min is obtained from this ordering by setting set-min$(a, b) = a \Leftrightarrow a \leq b$. Another possible ordering, using node evaluation as a heuristic measure of proof effort [1], is given in Sect. 5. Set operations are extended to $\{\}_\infty$ in the natural way, by $s \cup \{\}_\infty = \{\}_\infty$ and set-min$(s, \{\}_\infty) = s$ for all finite sets $s$, and by $\{\}_\infty \cup \{\}_\infty = \{\}_\infty$ and set-min$(\{\}_\infty, \{\}_\infty) = \{\}_\infty$.

*Example 2.* See Fig. 2. Assume that the nodes of Example 1 are ordered $A < B < C < D < E < F < G$, and that tie-breaks among sets of the same size are resolved by a lexicographical ordering of the sorted lists of elements, so that for example $\{F, E, D\} = \{D, E, F\} < \{D, E, G\}$. Then the computation of proof and disproof sets proceeds as follows:

1. $pset(F) = dset(F) = \{F\}$, $pset(G) = dset(G) = \{G\}$.
2. $pset(B) = $ set-min$(pset(F)) = \{F\}$. $dset(B) = \bigcup(dset(F)) = \{F\}$.
3. In the same way, $pset(C) = pset(D) = \{F\}$ and $dset(C) = dset(D) = \{F\}$.
4. $pset(E) = $ set-min$(\{F\}, \{G\}) = \{F\}$ and $dset(E) = \{F\} \cup \{G\} = \{F, G\}$.
5. $pset(A) = \bigcup(pset(B), pset(C), pset(D), pset(E)) = \{F\}$.
6. $dset(A) = $ set-min$(dset(B) \ldots dset(E)) = $ set-min$(\{F\}, \{F, G\}) = \{F\}$.

By using proof sets instead of proof numbers, the root's proof set has cardinality 1, whereas the proof number computed by PNS was 4. So with proof sets, the promising node $F$ is likely to be expanded much earlier.

## 2.2 Existence of a Most-Promising Node

In proof-number search on trees, a *most-proving node* always exists, and it can easily be identified. (Dis)proving a most-proving node reduces (dis)proof number of the root by at least one.

Schijf [2] proves the existence of a most-proving node in a DAG. However, it is not easy to identify such a node. Schijf develops theoretically correct but impractically slow algorithms to identify a most-proving node. On the other hand, since PSS uses a locally greedy heuristic to select a minimal set, it can not always identify a most-proving node. However, PSS computes a *most-promising node (mpn)* which lies in the intersection of the proof and disproof sets computed for the root node by an efficient PNS-like backup scheme. The following theorem states that such a most-promising node always exists.

**Theorem 1.** *Given a DAG $G$, compute proof and disproof sets using PSS. Then for each unproven node $n$, the intersection of proof and disproof set, $pset(n) \cap dset(n)$, is nonempty.*

The induction proof follows the lines of the analogous theorem for proof numbers in [2]:

1. The theorem holds for all unproven leaf nodes $n$ since $pset(n) = dset(n) = \{n\}$.
2. Let $n$ be an unproven AND node with unproven children $\{n_1, \ldots, n_k\}, k > 0$. Assume the induction hypothesis holds for all these children. Let $n_i$ be a node such that $dset(n) = dset(n_i)$. Since $pset(n_i) \cap dset(n_i)$ is nonempty by the induction assumption, let $x$ be a node in the intersection. Then $x \in dset(n)$ and since $pset(n) \supseteq pset(n_i)$, it follows that $x \in pset(n)$.
3. The proof for OR nodes is obtained by interchanging the roles of *pset* and *dset* in step 2.

## 2.3   Dominance of Proof-Set Search over Proof-Number Search

As a more informed search method, proof-set search dominates proof-number search in the following sense:

**Theorem 2.** *For each node $n$ in a DAG $G$, the size of the proof (disproof) set computed by PSS is not larger than the proof (disproof) number of $n$ computed by proof-number search on $G$.*

$$|pset(n)| \leq pn(n)$$

$$|dset(n)| \leq dn(n)$$

We omit a proof here, since Theorem 2 is a special case of the more general Theorem 6, which will be proven in Sect. 5.

Note that the result holds only for PSS and PNS operating on the same DAG. The theorem cannot compare nodes in the two different DAGs which are generated when PNS and PSS respectively are used as the algorithm to select most-promising nodes for expansion. Since these two DAGs are generated by different mechanisms, their shape and their node sets may be completely different, and are therefore not comparable.

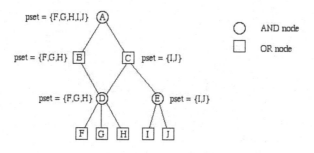

**Fig. 3.** PSS fails to find a smallest proof set.

## 2.4   Proof-Set Search Does Not Always Select a Smallest Proof Set

As mentioned in Sect. 2.2, PSS uses a locally greedy heuristic to select a minimum set among all unproven children. A local method cannot always choose a set that will perform best when taking unions with other sets further up in the DAG. In the worst case, PSS can do no better than PNS. In the following two examples, PSS fails to find a smallest proof set.

*Example 3.* In Fig. 3, it is easy to see that $\{F, G, H\}$ is the minimal proof set for the root $A$. This set is necessary to prove node $D$, which in turn is needed to prove $B$, which is required to prove $A$. On the other hand, proving $D$ also proves the only other child of $A$, node $C$, so a complete proof of $A$ is obtained by proving $\{F, G, H\}$.

At node $C$, the locally greedy selection of PSS turns out badly: given the proof sets of $C$'s children, $pset(D) = \{F, G, H\}$ and $pset(E) = \{I, J\}$, PSS selects the smaller set $\{I, J\}$ as the minimal proof set for $C$. Because of this choice, the proof set of $A$ becomes $pset(A) = pset(B) \cup pset(C) = \{F, G, H, I, J\}$, which is almost twice as large as the optimum.

*Example 4.* The ratio between the real optimum and the set computed by PSS can be made arbitrarily large by repeating a similar construction. In Fig. 4, the DAG of Fig. 3 has been extended to the top and right by nodes $S \ldots Z$. $\{F, G, H\}$ is still a smallest proof set for node $A$. Applying the same argument as above for nodes $S - T - U - A$, it can be seen that $\{F, G, H\}$ is also a smallest proof set for node $S$. However, as in Example 3 PSS computes a proof set of size five for node $A$, and therefore selects the right branch at node $U$, with a four-element proof set $\{W, X, Y, Z\}$. The proof set backed up to the root contains all nine leaf nodes, whereas the smallest set has only three nodes. Repeating this construction $n$ times yields a DAG for which PSS computes a proof set of size $2^{n+1} + 1$ containing all leaf nodes, while the size of the smallest proof set is 3.

When computing proof sets incrementally, it is possible to take the proof sets of other siblings into account for the minimum selection. Section 5.2 discusses such a variation of the algorithm, which tries to improve the likelihood of selecting a set that works together well with sets from other siblings.

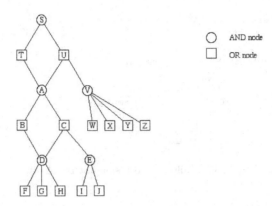

**Fig. 4.** Example 4: extending Example 3.

# 3    Algorithmic Aspects of Proof-Set Search

This section describes some algorithmic aspects of proof-set search, especially in areas where it differs from proof-number search, such as the selection of a most-promising node, different resource requirements and the representation of sets of nodes. It also discusses the problem of multiple updates of the same node. This problem already exists in PNS, but is more severe for PSS since the update cost per node is larger.

## 3.1    Selecting a Most-Promising Node

Selecting a most-promising node in proof-set search is extremely simple. Since PSS represents proof and disproof sets directly, it is not necessary to traverse the DAG as in PNS. Theorem 1 guarantees that the intersection of proof and disproof sets of the root is nonempty, and any node in the intersection is a suitable most-promising node.

## 3.2    Ancestor Updating Algorithm

This subsection discusses some problems of value propagation in DAGs. It applies to proof-number search as well as PSS. An ancestor updating algorithm must update all children of a node before the node itself can be computed. In a tree, a simple bottom-up computation suffices. On the other extreme, in a directed *cyclic* graph (DCG) a 'right' ordering of nodes does not exist because of cyclic dependencies. Special methods have been developed for proof-number search in DCGs [4], but it is unclear how they relate to PSS. This is a topic for future research, see the discussion in Sect. 6.1.

In a DAG, the right order of updates can be assured by a topological sorting of nodes. However, dynamically maintaining such a sorted order may be expensive, and in practice it may be preferable to accept multiple updates of some nodes instead.

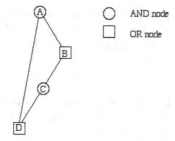

**Fig. 5.** Queue update algorithm causes multiple updates of the same node.

The simplest updating algorithm for DAGs, modeled after that for trees, starts with the just developed node *mpn*, then adds its predecessors to a queue [2]. For node sets, such an algorithm can be written as follows:

```
UpdateAncestorSets(mpn)
{  ListOf<Node> updateQ = [mpn]; // start with just expanded mpn
   while (updateQ.NonEmpty())
   {
       node = updateQ.Pop(); // extract first node from queue
       NodeSet oldPS = node->PS(), oldDS = node->DS();
       node->SetProofAndDisproofSets(); // backup from children
       if ((oldPS == node->PS()) && (oldDS == node->DS()))
       {} // unchanged, no need to propagate
       else // update parents
           updateQ.InsertAll(node->Parents());
           // insert new parents in queue
   }
} // UpdateAncestorSets
```

In the general case, this method does not guarantee a perfect order of updates. Some nodes might be updated more than once, as the following example shows.

*Example 5.* In the DAG of Fig. 5, there is a direct move from $A - D$, but there is also a longer path $A - B - C - D$ between the same nodes. Assume that parents of a node are added to the queue in left-to-right order. Then after updating $D$, $A$ is appended to the queue before $C$, so $A$ will be updated first. However, this is a useless update, since $A$ is also an ancestor of $C$ via $B$. After updating $C$ and $B$, $A$ is re-added to the queue and updated once more.

As the example shows, the queue backup algorithm can cause multiple updates of nodes in a DAG, which slows down the computation. There are two choices: either accept the inefficiency caused by multiple updates of the same node, or keep the nodes sorted in a priority queue, using a topological sorting of the DAG. This can be accomplished by an algorithm such as *Topological_Order* in [5, p.137], which enumerates nodes from the root downwards. We have not yet implemented such an ordering method. It is unclear whether an efficient incremental version of the algorithm exists, which can update the ordering after each node expansion. The problem is at least as hard as cycle detection in directed graphs, since computing an ordering is possible exactly for cycle-free graphs, and the algorithm must be able to complain if no ordering is possible because of a cycle.

**A Sufficient Condition for Optimality of the Queue Backup Algorithm.** The ancestor relation defines a partial ordering of the nodes in the DAG. If this partial order has a special structure, the optimality of the queue backup algorithm is assured. A *rank function* $r$ [6, p.99] on a partially ordered set is a function mapping elements to integers such that $r(y) = r(x) + 1$ whenever $y$ *covers* $x$ (immediately follows $x$) in the partial order. In our case, that means that $r(c) = r(n) + 1$ for all children $c$ of a node $n$. For example, for all the DAGs in Figs. 1 - 4 a rank function exists, while for the DAG in Fig. 5 there is none.

**Lemma 1.** *Consider the partially ordered set $P(G, Anc)$ given by the nodes in a DAG $G$ and the ancestor relation $Anc$ on $G$. If a rank function $r$ exists for $P(G, Anc)$, then the queue backup algorithm is optimal: it computes each node value at most once.*

Proof of Lemma 1: The queue backup algorithm processes nodes in order of their rank: if $x$ is inserted into the queue before $y$, then $r(x) \geq r(y)$. Actually, the following stronger statement will be proven: At each stage of the algorithm, the ranks of nodes in the queue are in monotonically decreasing order and assume at most two distinct values $v$ and $v - 1$. In other words, if the queue $q = [n_1, \ldots, n_k]$ contains $k > 0$ elements, then there exists $j$, $1 \leq j \leq k$, such that $r(n_i) = r(n_1)$ for $1 \leq i \leq j$ and $r(n_i) = r(n_1) - 1$ for $j + 1 \leq i \leq k$. It is easy to see that this property is an invariant maintained by the algorithm: Initially, the queue contains only a single element. Removing the element $n_1$ of rank $r(n_1)$ from the head of the queue maintains the invariant. The only nodes appended to the end of the queue by the algorithm are parents of $n_1$, which all have rank $r(n_1) - 1$ by definition of $r$.

Examples of games for which a rank function exists are all those where a move adds exactly one stone to the game state and does not remove anything, such as Othello (except for forced pass moves), connect-4, qubic, gomoku or Tic-Tac-Toe. The number of stones on the board is a rank function for positions in such games. In contrast, games with loops such as chess, shogi or Go do not have a rank function, unless the whole move history is taken into account for defining a position.

**Comparing the Ancestor Updating Algorithms of PNS and PSS.** Both PNS and PSS can stop propagating values to ancestors as soon as a node's value does not change. However, updates in PSS are certain to propagate all the way to the root, since the root's proof and disproof sets contain the just expanded *mpn*, which is no longer a leaf node. Usually, PSS will have to update more ancestors than PNS, since it distinguishes between sets of the same size with different elements. However, because of transpositions, the opposite case can also happen. If a node which is already contained in a proof set is re-added along a new path, the proof number increases but the proof set remains unchanged.

## 4    Truncated Node Sets for Proof-Set Search

This section describes a *truncated node set* data type with bounded memory requirements per set, and uses it in the algorithm $PSS_{P,D}$, PSS with (P,D)-truncated node sets. We prove theorems that characterize both PNS and PSS as extreme cases of $PSS_{P,D}$.

Set union and assignment operations on large node sets are expensive, both in terms of memory and computation time. The new data structure of a $K$-*truncated node set* provides a compromise between the two extremes of using a number and using a node set of unbounded size. A truncated node set stores at most $K$ nodes explicitly. In addition, it stores an upper bound on the overall set size, in the same sense that proof numbers represent an upper bound on the size of proof sets. In this sense, proof numbers can be regarded as 0-truncated node sets, which store only a bound but no elements.

**Definition 1.** *Let $K$ be a nonnegative integer, and let set-min be a function computing the minimum of two sets based on a total order of nodes. A $K$-truncated node set $s$ is a pair $(rep(s), bound(s))$, where $rep(s)$ is a set of nodes of cardinality at most $K$ and $bound(s)$ is a nonnegative integer. In addition, a node set has the following properties:*

- $bound(s) < K \Leftrightarrow |rep(s)| = bound(s)$
- $bound(s) \geq K \Leftrightarrow |rep(s)| = K$

**Definition 2.** *The operations of minimum selection and set union for truncated node sets are defined as follows:*

- $bound(a \cup b) = bound(a) + bound(b) - |rep(a) \cap rep(b)|.$
- $rep(a \cup b) = x$, *where $x$ is the smallest $K$-element subset of $rep(a) \cup rep(b)$ according to set-min.*
- $min(a, b) = a \Leftrightarrow bound(a) < bound(b) \vee (bound(a) = bound(b) \wedge set - min(rep(a), rep(b)) = rep(a)).$

In other words:

- Minimum selection: If the bounds of two sets are different, the set with the smaller bound is the minimum. Otherwise, the explicitly represented parts of the sets are compared as if they were unbounded sets.
- Set union: The truncated set union computes the union of the explicitly represented sets, and stores the first $K$ elements according to some total ordering of nodes, plus the best-possible bound for the size of the union.
- Initialization by a single element: A truncated node set $s$ is initialized to store a single node $n$ as follows: $s = (\{n\}, 1)$ if $K > 0$, $s = (\{\}, 1)$ if $K = 0$.

*Example 6.* Let $K = 8$, let nodes be represented by letters ordered alphabetically, let $s_1 = \{A, D, E, H, K, L, M, Q\}_{16}$, $s_2 = \{C, D, E, F, H, M, P, Q\}_{13}$. For each truncated set, $K$ elements are given explicitly, and the subscript represents the bound on the set size. $s_1$ represents a set of at most 16 elements, including the eight listed. The truncated set union $s_1 \cup s_2 = \{A, C, D, E, F, H, K, L\}_{24}$ is computed by truncating $rep(s_1) \cup rep(s_2) = \{A, C, D, E, F, H, K, L, M, P, Q\}$ to the smallest $K = 8$ elements. The bound on the set union size is computed by adding the bounds, then subtracting the double-counted elements in $rep(s_1) \cap rep(s_2)$, $16 + 13 - 5 = 24$.

As before, infinite size sets $\{\}_\infty$ are added along with rules for computing minima and unions involving such sets.

## 4.1   Some Properties of Truncated Node Sets

The next lemma formalizes the intuitively clear fact that larger truncation thresholds result in tighter bounds for set unions.

**Lemma 2.** *Given two integers $K > L$, and sets of nodes $s_1 \ldots s_n$, compute both the $K$−truncated and the $L$−truncated union of $s_1 \cup s_2 \cup \ldots \cup s_n$, using the truncated set method with the same node ordering and the same sequence of two-set union operations. Then the bound on the $K$−truncated union is not larger than that on the $L$−truncated union. Furthermore, the explicitly represented set of the $L$−truncated union is a subset of the explicitly represented set of the $K$−truncated union.*

The proof is straightforward from the definition of truncated set union and is omitted here to save space. It is worth noting that since truncating sets loses information, some properties of the usual set union operations are lost. For example, the absorption law $a \cup a = a$ does no longer hold, since two sets that look identical may contain different nonrepresented nodes.

*Example 7.* In the extreme case of 0-truncated sets, computing the truncated set union is equivalent to adding the bounds for both sets, as in $\{\}_5 + \{\}_5 = \{\}_{10}$.

*Example 8.* Let $a = a_1 = a_2 = \{A, B\}_5$ be 2-truncated sets. Then it would be wrong to set $a_1 \cup a_2 = a$, since the two sets might contain up to three different elements. The correct result is $a_1 \cup a_2 = \{A, B\}_{5+5-2} = \{A, B\}_8$.

## 4.2   Proof-Set Search with Truncated Node Sets

Given two integers $P$ and $D$, the algorithm *proof-set search with (P,D)-truncated node sets*, $PSS_{P,D}$, is a modified version of PSS where all proof sets are replaced by $P$-truncated node sets, and all disproof sets by $D$-truncated node sets.

**Selecting a Most-Promising Node.** Selection of a most-promising node in PSS relies on Theorem 1 of Sect. 2.2, which guarantees that for each unproven node $n$, $pset(n) \cap dset(n) \neq \emptyset$. With truncated node sets, a most-promising node may not always be found immediately, since all the elements in the intersection might have been cut off by truncation. The algorithm for selecting a most-promising node with truncated node sets is an intermediate form of the respective algorithms in PNS and PSS.

```
SelectMPN() // find most-promising node with truncated node sets
{
    node = root; mpn = NIL;
    while (IsInteriorNode(node))
    {
        // CommonNode(a,b) returns NIL if no common node
        // is found in rep(a) and rep(b).
        mpn = CommonNode(pset(node), dset(node));
        if (mpn != NIL) // found a node in intersection, done
            return mpn;
        SetType t = 'disproof' if 'node' is AND node,
                    'proof' if 'node' is OR node;
        node = child with same set of type t as node;
    }
    return node; // reached a leaf node, done.
} // SelectMPN
```

## Characterizing PSS and PNS as Special Cases of $PSS_{P,D}$.

**Theorem 3.** $PSS_{\infty,\infty}$ is the same algorithm as standard PSS.

**Theorem 4.** $PSS_{0,0}$ is the same algorithm as proof-number search.

The proofs follow immediately from the facts that an $\infty$-truncated node set is an ordinary untruncated node set and a 0-truncated node set is equivalent to a proof number. Two other special cases are interesting, and may be appropriate parameter choices if search behavior is highly biased towards only proofs or only disproofs: $PSS_{\infty,0}$ combines proof sets with disproof numbers, while $PSS_{0,\infty}$ uses proof numbers together with disproof sets.

A nice property of $PSS_{P,D}$ is that if $P < \infty$ and $D < \infty$, then the required memory remains bounded by a constant factor of what PNS would use on the same DAG.

## Dominance Theorem of PSS with Truncated Node Sets.

**Theorem 5.** Let $G$ be a DAG and let $P \geq 0$ and $D \geq 0$ be integers. Then at each node $n \in G$:

$$bound(pset(n)) \leq pn(n)$$

$$bound(dset(n)) \leq dn(n)$$

The proof is by induction: the theorem holds for single nodes, and remains true when taking truncated set unions or selecting minima in interior nodes.

1. The theorem holds for all proven, disproven and unproven leaf nodes by definition of $PSS_{P,D}$ and PNS.
2. Assume $bound(pset(n_i)) \leq pn(n_i)$ for all children $n_i$ of node $n$.
3. Set union: By Definition 2,

$$bound(pset(n)) = bound(pset(n_1) \cup \ldots \cup pset(n_k))$$

$$\leq bound(pset(n_1)) + \ldots + bound(pset(n_k))$$

$$\leq pn(n_1) + \ldots + pn(n_k) = pn(n).$$

4. Minimum selection: By definition,

$$pn(n) = min(pn(n_1), \ldots, pn(n_k)) \text{ and}$$

$$bound(pset(n)) = min(bound(pset(n_1)), \ldots, bound(pset(n_k))).$$

Assume the minimal proof number is achieved in node $n_i$. Then

$$pn(n) = pn(n_i) \geq bound(pset(n_i))$$

$$\geq min(bound(pset(n_1)), \ldots, bound(pset(n_k))) = bound(pset(n)).$$

5. The proof for disproof sets is the same as for proof sets.

Theorem 5 generalizes Theorem 2 of Sect. 2.3, which dealt with the extreme case $PSS_{\infty,\infty}$. As in Theorem 2, the dominance holds only when comparing the computation of the algorithms on the same DAG $G$. It does not hold, and is not even meaningful, for the different DAGs generated by using $PSS_{P,D}$ and PNS respectively to generate the DAG.

## 5    Using PSS with a Heuristic Leaf Evaluation Function

In analogy to the refinement of PNS described in Sect. 1.1, PSS can utilize a heuristic node initialization function $h$ for new frontier nodes. For simplicity, we will use the same function symbol $h$ for both the proof and the disproof case. In practice, two different initialization functions will be used, since proofs and disproofs are opposite goals.

For a node set $s = \{n_1, \ldots, n_k\}$, define the heuristic weight by $h(s) = \sum h(n_i)$, and set $h(\{\}_\infty) = \infty$. In PSS, modify the minimum selection among sets as follows: set-min$(s_1, s_2) = s_1$ if $h(s_1) < h(s_2)$. If $h(s_1) = h(s_2)$, break the tie according to a secondary criterion such as lexicographical ordering. Combining heuristic initialization with truncated node sets is also relatively straightforward and will be described in Sect. 5.1. Let the proof and disproof number of $n$ computed by PNS with leaf node initialization function $h$ be denoted by $pn_h(n)$ and $dn_h(n)$ respectively.

**Theorem 6.** *Let G be a DAG, and let a positive heuristic leaf node initialization function h be defined for each game position represented by a node in G. Furthermore, extend h to sets of nodes by $h(s) = \sum_{n \in s} h(n)$. Then for each unproven node $n \in G$:*

$$h(pset(n)) \leq pn_h(n)$$
$$h(dset(n)) \leq dn_h(n)$$

Proof: by induction.

1. The theorem holds for all leaf nodes $n$ since $pset(n) = dset(n) = \{n\}$ and $h(pset(n)) = h(dset(n)) = h(n) = pn_h(n) = dn_h(n)$.
2. Let $n$ be an unproven AND node with (unproven) children $\{n_1, \ldots, n_k\}$. Assume that the induction hypothesis holds for all children $n_i$, so $h(pset(n_i)) \leq pn_h(n_i)$.

$$h(pset(n)) = h(pset(n_1) \cup pset(n_2) \cup \ldots \cup pset(n_k))$$
$$\leq h(pset(n_1)) + h(pset(n_2)) + \ldots + h(pset(n_k))$$
$$\leq pn_h(n_1) + pn_h(n_2) + \ldots + pn_h(n_k) = pn_h(n).$$

3. Let $n$ be an OR node with children $\{n_1, \ldots, n_k\}$, and again assume the induction hypothesis holds for the children.

$$h(pset(n)) = \min(h(pset(n_1)), h(pset(n_2)), \ldots, h(pset(n_k)))$$
$$\leq \min(pn_h(n_1), pn_h(n_2), \ldots, pn_h(n_k)) = pn_h(n).$$

4. The claim for disproof sets is proved by swapping the AND with the OR case in steps 2 and 3.

Setting $h(n) = 1$ for all $n$ results in another proof of Theorem 2 in Sect. 2.3, since for every set $s$, $h(s) = \sum_{n \in s} 1 = |s|$.

## 5.1 Combining Truncated Node Sets with Heuristic Leaf Evaluation Functions

The two generalizations of PSS by truncated node sets and heuristic leaf initialization can be combined by reinterpreting the bound on the set size as a bound on the set evaluation: The bound of a union of two sets is then defined by $bound(a \cup b) = bound(a) + bound(b) - \sum_{x \in I} h(x)$, $I = rep(a) \cap rep(b)$.

*Example 9.* Given the following nodes, with their heuristic evaluation written as a subscript: $A_{10}, B_{15}, C_7, D_{16}, E_{18}, F_{39}$. Consider a 4-truncated node set representation, with the subscript of the whole set showing the evaluation bound for the whole set. Examples of exactly representable sets are $s_1 = A_{10} \cup B_{15} \cup C_7 = \{C_7, A_{10}, B_{15}\}_{32}$ and $s_2 = A_{10} \cup D_{16} \cup E_{18} = \{A_{10}, D_{16}, E_{18}\}_{44}$. Taking the union leads to set truncation, $s_1 \cup s_2 = \{C_7, A_{10}, B_{15}, D_{16}\}_{66}$. Different sets may share the same truncated set but have different bounds. For example, $s_1 \cup \{B_{15}, D_{16}, F_{39}\}_{70} = \{C_7, A_{10}, B_{15}, D_{16}\}_{87}$.

The following lemma and theorem are easy generalizations of Lemma 2 and Theorem 5 respectively. Proofs are omitted for lack of space.

**Lemma 3.** *Let $K > L \geq 0$ be integers, let $s_i$ be (untruncated) node sets, and let $h$ be a heuristic node evaluation function. Compute the $K$-truncated set union $u_K$ and the $L$-truncated set union $u_L$ using a set ordering based on $h$, the same secondary sorting criterion, and the same sequence of two-set union operations. Then $bound(u_K) \leq bound(u_L)$.*

**Theorem 7.** *Let $K$ be an integer, let $h$ be a heuristic node evaluation function, and let $G$ be a DAG. For each node $n \in G$, compute proof and disproof sets using $PSS_{K,K}$, and compute proof and disproof numbers using PNS with node initialization by $h$. Then for each unproven node $n$ in $G$:*

$$bound(pset(n)) \leq pn_h(n)$$

$$bound(dset(n)) \leq dn_h(n).$$

### 5.2    Favoring Nodes from a Given Set

Assume a most-promising OR node $n$ is expanded by PSS, and that before the expansion, the proof set of the root already contains other nodes $P = \{p_1, \ldots, p_k\}$. When computing the new minimal proof sets after expanding $n$, it is probably better to choose nodes which are already contained in $P$, since they will not increase the size of sets further up in the DAG. The node evaluation can be modified to discount such nodes by a factor $\epsilon$, $0 \leq \epsilon < 1$: $h'(x) = \epsilon h(x)$ if $x \in P, h'(x) = h(x)$ if $x \notin P$.

A problem with this approach is that the evaluation of node sets becomes context-dependent. Also, care should be taken to always prefer a proven node $x$ with $h(x) = 0$ over an unproven but completely discounted node $y$ with $h(y) \neq 0$ but $h'(y) = 0$.

*Example 10.* In the DAG on the left side of Fig. 6, node $C$ is the only most-promising node and is expanded next. The figure on the right shows the situation at the time of recomputing the proof set of $C$ after the expansion. (Proof sets of node $D$ and below are not affected by expanding $C$.)

At node $C$, standard PSS would select the small proof set $\{E\}$ as its minimal proof set. However, the modified algorithm would discount the values of $F$, $G$ and $H$, since they are contained in the proof set of the root $A$, and therefore select $D$ over $E$. This way, the proof set of $A$ shrinks because of the expansion of $C$, $pset(A) = \{F, G, H\}$.

## 6    Future Work and Summary

Future work on PSS includes extending it to cyclic game graphs and testing it in a variety of applications.

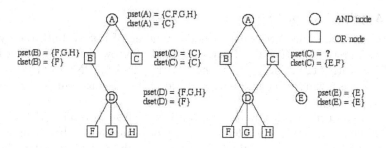

**Fig. 6.** Discounting existing nodes.

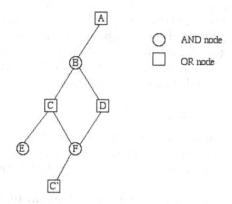

**Fig. 7.** Directed cyclic graph with cycle $C - F - C$.

## 6.1   PSS for Directed Cyclic Graphs (DCG)

Directed cyclic graphs, or DCGs, cause problems for proof-based search procedures because of cyclical dependencies, often called *graph history interaction* in this context. Breuker et al. develop one solution to this problem with their *base-twin algorithm* [4].

It is presently unknown whether PSS is well-defined and how it works on general DCGs. It seems necessary to adapt the update and propagation rules, since now the same node can appear both as leaf and as interior node in the DCG. A promising sign is that PSS has no trouble solving the following example, taken from Fig. 5 of [3]:

*Example 11.* $C'$ is a transposition of $C$, leading to a cycle $C - F - C$. Let's treat the graph as a DAG and update proof sets, assuming nodes are ordered alphabetically for minimum selection. Initially, $pset(C') = \{C'\} = \{C\}$, $pset(E) = \{E\}$, $pset(F) = \{C\}$, $pset(C) = \min(pset(E), pset(C)) = \{C\}$, $pset(D) = \{C\}$, $pset(B) = pset(C) \cup pset(D) = \{C\}$, $pset(A) = \{C\}$. After proving $E$, $pset(E) = \emptyset$, $pset(C) = \min(pset(E), pset(C)) = \emptyset$. Now $C$ is proved, and can be propagated through the DCG, leading successively to proofs of $F$, $D$, $B$ and $A$.

## 6.2    Applications of PSS

Applications should lead to a better understanding of PSS in both theory and practice. A first test on Tic-Tac-Toe resulted in PSS growing a 10% smaller DAG than PNS, with 1114 nodes against the 1237 required by PNS to disprove that Tic-Tac-Toe is a win for the first player. However, large gains cannot be expected in this toy example.

The original motivation to develop PSS came from a report [7] that some *tsume shogi* (shogi mating) problems are hard for proof-number based algorithms because of transpositions. In first experiments with PSS in this domain, the method proved viable on moderately large DAGs with up to several hundred thousand nodes when using truncated node sets with $P = D = 20$. Preliminary results seem to indicate a correlation between the frequency of transpositions and the performance of PSS. However, a proper study remains as future work.

Another promising application area are subproblems in the game of Go, such as life and death puzzles or tactical capturing problems. Both shogi and Go provide a rich and very challenging set of test cases.

## 6.3    More Research Topics

**Comparison with Nagai's method.** In his PhD thesis [8], Nagai describes an alternative method that can avoid the double-counting of leaf nodes in a DAG. The method should be compared with PSS both theoretically and experimentally.

**Investigate the performance of the queue backup method.** What is the average and worst-case performance of the queue backup method on different types of DAGs? Is it sufficient for the DAGs encountered in practice? Are there DAGs for which the performance is unacceptably bad?

**Find necessary conditions for optimality of queue backup.** Lemma 1 in Sect. 3.2 proves that the existence of a rank function for the ancestor relation is sufficient to ensure the optimality of the queue backup algorithm. Are there other, more general sufficient conditions? What are necessary conditions for optimality? Such a criterion might be based on characterizing forbidden subgraphs, such as the one in Fig. 5.

**Efficient data structures for large node sets.** Our implementation of node sets using sorted lists is easy to program, but becomes slow for large sets. Are there applications where it is essential to deal with large sets, and if so, are there more efficient data structures?

**Choice of truncation values.** Which values of $P$ and $D$ provide a good trade-off between memory and accuracy for $PSS_{P,D}$? How are the values related to the problem type? Is it beneficial to dynamically adapt $P$ and $D$ during the search, or use different values in different regions of the DAG?

**Multi-level schemes.** Are schemes such as $pn^2$-search [9] effective for PSS?

## 6.4    Summary

*Proof-set search*, or *PSS*, is a new search method which addresses the problems caused by overestimating the size of the minimal proof set in DAGs. Like PNS,

PSS uses the simple children-to-parents propagation scheme for DAGs, but unlike PNS, it backs up proof sets instead of proof numbers. The sets computed by PSS provide better approximations than is possible with only proof numbers on the same DAG. However, more memory and more computation is needed to store and manipulate node sets instead of numbers. The trade-off between the advantages of a more focused search and the disadvantages of using more memory per node and a more expensive backup procedure need further investigation. Since overestimating proof numbers in a DAG can lead search into a completely wrong direction for a long time, any improvement in the node expansion strategy can potentially achieve large savings in search efficiency, especially on hard problems.

# References

1. Allis, V.: Searching for Solutions in Games and Artificial Intelligence. PhD thesis, University of Limburg, Maastricht (1994)
2. Schijf, M.: Proof-number search and transpositions. Master's thesis, University of Leiden (1993)
3. Schijf, M., Allis, V., Uiterwijk, J.: Proof-number search and transpositions. International Computer Chess Association Journal 17 (1994) 63–74
4. Breuker, D., van den Herik, J., Uiterwijk, J., Allis, V.: A solution to the GHI problem for best-first search. In van den Herik, J., Iida, H., eds.: First International Conference on Computers and Games (CG 1998). Volume 1558 of Lecture Notes in Computer Science. Springer-Verlag (1999) 25–49
5. McHugh, J.: Algorithmic Graph Theory. Prentice-Hall (1990)
6. Stanley, R.: Enumerative Combinatorics Vol. 1. Number 49 in Cambridge Studies in Advanced Mathematics. Cambridge University Press (1997)
7. Kishimoto, A.: Seminar presentation. Electrotechnical Laboratory, Tsukuba, Japan (1999)
8. Nagai, A.: DF-PN Algorithm for Searching AND/OR Trees and Its Applications. PhD thesis, University of Tokyo (2001)
9. Breuker, D., Uiterwijk, J., van den Herik, J.: The $PN^2$-search algorithm. In van den Herik, J., Monien, B., eds.: Advances in Computer Games 9. Universiteit Maastricht (2001) 115–132

# A Comparison of Algorithms
# for Multi-player Games

Nathan Sturtevant

Computer Science Department, University of California, Los Angeles
nathanst@cs.ucla.edu

**Abstract.** The $max^n$ algorithm for playing multi-player games is flexible, but there are only limited techniques for pruning $max^n$ game trees. This paper presents other theoretical limitations of the $max^n$ algorithm, namely that tie-breaking strategies are crucial to $max^n$, and that zero-window search is not possible in $max^n$ game trees. We also present quantitative results derived from playing $max^n$ and the paranoid algorithm (Sturtevant and Korf, 2000) against each other on various multi-player game domains, showing that paranoid widely outperforms $max^n$ in Chinese checkers, by a lesser amount in Hearts and that they are evenly matched in Spades. We also confirm the expected results for the asymptotic branching factor improvements of the paranoid algorithm over $max^n$.

## 1 Introduction and Overview

Artificial Intelligence researchers have been quite successful in the field of two-player games, with well-publicized work in chess [1] and checkers [2], and are producing increasingly competitive programs in games such as Bridge [3].

If we wish to have similar success in multi-player games, there is much research yet to be done. Many of the issues surrounding two-player games may be solved, but there are many more unanswered questions in multi-player games. The most basic of these is the question of which algorithm should be used for playing multi-player games.

Unfortunately, the question is not as simple as just which algorithm to use. Every algorithm is associated with other techniques (such as alpha-beta pruning or transposition tables) by which it can be enhanced. There are too many different techniques to cover here, but we have chosen a few algorithms and techniques as a starting point.

This paper considers the $max^n$ [4] and paranoid [5] algorithms. In this paper we introduce and discuss theoretical limitations with each algorithm, and then present our results from playing these two algorithms against each other in various domains. Our results indicate that the paranoid algorithm is worth considering in a multi-player game, however the additional depth of search offered by the paranoid algorithm may not always result in better play. These results address games solely from a perfect-information standpoint, leaving the question of imperfect information for future research.

J. Schaeffer et al. (Eds.): CG 2002, LNCS 2883, pp. 108–122, 2003.
© Springer-Verlag Berlin Heidelberg 2003

One popular multi-player game we will not address here is poker. Billings et al. describe work being done on this domain [6]. However, their focus is directed towards opponent modeling and simulations (limited search), while our focus is more on search.

## 2 Multi-player Games: Hearts, Spades, and Chinese Checkers

To help make the concepts here more concrete, we chose two card games, Hearts and Spades, and the board game Chinese checkers to highlight the various algorithms presented.

Hearts and Spades are both trick-based card games. Cards are dealt out to each player before the game begins[1]. The first player plays (*leads*) a card face-up on the table, and the other players follow in order, playing the same suit as lead if possible. When all players have played, the player who played the highest card in the suit that was led "wins" or "takes" the trick. He then places the played cards face down in his discard pile, and leads the next trick. This continues until all cards have been played.

Hearts is usually played with four players, but there are variations for playing with two or more players. The goal of Hearts is to take as few points as possible. A player takes points when he takes a trick which contains point cards. Each card in the suit of hearts is worth one point, and the queen of spades is worth 13. At the end of the game, the sum of all scores is always 26, and each player can score between 0 and 26. If a player takes all 26 points, or "shoots the moon," he instead gets 0 points, and the other players all get 26 points each. These fundamental mechanics of the game are unchanged regardless of the number of players.

Spades can be played with 2-4 players. Before the game begins, each player predicts how many tricks they think they are going to take, and they then get a score based on how many tricks they actually do take. With 4 players, the players opposite each other play as a team, collectively trying to make their bids, while in the 3-player version each player plays for themselves. More in-depth descriptions of these and many other multi-player games can be found in Hoyle *et al.* [7].

Chinese checkers is a perfect information game for 2-6 players. A Chinese checkers board is shown in Fig. 1. The goal of the game is to get 10 pegs or marbles from one's starting position to one's ending position as quickly as possible. These positions are always directly across from each other on the board. Pegs move by stepping to an adjacent position on the board or by jumping over adjacent pegs. One can jump over any player's pegs, or chain together several jumps, but pegs are not removed from the board after a jump.

---

[1] There are rules for trading cards between players, but they have little bearing on the work presented here.

**Fig. 1.** A Chinese checkers board.

## 2.1   Imperfect-Information Games

The paranoid and $max^n$ algorithms, which we will cover next, are designed for perfect-information games such as Chinese checkers. In order to use them with imperfect-information games such as Spades or Hearts, we must either modify the algorithms or modify the nature of the games we want to play. Both approaches have been used successfully in two-player games, but it remains to be seen how they can be applied successfully in multi-player games.

If, in a card game, we could see our opponents' cards, we would be able to use standard search algorithm to play the game. While in most games we don't know the exact cards our opponent holds, we do know the probability of our opponent holding any particular hand. Thus, we can create a hand that should be similar to what our opponent holds, and use a perfect-information algorithm to play against it.

The full expansion of this idea uses Monte-Carlo sampling. Instead of generating just a single hand, we generate a set of hands that are representative of the actual hand we expect our opponent to have. We then solve each of these hands using the standard minimax algorithm. When we have completed the analysis of each hand, we combine and analyze the results from each hand to produce our next play. As the play continues we update our models to reflect the plays made by our opponent. This is one of the techniques used to create a strong Bridge program [3].

## 3   Multi-player Game Algorithms

We review two multi-player game algorithms and their properties as a background for the theory and results found in later sections.

### 3.1   Max$^n$

The $max^n$ algorithm [4] can be used to play games with any number of players. For two-player games, $max^n$ simply computes the minimax value of a tree.

In a $max^n$ tree with $n$ players, the leaves of the tree are $n$-tuples, where the $i$th element in the tuple is the $i$th player's score. At the interior nodes in the

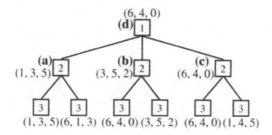

**Fig. 2.** A 3-player $\max^n$ game tree.

game tree, the $\max^n$ value of a node where player $i$ is to move is the child of that node for which the $i$th component is maximum. This can be seen in Fig. 2. In this tree there are three players. At node (a), Player 2 is to move. Player 2 can get a score of 3 by moving to the left, and a score of 1 by moving to the right. So, Player 2 will choose the left branch, and the $\max^n$ value of node (a) is (1, 3, 5). Player 2 acts similarly at node (b) selecting the right branch, and at node (c) breaks the tie to the left, selecting the left branch. At node (d), Player 1 chooses the move at node (c), because 6 is greater than the 1 or 3 available at nodes (a) and (b).

One form of pruning, shallow pruning, is possible in a $\max^n$ tree. Shallow pruning refers to cases where a bound on a node is used to prune at the child of that node. To prune, we need at least a lower bound on each player's score, and an upper bound on the sum of all players scores (for a full discussion, see [5, 8]). This work shows that theoretically there are no asymptotic gains due to shallow pruning in $\max^n$. This contrasts with the large gains available from using alpha-beta pruning with minimax. Another type of pruning, deep pruning, is not possible in $\max^n$.

If a monotonic heuristic is present in a game, it can also be used to prune a $\max^n$ tree. The full details of how this occurs is contained in [5]. An example of a monotonic heuristic is the number of tricks taken in Spades. Once a trick has been taken, it cannot be lost. This guarantee can provide a lower bound on a player's score, and an upper bound on one's opponents scores.

In practice, almost no pruning occurs when using $\max^n$ to play Chinese checkers. In Hearts, monotonic heuristic bounds on the scores can provide some pruning. In Spades, both monotonic heuristics and shallow pruning can be combined for more significant amounts of pruning.

## 3.2  Paranoid Algorithm

The lack of pruning in $\max^n$ has motivated research into other methods which might be able to search faster or deeper into a multi-player game tree. The paranoid algorithm does this by reducing a $n$-player game to a 2-player game. It assumes that a coalition of $n$-1 players have formed to play against the remaining player. We demonstrate this in Fig. 3. In this tree Players 2 and 3 ignore their

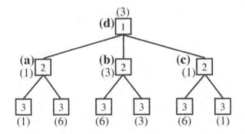

**Fig. 3.** The paranoid version of the game tree in Fig. 2.

own scores, and simply try to minimize Player 1's score. At nodes (a) and (b), Player 2 makes the same choice as in Fig. 1, but at node (c), Player 2 chooses the right branch. Then, at the root, (d), Player 1 chooses to move towards node (b) where he can get a score of 3.

If we search an $n$-player game tree with branching factor $b$ to depth $d$, the paranoid algorithm will, in the best case, expand $b^{d(n-1)/n}$ nodes [5]. This is the general version of the best case bound for a two-player game, $b^{d/2}$ [9]. So, making the assumption that our opponents have formed a coalition against us should allow us to search deeper into a game tree, and therefore produce better play. Obviously as the number of players grows, the added efficiency of paranoid will drop. But, most multi-player games are played with 3-6 players, in which case paranoid can search 20-50% deeper.

## 4    Theoretical Properties of Max$^n$

Before we delve into the theoretical properties of multi-player algorithms, we return briefly to two-player games. For those familiar with game theory, one reason the minimax algorithm is so powerful is because it calculates an equilibrium point and strategy for a given game tree. This strategy guarantees, among other things, some payoff $p$, regardless the strategy of the opponent. This statement is quite strong, and it allows us to, for the most part, ignore our opponents strategy. As we will see, we cannot make such strong statements about max$^n$.

### 4.1    Equilibrium Points in Max$^n$

It has been shown that, in a multi-player game, max$^n$ computes an equilibrium point [4]. But, the concept of an equilibrium point in a multi-player game is much weaker than that in a two-player game. In a multi-player game there are multiple equilibrium points that may have completely different equilibrium values and strategies. For those unfamiliar with game theory, it will suffice to understand that while there is a single minimax value for a two-player game tree, there are multiple possible max$^n$ values for a multi-player game tree, which can all be different.

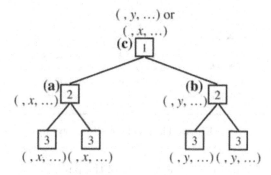

**Fig. 4.** Tie breaking in a $\text{max}^n$ game tree.

An example of this can be seen in Fig. 2 at node (c). At this node Player 2 can make any choice, as either will lead to the same score for Player 2. But, if Player 2 chooses $(1, 4, 5)$ as the $\text{max}^n$ value of node (c), Player 1 will choose the result from node (b), $(3, 5, 2)$ to be the $\text{max}^n$ value of the tree. These are both valid, but different, equilibrium points in the tree. In a two-player game, since tie-breaking cannot affect the minimax value of the tree, ties are broken in favor of the left-most branch, allowing any additional ties to be pruned. However, in a multi-player game we cannot do this.

**Lemma 1.** *In a multi-player game, changing the tie-breaking rule may arbitrarily affect the $\text{max}^n$ value of the tree.*

*Proof.* Figure 4 contains a generic $\text{max}^n$ tree. We have represented Player 2's possible scores by $x$ and $y$. The scores for the other players can obviously be arbitrarily affected by the way we break the ties for Player 2. By adjusting the tie breaking, we can also change whether Player 1 moves to the left or right from the root, and so can affect whether Player 2's score will be $x$ or $y$.

This result doesn't mean that $\text{max}^n$ is a worthless algorithm. A game tree with no ties will have a single $\text{max}^n$ value. In addition, each possible $\text{max}^n$ value that results from a particular tie-breaking rule, will play reasonably, given that all players use that tie-breaking rule.

The choice of a tie-breaking rule amounts to a strategy for play. In Hearts, for instance, good players will often save the queen of spades to play on the player with the best score. Thus, we must consider our opponents' strategy.

We illustrate the implications of lemma 1 in Fig. 5. Each player holds 2 cards, as indicated, and three possible outcomes of the play are shown. The winning card of each trick is underlined. If cards are played from left to right in your hand by default, Player 1 can lead the A♠, and Player 2 will not drop the Q♠, as in play (a). However, if Player 2 breaks ties differently, this could be a dangerous move, resulting in Player 1 taking the Q♠, as in (b). But, if Player 2 leads the 2♣, as in (c), Player 3 will be forced to take the Q♠.

The tie-breaking rule we have found most effective in such situations has been to assume that our opponents are going to try to minimize our score when

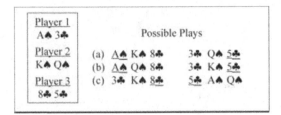

**Fig. 5.** Tie breaking situation.

they break ties. This obviously has a flavor of the paranoid algorithm, and it will cause us to try and avoid situations where one player can arbitrarily change our score. From our own experience, we believe that this is similar to what humans usually do when playing Hearts.

## 4.2   Zero-Window Search

Zero-window search originates from two-player games [10]. The idea behind zero-window search is to turn a game tree with a range of evaluations into a tree with only two evaluations. This is done by choosing some value $v$, and treating a leaf node as a win if its evaluation is $> v$, and as a loss if it is $\geq v$. Combining this approach with a binary search will suffice to find the minimax value of a game tree to any precision. This assumption results in highly optimized searches that can prune away most of the game tree. Zero-window search is one of the techniques that helps make partition search efficient [11].

While there are limitations on pruning during the calculation of the $\max^n$ value of a tree, it is not immediately obvious that we cannot somehow prune more if we just try to calculate the bound on the $\max^n$ value of a tree, instead of the actual $\max^n$ value.

So, we could attempt to search a $\max^n$ game tree to determine whether the $n$th player will be able to get at least a score of $v$. Supposing at the root of a tree Player 1 gets $v$ points. We can obviously stop at that point knowing that Player 1 will get $\leq 3$ points. This is a bound derived from shallow $\max^n$ pruning, and at most it will reduce our search size from $b^d$ to $b^{d-1}$. In addition, this will not combine well with a technique like partition search. We would like to know if this idea can be applied to every node in the tree.

Suppose we want to consider all scores $> 2$ as a win. We illustrate this in Fig. 6. Since Player 2 can get a score of 5 by moving to the left at node (a), Player 2 will prune the right child of (a), and return a $\max^n$ value of $(w, w, l)$, where $w$ represents a win and $l$ represents a loss. However, at node (b), Player 1 would then infer that he could win by moving towards node (a). But, the exact $\max^n$ value of (a) is $(1, 8, 1)$, and so in the real game tree, Player 1 should prefer node (c) over node (a).

So, the only bounds we can get on the $\max^n$ value of a tree are those that come from a search with shallow pruning, which, given no additional constraints, is already optimal.

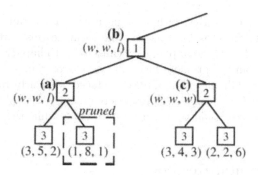

**Fig. 6.** Finding bounds in a $max^n$ game tree.

Korf shows that "Every directional algorithm that computes the $max^n$ value of a game tree with more than two players must evaluate every terminal node evaluated by shallow pruning under the same ordering" [8]. We can now expand this statement:

**Theorem 1.** *Given no additional constraints, every directional algorithm that computes either the $max^n$ value or a bound on the $max^n$ value of a multi-player game tree must evaluate every terminal node evaluated by $max^n$ with shallow pruning under the same ordering.*

*Proof.* Korf has already shown that shallow pruning is optimal in its computation of the $max^n$ value of a tree [8]. If we replace backed-up $max^n$ values in a game tree with just bounded $max^n$ values, there may be ties between moves that were not ties in the original $max^n$ tree, such as at node (a) in Fig. 6. By definition, all possible moves when breaking ties must lead to equilibrium points for the sub-tree below the node. Since some of these moves may not have been equilibrium points in the original tree, any bound computed based on these moves may not be consistent with the possible $max^n$ values for the original tree.

## 5   Equilibrium Points in the Paranoid Algorithm

While the paranoid algorithm may be a pessimistic approach to game playing, it is not entirely unreasonable, and it offers a theoretical guarantee not offered by $max^n$.

Equilibrium points calculated by the paranoid algorithm are theoretically equivalent to those calculated by minimax. This means that a paranoid game tree has a single paranoid value which is the guaranteed lower bound on one's score.

This may, however, be lower than the actual score achievable in the game, because it is unlikely that your opponents have truly formed a coalition against you. This leads to moves which are sub-optimal, because of the unreasonable assumptions being made. In Hearts, for instance, paranoid may find there is no way to avoid taking a trick with the queen of spades. But, instead of forcing the opponents to work as hard as possible to make this happen, the paranoid algorithm may select a line of play that causes this to happen immediately. This is no different from the problem faced in two-player games. In [2], for instance, they discuss the difficulty that CHINOOK faced with arbitrating between two lines of play that both lead to a draw, where on one path a slight mistake by the opposing player could lead to a win by CHINOOK, while the other leads clearly to a draw.

## 6    Experimental Results

Given these theoretical results, we would like to see what we can do in practice to play multi-player games well. To this end, we have run a number of experiments on various games to see how $\max^n$ and paranoid perform against each other. We have written a game engine that contains a number of algorithms and techniques such as the paranoid algorithm, $\max^n$, zero-window search, transposition tables and monotonic heuristic pruning. New games can easily be defined and plugged in to the existing architecture without changing the underlying algorithms.

We first present the general outline of our experiments applicable to all the games, and then we will present the more specific details along with the results.

Our game engine supports an iterative deepening search process, but each particular game can choose the levels at which the iterations take place. Search can be bounded by time, nodes expanded, or search depth. In our experiments, we gave the algorithms a node limit for searching. When the cumulative total of all iterative searches expands more nodes than the limit, the search is terminated, and the result from the last completed iteration is returned.

We are using 3, 4, and 6-player games to compare 2 different algorithms. In a 3-player game, there are $2^3 = 8$ different ways we could assign the algorithm used for each player. However, we are not interested in games that contain exclusively $\max^n$ or exclusively paranoid players, leaving 6 ways to assign each player to an algorithm. These options are shown in Table 1. So, we ran most experiments 6 times, one time with each distribution in Table 1. For card games, that means that the same hand is played 6 times, once with each possible arrangement of cards. For Chinese checkers, this varies who goes first, and what player type goes before and after you.

Similarly, in a 4-player game there are $2^4 = 16-2 = 14$ ways to assign player types, and in a 6-player game there are $2^6 = 64-2 = 62$ ways to assign player types.

For $\max^n$, we implemented a tie-breaking rule that assumes our opponents will break ties to give the player at the root the worst possible score. Without this tie-breaking rule, $\max^n$ plays much worse.

### 6.1    Chinese Checkers

To simplify our experiments in Chinese checkers, we used a slightly smaller board than is normally used. On the smaller board each player has 6 pieces instead of

**Table 1.** The six ways to assign paranoid and $\max^n$ player types to a 3-player game.

|   | Player 1 | Player 2 | Player 3 |
|---|----------|----------|----------|
| 1 | $\max^n$ | $\max^n$ | paranoid |
| 2 | $\max^n$ | paranoid | $\max^n$ |
| 3 | $\max^n$ | paranoid | paranoid |
| 4 | paranoid | $\max^n$ | $\max^n$ |
| 5 | paranoid | $\max^n$ | paranoid |
| 6 | paranoid | paranoid | $\max^n$ |

the normal 10 pieces. Despite this smaller board, a player will have, on average, about 25 possible moves (in the 3-player game), with over 50 moves available in some cases.

Besides reducing the branching factor, this smaller board also allowed us to create a look-up table of all possible combinations of a single players pieces on the board, and an exact evaluation of how many moves it would take to move from that state to the goal. The table is the solution to the single-agent problem of how to move your pieces across the board as quickly as possible. This makes a useful evaluation for the two-player version of Chinese checkers. However, as additional players are added to the game, this information becomes less useful, as it doesn't take into account the positions of one's opponents on the board. It does have other uses, such as measuring the number of moves a player would need to win at the end of the game.

Because only one player can win the game, Chinese checkers is a zero-sum, or constant-sum game. However, within the game, the heuristic evaluation is not constant-sum. Our heuristic evaluation was based on the distance of one's pieces from the goal. This means that we cannot use any simple techniques to prune the $\max^n$ tree. This combined with the large branching factor in the game makes $\max^n$ play Chinese checkers rather poorly.

In our 3-player experiments we played 600 games between the $\max^n$ and paranoid algorithms. To avoid having the players repeat the same order of moves in every game, some ties at the root of the search tree were broken randomly. We searched the game tree iteratively, searching one level deeper in each successive iteration.

We report our first results at the top of Table 2. We played 600 games, 100 with each possible configuration of players. If the two algorithms played evenly, they would each win 50% of the games, however the paranoid algorithm won over 60% of the games it played.

Another way to evaluate the difference between the algorithms is to look at the state of the board at the end of the game and measure how many moves it would have taken for each player to finish the game from that state. When tabulating these results, we've removed the player who won the game, who was 0 moves away from winning. The paranoid player was, on average, 1.4 moves ahead of the $\max^n$ player.

**Table 2.** Chinese checkers statistics for $\max^n$ and paranoid.

| Experiment | Statistic | Paranoid | $\text{Max}^n$ |
|---|---|---|---|
| 3-player 250k nodes | games won | 60.6% | 39.4% |
|  | moves away | 3.52 | 4.92 |
|  | search depth | 4.9 | 3.1 |
| 4-player 250k nodes | games won | 59.3% | 40.7% |
|  | moves away | 4.23 | 4.73 |
|  | search depth | 4.0 | 3.2 |
| 6-player 250k nodes | games won | 58.2% | 41.8% |
|  | moves away | 4.93 | 5.49 |
|  | search depth | 4.6 | 3.85 |

Finally, we can see the effect the paranoid algorithm has on the search depth. The paranoid player could search ahead 4.9 moves on average, while the $\max^n$ player could only look ahead 3.1 moves. This matches the theoretical predictions made in section 3.2; Paranoid is able to look ahead about 50% farther than $\max^n$.

We took the same measurements for the 4-player version of Chinese checkers. With 4 players, there are 14 configurations of players on the board. We played 50 games with each configuration, for a total of 700 games. The results are in the middle of Table 2. Paranoid won 59.3% of the games, nearly the same percentage as in the 3-player game. In a 4-player game, paranoid should be able to search 33% farther than $\max^n$, which these results confirm, with paranoid searching, on average, 4-ply into the tree, while $\max^n$ was able to search 3.2-ply on average. Finally, the paranoid players that didn't win were 4.23 moves away from winning at the end of the game, while the $\max^n$ players were 4.73 moves away. In the 4-player game some players share start and end sectors, meaning that a player can block another player's goal area, preventing them from winning the game. This gave $\max^n$ a chance to get closer to the goal state before the game ended.

In the 6-player game, we again see similar results. We played 20 rounds on each of 64 configurations, for 1280 total games. Paranoid won 58.2% of the games, on average 4.93 moves away from the goal state at the end of the game, while $\max^n$ was 5.49 moves away on average. In the 6-player game, we expect paranoid to search 20% deeper than $\max^n$, and that is the case, with $\max^n$ searching 3.85 moves deep on average and paranoid searching 4.6 moves on average.

The $\max^n$ algorithm has an extremely limited search, often not even enough to look ahead from its first move to its second. This means that, although $\max^n$ can, in theory, use transposition tables in Chinese checkers, it is unable to in practice because it cannot even search deep enough to cause a transposition to occur.

Because of this, we conducted another experiment with the 3-player games. In this experiment we again played 600 total games, limiting the branching factor for each algorithm, so that only the 6 best moves were considered at each branch. We chose to limit the branching factor to 6 moves because this will allow reasonable depth searches without an unreasonable limitation on the possible moves. If we limited the branching factor to just 2 moves, there wouldn't be enough variation in moves to distinguish the two algorithms.

**Table 3.** 3-player Chinese checkers statistics for $max^n$ and paranoid.

| Experiment | Statistic | Paranoid | $Max^n$ |
|---|---|---|---|
| 250k nodes, | games won | 71.4% | 28.6% |
| fixed | moves away | 2.47 | 4.4 |
| branching factor | search depth | 8.2 | 5.8 |
| fixed depth | games won | 56.5% | 43.5% |
| search | moves away | 3.81 | 4.24 |

The results from these experiments are found in Table 3. Under these conditions, we found that paranoid did even better than $max^n$, winning 71.4% of all the games even though $max^n$ was able to search much deeper than in previous experiments. The paranoid algorithm could search 8.2 moves deep as opposed to 5.8 for $max^n$. At the end of the game, paranoid was, on average, only 2.47 moves away from finishing, as opposed to 4.4 for $max^n$.

Finally, we played the algorithms against each other with a fixed depth search. In this experiment, both algorithms were allowed to search 4-ply into the tree, regardless of node expansions. In these experiments the paranoid algorithm again was able to outperform the $max^n$ algorithm, albeit by lesser margins. Paranoid won 56.5% of the games played, and was 3.81 moves away at the end of the game, as opposed to 4.24 moves for $max^n$.

These results show that the paranoid algorithm is winning in Chinese checkers both because it can search deeper, and because its analysis produces better play. We would expect similar results for similar boards games.

## 6.2   Perfect Information Card Games

For the card games Hearts and Spades we deal a single hand and then play that same hand six times in order to vary all combinations of players and cards. If $max^n$ and paranoid play at equal strength, they will have equal scores after playing the hand 6 times. For both games we used a node limit of 250,000 nodes per play. These games were played openly allowing all players to see all cards.

For the 3-player games of Hearts and Spades we played 100 hands, 6 times each. In Hearts we also run experiments with the 4-player version of the game. For the 4-player game we also used 100 hands, played 14 times each for arrangement of players, for 1400 total games. Our search was iterative, as in Chinese checkers. But, since points are only awarded when a trick is taken, we didn't search to depths which ended in the middle of a trick. We used a hand-crafted heuristic to determine the order that nodes were considered within the tree. This heuristic considered things like when to drop trump in Spades, and how to avoid taking the Queen of Spades in Hearts.

## 6.3   Hearts

The top of Table 4 contains the results for Hearts. Over these games, the paranoid player had an average score of 8.1 points, while the $\text{max}^n$ player had an average score of 8.9 points. The standard deviation of the scores was 1.3, so these results are close, but paranoid has a definite advantage. There are about 26 points available in the game (in the 3-player version, one card is taken out of the deck randomly), so if the algorithms played with equal strength, they would have averaged about 8.5 points each. The paranoid algorithm could search to depth 15.2 on average, while the $\text{max}^n$ algorithm could only search to depth 11.0 on average. The paranoid algorithm is searching close to 50% farther in the tree than $\text{max}^n$, as expected for a 3-player game.

**Table 4.** Games won in Spades and Hearts by $\text{max}^n$ and paranoid.

| Experiment | Statistic | Paranoid | $\text{Max}^n$ |
|---|---|---|---|
| 3-player | average score | 8.1 | 8.9 |
| Hearts | search depth | 15.2 | 11.0 |
| 250k nodes | vs. heuristic | 5.6 | 5.6 |
| 4-player | average score | 6.45 | 6.55 |
| Hearts | search depth | 14.3 | 11.2 |
| 250k nodes | vs. heuristic | 4.3 | 4.2 |
| Spades | average score | 5.67 | 5.67 |
| 250k nodes | search depth | 15.4 | 10.6 |
|  | vs. heuristic | 6.06 | 6.12 |

To compare both algorithms against a different standard, we also set up a different experiment that pitted each algorithm against a player that did no search, but just used the node-ordering function to pick the next move. $\text{Max}^n$ and paranoid were allowed to search just 6-ply (2 tricks) into the game tree, but only took 5.6 points in an average game. This confirms that the search is doing useful computation and discovering interesting lines of play above and beyond what a simple heuristic can do.

In the 4-player game the algorithms are more closely matched. Paranoid did just slightly better, averaging 6.45 points per hand as opposed to 6.55 for $\text{max}^n$. The standard deviation per round was 0.67 points. Paranoid was able to search depth 14.3 on average, about 33% farther than $\text{max}^n$, which could search depth 11.2. In this case, it seems that the extra search depth allowed by paranoid is being offset by the (incorrect) assumption that our opponents have formed a coalition against us. Both players were again able to easily outperform the heuristic-based player, scoring, on average, just over 4 points per game.

## 6.4   Spades

In the actual game of Spades, players bid on how many tricks they are going to take, and they then play out the game, attempting to take exactly that number of tricks. We have experimented with the first phase of that process, attempting to ascertain how many tricks can be taken in a given hand, and we do this by playing out a game, trying to take as many tricks as possible.

The bottom of Table 4 contains the results. Over these games, both players had an average score of 5.67 points, which is what we expect for algorithms of equal strength, given the 17 points (tricks) in the 3-player game. However, the paranoid algorithm is searching 15.4-ply deep, on average, while $max^n$ is only searching 10.6-ply deep. This means that paranoid can look ahead nearly 2 tricks more than $max^n$, about 50% deeper.

This is an interesting result because the paranoid algorithm is able to search deeper than $max^n$ by a wider margin than in it can in Hearts, but it is still not able to outperform $max^n$. However, looking at the results from play against a heuristic player makes this more clear. When $max^n$ and paranoid played a heuristic based player that does no search, they were barely able to outperform it. This indicates one of two things: Either the process of estimating how many tricks can be taken in a three-player game is very simplistic, or neither algorithm is able to use its search to devise good strategies for play.

We expect that both are happening. In games like Bridge, a lot of work is spent coordinating between the partners in the game. With no partners in 3-player spades, there is little coordination to be done, and the tricks are generally won in a greedy fashion. Our own experience playing against the algorithms ourselves also lead us to this conclusion. Often times the only decision that can be made is how to break ties in giving points to the other players.

Note that these results are only for the bidding portion of the game. The scoring of the actual game play is somewhat different, with players trying to make their bid exactly. This changes the rules significantly, and may give one algorithm an edge.

# 7   Conclusion and Future Work

Our results show that the paranoid algorithm should be considered as a possible multi-player algorithm, as it is able to easily outperform the $max^n$ algorithm in checkers and slightly less so in Hearts. However, our current results from Hearts and Spades indicate that the $max^n$ algorithm is more competitive than we expected in these domains.

As a general classification, we can conclude that in games where $max^n$ is compelled to do a brute-force search, the paranoid algorithm is a better algorithm to use, while games in which the $max^n$ algorithm can prune result in more comparable performance. Another factor that differs between Chinese checkers and card games is that in card games your opponents can easily team up and work together. This is more difficult in Chinese checkers, because if they block one piece you can simply move another, and the limited search depth prevents the formation of more complicated schemes. This makes the paranoid assumption more reasonable.

There are several areas into which we are directing our current research. First, we need to consider the development of additional algorithms or pruning techniques that will result in improved performance. We are also working on optimizing the techniques and algorithms we have already implemented. Next, and more importantly, the issue of imperfect information needs to be addressed. Paranoid and $max^n$ played similarly in card games, which are always games of imperfect information. As there is no clear edge for either algorithm, the point that will differentiate these algorithms is how well they can be adapted for games of imperfect information. Finally, we need to concern ourselves not just with play against different algorithms, but with play against humans. We are in the beginning stages of attempting to write a Hearts program that is stronger than any existing program, and one that also plays competitively against humans.

## Acknowledgments

We wish to thank Rich Korf for his guidance in developing this work, Haiyun Luo and Xiaoqiao Meng for the use of their extra CPU time to run many of our experiments, and the reviewers for their valuable comments and suggestions.

## References

1. Hsu, F.: Behind DEEP BLUE. Princeton University Press (2002)
2. Schaeffer, J., Culberson, J., Treloar, N., Knight, B., Lu, P., Szafron, D.: A world championship caliber checkers program. Artificial Intelligence **53** (1992) 273–290
3. Ginsberg, M.: GIB: Imperfect information in a computationally challenging game. Journal of Artificial Intelligence Research **14** (2001) 303–358
4. Luckhardt, C., Irani, K.: An algorithmic solution of N-person games. In: Fifth National Conference of the American Association for Artificial Intelligence (AAAI-86), AAAI Press (1986) 158–162
5. Sturtevant, N., Korf, R.: On pruning techniques for multi-player games. In: Sixteenth National Conference of the American Association for Artificial Intelligence (AAAI-00), AAAI Press (2000) 201–207
6. Billings, D., Davidson, A., Schaeffer, J., Szafron, D.: The challenge of poker. Artificial Intelligence **134** (2002) 201–240
7. Hoyle, E., Frey, R., Morehead, A., Mott-Smith, G.: The Authoritative Guide to the Official Rules of All Popular Games of Skill and Chance. Doubleday (1991)
8. Korf, R.: Multiplayer alpha-beta pruning. Artificial Intelligence **48** (1991) 99–111
9. Knuth, D., Moore, R.: An analysis of alpha-beta pruning. Artificial Intelligence **6** (1975) 293–326
10. Pearl, J.: Asymptotic properties of minimax trees and game-searching procedures. Artificial Intelligence **14** (1980) 113–138
11. Ginsberg, M.: Partition search. In: Thirteenth National Conference of the American Association for Artificial Intelligence (AAAI-96), AAAI Press (1996) 228–33

# Selective Search in an Amazons Program

Henry Avetisyan and Richard J. Lorentz

Department of Computer Science, California State University
henry.avetisyan@cp.net, lorentz@csun.edu

**Abstract.** Throughout at least the first two thirds of an Amazons game the number of legal moves is so great that doing a full width search is impractical. This means we must resort to some sort of selective search and in this paper we study a number of algorithms to perform forward pruning. In particular we describe techniques for selecting from the set of all legal moves a small subset of moves for look-ahead evaluation, eliminating all other moves from consideration. We then study the effects of these techniques experimentally and show which have the most potential for producing the best Amazons-playing program.

## 1 Introduction

Most game playing programs that play as well as or better than the best human players are strong because they have the luxury of being able to perform full width searches down many levels of the search tree. Examples of such programs are DEEP BLUE in chess [1], CHINOOK in checkers [2], and LOGISTELLO in Othello [3].

However, there are a number of important and interesting games where this cannot be done because the number of available moves at each ply is too large. Probably the best known of these games is Go, where the best programs are playing at the level of an intermediate amateur [4] (well under 1-dan) and far below that of the best humans. Another example is shogi [5], where the best programs are playing at the level of a strong amateur [6], but far from world-class strength.

A third example is Amazons. Amazons is a fairly new game, invented in 1988 by Walter Zamkauskas of Argentina and is a trademark of Ediciones de Mente. In the early stages of a game of Amazons there are hundreds of legal moves from which to choose. Hence, one of the greatest challenges of writing a program to play this game well is deciding which, of the many available moves, deserves to be considered for look-ahead evaluation and possible selection as the best move. Also, since the game is quite new, there is no collection of literature and experience to help guide the player (and the programmer) in choosing good moves. There are no opening books. There are no well-understood middle game strategies. Even the endgame, though seemingly simpler than the other phases of the game, still poses its share of difficulties. There are "defective shapes" [7] to avoid and in the general case Amazon endgames have actually been proven to be NP-Complete [8].

J. Schaeffer et al. (Eds.): CG 2002, LNCS 2883, pp. 123–141, 2003.
© Springer-Verlag Berlin Heidelberg 2003

Our work here focuses mainly on the parts of the game where there are still many moves available and we must decide which of these moves require further study, that is, which moves will be selected for look-ahead evaluation. Briefly, our experiments indicate that in general 50% or more of the possible moves need never undergo a time consuming evaluation and many other moves can be eliminated through a pre-processing step that utilizes a faster but less precise evaluation function and so these time savings can be used for deeper look-ahead evaluation.

In Sect. 2 we discuss the game of Amazons including the rules, the number of legal moves available as the game progresses, and possible evaluation functions. In Sect. 3 we explain the different methods employed in selecting candidate moves and how well each of these methods performed experimentally. In the final section we attempt to explain the experimental results and discuss ideas for future work.

## 2    The Game of Amazons

Amazons is usually played on a $10 \times 10$ board where each player is given 4 amazons initially placed as shown in Fig. 1(a). Although the game may be played on other sized boards and with different numbers and initial placements of amazons, it is this version of the game that is most often played and so will be the only version that will be discussed here.

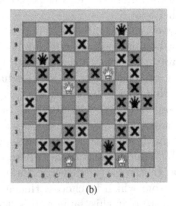

(a)                                              (b)

**Fig. 1.** The initial position and a typical position 30 moves into the game.

A move comprises two steps: first an amazon is moved like a chess queen through unoccupied squares to an unoccupied square. Then from this new location the amazon "throws an arrow" again across unoccupied squares to an unoccupied square where it stays for the remainder of the game. The first player who cannot make a legal move loses. We will refer to the different sides as white and black, where white moves first and appears at the bottom of the original position in our diagrams.

For example, from the initial position white may move the amazon from square D1 to square D7 and then throw an arrow to square G7. This move is denoted D1-D7(G7). Similarly D6-G9(G10) is a legal move in Fig. 1(b).

The feature of this game that makes it so interesting is that throughout most of the game there are many legal moves available. For example, from the initial position there are 2176 legal moves. Even in Fig. 1(b), after 30 moves have been made and the board is noticeably crowded with arrows, white still has 274 legal moves. Because of these extremely high numbers we must be quite selective of our moves before we can begin doing look-ahead.

We collected 100 games played by our program, INVADER, and recorded the number of legal moves at each turn. It turns out that the number of legal moves for white is always a bit higher than for black, so it is clearer to show the results separately, which we do in Figs. 2 and 3.

Note that even as late as move 40 both white and black, on average, still have more than 100 legal moves. But for most Amazons games, by move 40 it is already determined who will win the game, so it is hopeless to expect to ever be able to do a full width search at any important stage of the game with current computer technology.

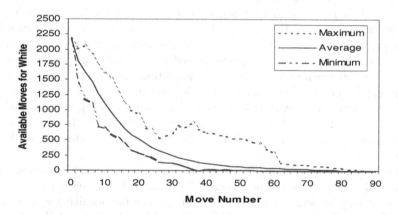

**Fig. 2.** The maximum, average, and minimum number of moves possible from 100 test games for the white player.

When programming any game, including Amazons, finding a good evaluation function is, of course, very important. Though not all Amazons programmers are completely forthright when discussing their evaluation functions, we suspect most use something similar to what has been called *minimum distance* [7]. The idea is that for each empty square on the board we determine in how few moves an amazon of each color can arrive at that square. For example, square A6 in Fig. 1(b) can be reached in a minimum of two moves by black (by the amazon on

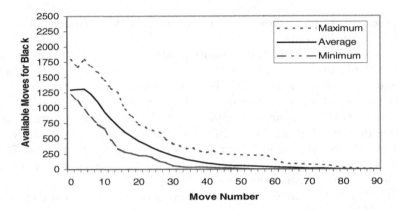

**Fig. 3.** The maximum, average, and minimum number of moves possible from the same 100 test games for the black player.

B8) and in a minimum of three moves by white (by either amazon D6 or G7) so for that square the evaluation gives the edge to black. In doing these calculations we ignore the arrow-throwing portion of the amazon move. Since an amazon can always throw an arrow to the square it just vacated, arrow throwing does not affect the minimum distance.

If some square cannot be reached at all by one color but can be reached by the other, such as J10 in Fig. 1(b), then we say that square is owned by the color that can reach it. The total board evaluation is then determined by various weighted sums of the squares owned by each color and the squares that each color can reach more quickly. The weighting can either simply count how many squares each player gets to first, as is suggested in [7], or take into account how many steps closer each side is to the squares it can reach first, as is done in [9]. We do the latter. In particular, if a square is the same distance from both players then the square is considered neutral and does not affect the evaluation value.

Minimum distance evaluation has proven to be quite good for the late middle game and beyond, but it is not so clear how useful it is for earlier stages of the game. In fact, experience now tells us that other factors like the mobility of the amazons plays a more important role in the early stages of the game. We suspect that there are other, even more important factors yet to be discovered. Nevertheless, INVADER is able to play a reasonably strong game even when the evaluation is based entirely on the minimum distance evaluation, so for the purposes of this research we assume all evaluations are done in this manner.

## 3   Forward Pruning

The main purpose of this study was to investigate algorithms for selecting candidate moves when there are hundreds, or even thousands, of legal moves available. Performing a full-width search is clearly too slow. On a 300 MHz Pentium II machine doing a full-width search with 30 seconds per move, INVADER was unable to complete a depth two search until the legal move count dropped under 300. Looking at Figs. 2 and 3 we see this does not usually occur until after move 20 so one would expect the quality of play to be very poor. Indeed, games played by INVADER set to do a full-width search were of such low quality that there is no need to discuss this further.

In the case of Amazons, deciding how well a particular algorithm is performing is not necessarily easy to judge for a number of reasons. First, as noted earlier, Amazons is a very new game, so there is no body of knowledge about the game that we can draw on. For more established games with centuries of history one can judge the quality of moves based on acquired knowledge of the game. Similarly, a standard technique for evaluating the quality of moves made by programs is to replay published games played by experts and then see how closely the program is able to mimic expert play. In the case of Amazons there are virtually no published games of any quality to draw on and there are very few known human experts. Certainly the authors do not consider themselves experts in the game.

Another difficulty is that even though INVADER is playing a fairly strong game, it is still sufficiently weak that it can make disturbingly bad moves at unexpected moments. Hence, even though Version A, say, might be making better moves than Version B on average, however we might be measuring "better," Version A might occasionally make catastrophic moves. It then becomes problematic as to which version we really prefer. Related to this difficulty is the fact that there are very few other Amazons programs available so all of our tests by necessity are based on results of self-play games and games against the authors.

Summarizing, we are basically working in a vacuum with very little knowledge about the game, so we must be careful that our conclusions about the quality of play are correct. Since our goal is to be able to select a good candidate move list as quickly as possible, we judge the relative merits of two versions by comparing: 1) the number of nodes that each version is able to visit when making a move; 2) the candidate moves actually selected; and 3) the quality of the game played, looking especially for particularly good or bad moves. Exactly how we measure these comparisons is described in detail in the next section.

## 3.1 The Basic Approach

Given that we must somehow choose candidate moves from a large list, the most obvious and direct approach is to simply use the evaluation function. At each ply we evaluate every possible move and then perform look-ahead on the top few. Exactly how many of the top moves to choose at each level is by no means clear. We have tried numerous settings, such as looking at the top 40 moves at the first level and then decreasing the number of moves at lower levels by 5 or 10 moves per ply. We have also tried selecting more moves at the top level and then dramatically dropping the number of moves at lower levels. In any case, we have not obtained convincing results as to what the proper numbers should be. It surely depends on many factors, including the particular algorithms being used. So, for the sake of consistency, for this and every other selection algorithm we study, we simply select the top 20 moves at each level. We have found this to be a reasonable setting for most cases. Also, we selected a time limit of 30 seconds per move allowing us to visit a fair number of nodes for each search, but still allowing us to finish a game in a reasonable amount of time, typically about 30 minutes. At 30 seconds per move we usually manage to search at least through depth 3, so the quality of the moves being made is adequate.

Since this basic approach by design must sort the various moves based on evaluation values, it has the additional benefit of increasing the chances of alpha-beta cutoffs. Alpha-beta is, of course, implemented as part of our basic approach, but we supplement it further by using the NegaScout algorithm [10].

## 3.2  Amazon-Piece-Based Efficient Move Selection

The first technique we study attempts to speed up the generation of candidate moves by ignoring certain legal moves. The moves that are actually considered are based solely on evaluations calculated after an amazon moves, but before an arrow is thrown. We will discuss speed-up attempts based on arrow positions in Sect. 3.3.

Specifically, this is what we do. Recall that an amazon move actually comprises two discrete steps: first an amazon moves to a new location, then it fires an arrow. The idea of the amazon piece (as opposed to arrow) approach to forward pruning is that the location of the amazon is important regardless of where she eventually throws her arrow. So, if we evaluate the board position based only on the location of the amazon move we get a good approximation of the value of any move that is completed from this first step. If the evaluation reports a low value for the first step of the amazon move we don't bother throwing any arrows from that location and so no moves associated with that amazon position are even considered. Among all the possible places an amazon may relocate we keep a certain fraction of those that evaluate the highest, then complete these moves in all possible ways, and evaluate these board positions to decide which moves make it to the final candidate list. Since the other moves are not completed many evaluations are saved. Exactly what fraction of the possible moves we actually keep is part of the experiment that is reported in this section.

We have to decide if generating a partial list of moves is an improvement over generating the full list. The main reason for only looking at some of the amazon moves is to reduce the number of board evaluations that need to be calculated since evaluation tends to be quite time consuming and we are eliminating hundreds of evaluations. This extra time can then be used for tree searching. But, having more time for tree searching is of no use if the quality of the candidate moves has degraded significantly. It is easy to count the number of nodes the program visits while choosing a move to make, so measuring this is completely quantifiable. But it is not so easy to measure the quality of the candidate move list.

We attempt to determine the quality of the candidate move lists by looking at two things. First, we assume that the basic approach from Sect. 3.1 that looks at every possible move does as good a job of selecting candidate moves as can be expected from an evaluation-based approach. Then we look at the candidate moves found by generating only a partial list and see how many of these moves appear in the candidate move list generated by looking at the full list of amazon moves. The more moves they have in common, the better the new approach must be doing. Looking at extreme cases, for example, the idea is that if every move in the candidate list created using a fraction of the total number of amazon moves also appears in the list created using all the moves, then no information has been lost but extra time has been provided for searching. On the other hand, if only one or two moves on the candidate list from the new approach also appear in the list from the basic approach, then the new version is probably looking at the wrong moves and so the extra time provided is really of no benefit.

The other thing we look at is the move actually made by the original approach and see if it appears in the candidate list from the partial move generation approach. The idea here is that if this move remains in the new list then the new approach has the opportunity to choose this move as well, but should it actually find a different move given its extra search time, so much the better.

We conducted an experiment where we generated moves from six different games. First we generated moves in the normal manner, where we used all amazon moves to select the candidate moves. Then from the same positions we had the program find moves where only one half, one quarter, and one eighth of the total amazon moves were used to generate the candidate list. We counted how many nodes the search was able to visit in each case and show the results in Fig. 4. The graph reports the average over the six games of the number of nodes visited using each of the different selection schemes. It should be noted that because of certain idiosyncrasies of our implementation the node counts reported are the numbers of internal nodes visited – leaf nodes are not counted. But since we are only interested in the relative counts between different versions, this is of no consequence.

This graph reveals a number of interesting and useful facts. First we notice that all four graphs seem to intersect at roughly the same point, around move number 35. What this says is that no matter what fraction of moves is selected, eventually the overhead of evaluating and sorting partial amazon moves to create the list outweighs the benefit it provides. In other words, after move 35 it is always best to simply evaluate every possible complete amazon move and choose the candidates from this list. The reason for this is clear. Looking back at Figs. 2 and 3 we see that by this stage of the game there are on average "only" about 250 legal moves but probably at least 40 possible locations an amazon may move to, so the cost of evaluating these amazon moves, sorting them, and then generating moves from this list is simply too much overhead.

The other main thing to be noticed is that by using partial lists we do indeed save a lot of time. Through the first 20 moves we see that by looking at only half of the amazon moves we are able to visit approximately 50% more nodes in the search. The gains are comparable as we reduce the fraction down to one quarter and one eighth. It turns out there is no point in selecting a fraction much below

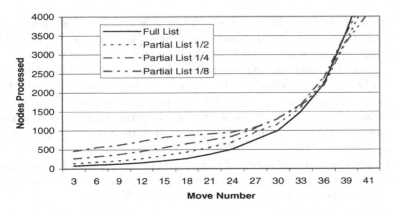

**Fig. 4.** Number of nodes searched keeping all, $\frac{1}{2}$, $\frac{1}{4}$, and $\frac{1}{8}$ of the total number of possible amazon locations.

one eighth because we insist that we always look at a minimum of 5 amazon moves per amazon and if the fraction is much larger, it will simply mean that we will always just look at the 5 best moves.

Now we must ask how good are the candidate moves calculated by the three partial lists? Using the same six games we counted how many moves from the original candidate move list, created by making every possible amazon move, appeared on the lists created by just looking at a fraction of the amazon moves. We only looked at the first 40 moves in each game since after move 40 we've already seen that partial move generation has no advantage. Also, to ensure all six games were different, except for the first game each game began with some fixed two move opening. This gave us 40 moves from the first game and 38 moves from the other 5 games for a total of 230 moves to look at for both of the players.

Table 1 shows the results expressed as percentages. For example, looking at the second column we see that by keeping half of the possible amazon moves, the candidate move list is off by no more than two moves 90% of the time. On the other hand, looking at the last two columns we see that by keeping one fourth and one eighth of the moves the candidate lists are off by two moves or less only 74% and 50% of the time, respectively.

We also counted how many times the move actually selected using the full amazon list did not appear on the other lists. In the case of the half list it only happened 9 times. This means that 96% of the time the program still could have selected the move the original version liked. But since it actually had more time available to decide, we conclude that it judged the move it made at least as good if not better than the one chosen the other way.

In the case of the one quarter list, there were 20 occurrences where the original program's best move did not appear at all on the candidate list meaning that 91% of the time the move was available. For the one eighth list it happened 32 times meaning 86% of the time the original best move could have been chosen.

**Table 1.** Number of candidate moves in common between the basic approach and the three versions of amazon-piece-based pruning.

| Match (out of 20) | Percentage of Moves Matched | | |
|---|---|---|---|
| | 1/2 List | 1/4 List | 1/8 List |
| 20 | 37.0 | 28.6 | 21.6 |
| 19 | 39.6 | 27.7 | 16.0 |
| 18 | 13.9 | 17.3 | 11.2 |
| 17 | 5.2 | 6.5 | 12.8 |
| 16 | 5.2 | 5.6 | 8.0 |
| 15 | 3.0 | 6.5 | 4.0 |
| 14 | | 1.7 | 3.2 |
| 13 | 0.4 | 2.6 | 4.4 |
| 12 | 0.4 | 1.7 | 3.2 |
| 11 | | | 5.2 |
| 10 | | 0.9 | 2.0 |
| 9 | | | 2.8 |
| 8 | | | 1.6 |
| 7 | | 0.4 | 1.2 |
| 6 | | | 2.4 |
| 5 | | 0.4 | |
| 4 | 0.4 | | |
| 3 | | | |
| 2 | | | 0.4 |
| 1 | | | |

Based on these results it seems clear that the small penalty for looking at only half of the list is easily outweighed by the 50% extra search time it provides. On the other hand, even though we gain about 250% extra time by limiting the lists to one eighth of the amazons, the cost seems too high since we lose the top move (as calculated by the original version) 14% of the time. Further, 50% of the time the candidate move list differs from the original by at least 3 moves, and 10% of the time not even half of the candidate move lists match.

It is not so clear what the situation is when looking at one quarter lists. In this case we try to do some additional analysis by looking at actual games. But we must treat such analysis with care. Our experience with actual games shows that self-play data can be misleading for a number of reasons. First, it is often difficult for us to determine the relative merits of moves, especially in the opening and middle games. Second, sometimes one program will tend to make what appear to be better moves for the majority of a game, but then make a particularly bad move at a critical juncture. In such a case, which version do we really prefer? Finally, since we are using a rather primitive evaluation function the results are limited to what can be determined by this evaluation. Nevertheless, we feel some useful information can be gained by looking at actual games.

Consider the two positions in Fig. 5. In position 5(a) where white is to move the full list version made the move D5-G5(G4). Clearly the intent is to try to establish territory on the right side of the board. Indeed black's access to that area is limited after that move. However, the one quarter version did not have this move on its candidate list. Instead it chose the move G2-I2(G4), apparently a weaker move because the territory on the right side is not as secure since black now has three entry points, one from the amazon on G8, another from the amazon on F4, and still another from the amazon on E2. Of course the analysis is not quite that simple since these moves do have an effect on the lower left side as well, but in this case that does not appear to be significant.

(a)

(b)

**Fig. 5.** Comparing moves made by the full list and the one quarter list players.

In position 5(b) where it is again white's move the full list version selected F4-H6(G6). Here it is clear the intent is to secure the upper right corner. However, again the one quarter list version did not have this move on its candidate list, so it selected F8-E8(G6) instead. Unfortunately in this case it actually seems to be a superior move since it secures the territory about as well as the other move, but the four white amazons appear much freer to move about the board and stake out more territory. Of course part of the problem is that the minimum distance evaluation does not have a good sense of what territory is, so from the minimum distance point of view, the two moves are quite similar and each version eventually chose its move based on what was seen during look-ahead. So it is really by pure chance that occasionally the one quarter and even the one eighth list versions manage to find good moves even in situations where the candidate list appears to have little in common with the full list version. Chances would seem to increase when there are many moves available, increasing the probability that a good move might sneak in, and indeed we do see this phenomenon more often in the early stages of the game.

Summarizing, using one half lists is clearly advantageous, using one eighth lists is probably not, and the benefits of one quarter lists is unclear. In the case of INVADER the extra time provided when using one quarter lists does not seem to be sufficient to compensate for the occasional lost moves, so for now we have chosen to implement one half lists. However this could easily change as new evaluation functions are developed.

## 3.3   Arrow-Based Efficient Move Selection

In the last section we pruned moves based solely on the location on the amazon. Here we try a complementary approach, where pruning is based on the location of the arrows that are thrown. The idea is that when an arrow is thrown to a particular square it will tend to have the same effect on the board evaluation regardless of the location of the amazon that threw it. That is, the value of an arrow on the board is independent of the location of the amazon that threw it. So, if for each empty square on the board we compute the board evaluation assuming an arrow is on that square, later when we are selecting our candidate list we can use these values instead of re-evaluating the board for every amazon move and arrow throw.

This is actually a three-step process. The first step is to compute evaluations for every empty square on the board by assuming an arrow has been placed on that square. The second step is to generate amazon moves and arrow throws for possible selection in the candidate move list. When generating amazon moves we use the technique described in the last section where we select one-half of the total amazon moves since that is clearly superior to looking at every possible amazon move. Then, when generating arrow throws for these amazon moves, instead of evaluating the board for each throw we simply take the value that has already been computed for an arrow corresponding to that throw. Even though this evaluation may not be very accurate since it does not take into account the fact that an amazon has moved, the value relative to the other arrow throws from this same position should be fine. So, if we keep a certain number, say five, of the best arrow throws for each possible amazon move we are still likely to get most of the good moves without having to do a separate evaluation for each position.

Since there are usually more than four legal amazon moves, the list created in the second step typically has more than the 20 moves on it. So, to end up with exactly 20 candidate moves the final step of this technique requires that we take all the moves that have been collected so far and do a full evaluation on each of them and use the result of this evaluation to then select the top 20 moves for the candidate list. But since this list is significantly smaller than the total number of legal moves, we still end up doing considerably fewer evaluations.

To study this algorithm we performed the same experiment as described in the last section. We generated moves from six different games in three different ways: (1) using half of the amazon moves and all of the arrow throws to generate the top twenty candidate moves; (2) using the top ten arrow throws for each amazon move; and (3) using the top five arrow throws for each amazon. Figure 6 shows the results in terms of the numbers of nodes we were able to process in each case.

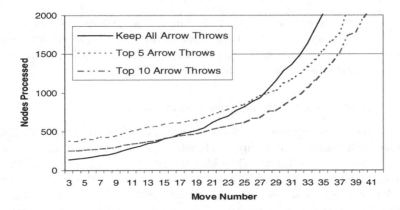

**Fig. 6.** Number of nodes searched keeping all possible arrow throws, the top five arrow throws, and the top ten arrow throws.

The most obvious difference between Figs. 6 and 4 is that in Fig. 6 the lines do not intersect near the same point. So for the case where we keep ten arrow throws we see that already by move 16 there is no time savings, hence the benefits of keeping ten arrows is already suspect even before we study the quality of the moves. Keeping more than ten arrows certainly will not pay off since the intersection will move even further to the left and the extra nodes searched will be even fewer. Keeping just five arrows, however, does show significant time savings in the early part of the game and does not actually cost extra time until around move 28, so this version seems to have the greater potential benefit.

We measure the quality of the candidate moves in the same way we did in the last section. We see how many moves in the candidate list created when all arrows are kept also appeared in the candidate lists of the two versions that kept only some of the arrows. Also we count how many times the move actually selected when keeping all arrows does not appear on the lists when only some of the arrow throws are kept.

Table 2 shows the results of the first measure expressed as percentages. We notice a big difference between keeping five and keeping ten arrow throws. In the case where ten arrows are kept we see a reasonable match between the candidate move lists. In two thirds of the cases no more than two moves from the candidate list where all arrows are considered were found missing from the candidate move list. However, in the case where only five arrows are kept, two thirds of the time as many as seven moves might be missing from the list. But even worse is the fact that only 14% of the time does the candidate list manage to miss two or fewer moves.

Counting how many times the move selected when all arrows are kept does not appear on the new candidate list is similarly revealing. In the case where ten arrows are kept, out of a total of 230 moves considered the chosen move does not appear on this candidate move list only 15 times. That is, 93% of the time it could have chosen the move selected by the original version. Of course, it did

**Table 2.** Number of candidate moves in common between the basic approach and the two versions of arrow-based pruning.

| Match (out of 20) | Percentage of Moves Matched | |
| --- | --- | --- |
| | Keep 10 Throws | Keep 5 Throws |
| 20 | 31.8 | 1.7 |
| 19 | 25.6 | 5.4 |
| 18 | 8.7 | 7.1 |
| 17 | 7.9 | 9.2 |
| 16 | 9.5 | 13.8 |
| 15 | 5.8 | 11.7 |
| 14 | 5.0 | 10.4 |
| 13 | 1.2 | 6.3 |
| 12 | 2.5 | 9.2 |
| 11 | 0.8 | 8.8 |
| 10 | 0.8 | 2.9 |
| 9 | | 5.4 |
| 8 | 0.4 | 2.9 |
| 7 | | 2.9 |
| 6 | | 0.8 |
| 5 | | 1.3 |
| 4 | | 0.4 |
| 3 | | |
| 2 | | |
| 1 | | |

not always make the same choice since it now also had more time to consider the move.

However, in the case where only five arrow throws are kept we find 50 occurrences where the previously chosen move does not appear on its candidate list, that is, only 78% of the time is the old best move even available for selection in this case.

Based on these results it seems clear that keeping only five arrow throws is not worth the time savings it provides. Even though it typically runs twice as fast during the first 10 moves and is still 50% faster up to about move 15, the cost in lost candidate moves is too great. However the situation is less clear when keeping ten throws. Here we typically get about a 50% time savings through the first ten moves and according to Fig. 6 we still get some savings up through about move 15. This doesn't get us very far into the game and, unfortunately, it is during a part of the game where the evaluation function is not its most useful, yet it still might be of some benefit. To help us settle the issue we looked at games played when keeping ten throws. Figure 7 shows two typical positions where the move selected by the version that keeps only ten throws differs from the move selected by the full version. It illustrates two different but telling problems.

In Fig. 7(a) the version that keeps all arrow throws made the move B2-C3(C2). This is clearly a strong move as it gives a huge piece of territory to white in the lower right corner. The version that keeps only ten throws, however, made the move D4-A7(E3). This is a noticeably weaker move since black can now gain access to that large corner area, but unfortunately there is a good reason for its choice and it reveals a weakness with this method. In this case an arrow by itself on either C2 or C3 does not do an adequate job of keeping black from encroaching on white's territory. Hence neither square evaluates an arrow very high, but when an amazon and an arrow combine to block both of the squares the strength of the move is obvious. In other words this approach is blind to the combined power of amazon and arrow. This is not so much of a problem for amazon-based-pruning because that is a sort of cumulative evaluation. It measures how much the amazon improves the position then how much more the arrow throw helps, whereas here the two effects are treated quite independently.

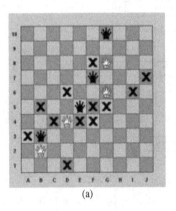

(a)                                                (b)

**Fig. 7.** Comparing moves made when keeping all arrow throws vs. keeping ten throws.

Figure 7(b) exposes another problem with this approach. From this figure the basic algorithm chose the move J7-G4(J7) while the pruning version selected C6-C2(C8). Why did it miss the first move, which seems to do a much better job of restricting white? The answer is that since the arrow is being thrown to the square that the amazon vacated, the arrow evaluation of that square is inappropriately low. In general, we currently evaluate the arrow for such a throw as the board value of the current position – as if nothing has changed. This, of course, is not correct, but it is not obvious how to correctly evaluate it. We could, of course, deal with this case separately, but there seem to be enough other weaknesses with this algorithm that we feel it will be difficult to make this approach tenable in any case. Hence, we currently feel this algorithm should not be implemented in an Amazons-playing program.

## 3.4    Evaluation-Based Efficient Move Selection

Strictly speaking this last approach is not a pruning algorithm but simply an attempt to speed up the candidate move selection process. The previous two approaches saved time by reducing the number of evaluations that needed to be performed. Now we investigate speeding up the evaluation. Even though this approach is highly dependent on the particular evaluation function being used such a study is still useful for two reasons: (1) it provides a concrete demonstration of the feasibility of the underlying concept, showing what sorts of gains might be expected, and (2) it suggests specific techniques that might be practical in an actual program. For example, the technique described in this section has been used to improve the performance of our program, INVADER.

In general the minimum distance evaluation, as described in Sect. 2, calculates the exact distance of every square from the white and from the black amazons, where this distance can be as small as one if the amazon can move to the square directly or as high as 10 or more in some complicated endgame situations. A reasonable approximation can be had by simply stopping the calculation after a certain distance, say two, and reporting all other squares as simply "more than two moves away." This is a particularly good approximation in the early stages of the game when most squares really are three or less moves away from some amazon of each color, meaning "more than two moves away" is actually equivalent to "exactly three moves away." Since this evaluation only calculates distances up to a distance of two, we call it *maximum of two*.

Given the fact that the maximum of two evaluation is significantly faster to compute than the more general minimum distance evaluation, how can we use this added efficiency without sacrificing much knowledge? We use an approach similar to what we described in the last section. First we evaluate all positions using maximum of two (after discarding half of the amazon moves using the minimum distance evaluation, as in Sect. 3.2), then we order all the moves based on this evaluation, select a certain number with the highest evaluation, and do a full minimum distance evaluation on this collection. Finally, we select the top 20 of these to be our candidate list. The most important variable in this case is the one that determines the size of the collection we perform the full evaluation on.

One overly hopeful idea is to simply select the top 20 highest evaluations so that we really do not need to perform any complete minimum distance evaluations at all, that is, the candidate list is chosen based solely on the maximum of two evaluations. Not surprisingly though, a maximum of two is sufficiently crude that one quickly sees that far too many good candidate moves are being lost, especially after the opening stages are over. Hence, we try three larger values: 40, 60, and 80 and for these values we report our results in the same way we did in the last two sections. Figure 8 shows how many nodes we were able to evaluate for each of these three cases.

Again, as in Fig. 6, we have a situation where the lines intersect at different points and the more conservative approaches provide gains for fewer moves into the game than the more aggressive ones. Still, even when we keep the top 80

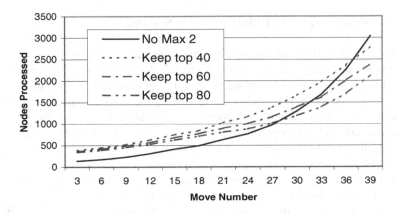

**Fig. 8.** Nodes searched when re-evaluating 40, 60, and 80 maximum two evaluations.

maximum two evaluations we still get sufficient time savings to visit about 50% more nodes up through about move 20 and gain some benefit almost to move 30.

Table 3 helps us analyze the quality of the moves. We notice that when reevaluating 60 moves our candidate lists differ by more than two moves nearly 50% of the time and when reevaluating only 40 moves we do much worse. Also, when counting how many times the move selected by the version that did not use the faster evaluation did not appear on the candidate move list we find that it happened 34 times when reevaluating 60 moves and 40 times when reevaluating 40. In other words, the move considered best by the normal version was available only 85% and 83% of the time, respectively. Because of these low values, we conclude that we need to reevaluate more than 60 moves.

Looking at the last column of the table we see that 72% of the time the lists differ by no more than two moves. Again, we counted the number of times the best move from the slower version appeared on the candidate move list and found it happened over 90% of the time. These numbers indicate that it might indeed be worthwhile to implement this algorithm, so we study games to gain some more insight.

When there are fewer arrows on the board in the beginning of the game the maximum of two evaluation is a better approximation of minimum distance than later in the game when there are many arrows on the board, because additional arrows make it more likely that there will be longer paths to squares. Hence, our first thought was that we are more likely to see different, and worse, moves made by the maximum of two version at later stages of the game, when it does a worse job of approximating minimum distance. This proved to be the opposite of what actually happens. Most of the bad moves made by maximum of two tend to be at the beginning of the game and when the moves differ from minimum distance later in the game, say around move 20, they tend not to be so bad. The two examples in Fig. 9 help explain this phenomenon.

**Table 3.** Number of candidate moves in common between the basic approach and re-evaluating some of the "maximum two" evaluations.

| Match (out of 20) | Percentage of Moves Matched | | |
|:---:|:---:|:---:|:---:|
| | Re-eval. 40 moves | Re-eval. 60 moves | Re-eval. 80 moves |
| 20 | 10.9 | 20.4 | 29.6 |
| 19 | 12.2 | 19.6 | 28.3 |
| 18 | 11.3 | 13.5 | 13.9 |
| 17 | 10.4 | 10.0 | 7.4 |
| 16 | 10.4 | 7.0 | 3.5 |
| 15 | 10.0 | 6.1 | 2.6 |
| 14 | 7.0 | 5.2 | 5.2 |
| 13 | 6.1 | 5.7 | 4.3 |
| 12 | 3.9 | 4.8 | 0.4 |
| 11 | 5.2 | 2.6 | 1.3 |
| 10 | 5.2 | 0.9 | |
| 9 | 1.7 | 0.9 | 0.4 |
| 8 | 1.3 | | 0.4 |
| 7 | 0.9 | 2.2 | 1.7 |
| 6 | 1.3 | | 0.4 |
| 5 | 0.4 | | 0.4 |
| 4 | 0.4 | 0.9 | |
| 3 | | 0.4 | |
| 2 | 0.9 | | |
| 1 | 0.4 | | |

In Fig. 9(a), the move made without using the maximum of two evaluation was H6-G6(F5), clearly a strong move, staking out a great deal of territory in the lower right. When using maximum of two, however, the move selected was H6-H3(F3), a noticeably worse move. The reason this happened is that even though maximum of two is a good approximation, it is still not exact, and in this case the better move, H6-G6(F5), evaluated a bit lower than it did with minimum distance. It was not much lower, but the fact that there are so many legal moves in this position (1,206) it was enough to drop the move from third to about $125^{th}$ best. When there are so many legal moves a small change in value can have a dramatic change in the ranking. Since the move was never even considered, its true value, which is discovered during look-ahead, is never revealed.

On the other hand, when there are not so many legal moves later in the game, even though maximum of two is a worse approximation, keeping 80 moves is a sufficiently high fraction of the total number of moves that most good moves are likely to be caught. And even if the best move is somehow lost, that usually means there are a number of other moves that appear to be about as good. For example, in Fig. 9(b) pure minimum distance made the move E2-C2(H2) while with maximum of two the move made was C3-C4(F7). Though the minimum

(a)                                          (b)

**Fig. 9.** Comparing moves made when using the normal evaluation function versus a faster approximation.

distance move is easy to understand (gaining territory on the lower part of the board) the maximum of two move also seems quite reasonable as it appears to claim at least as much territory in the center of the board. Typically we find the move made by maximum of two to be about as strong as the move made by minimum distance.

Approximating the evaluation with a faster, simpler version, saving full evaluation for a selected subset not only works well here, but the general approach provides a model for improving performance regardless of the evaluation function.

## 4    Conclusions and Future Work

Since a typical Amazons position has hundreds of legal moves we suggest three different methods for speeding up the candidate move selection process. The first method discards some amazon moves before an arrow is thrown. This technique is shown to be very effective and should be part of any amazon playing program.

The second technique speeds up the evaluation of all the legal moves by first calculating the value of various arrow throws independent of amazon moves and then using these values to help select the candidate list. Although the idea has intuitive appeal, the reality is that it does not improve the performance of the program. Perhaps when new evaluation functions are discovered that more accurately measure the value of a board position this method can be resurrected.

The last approach uses a simpler and faster approximation of the evaluation function to help prune moves that are unlikely to evaluate high. Though the results are not as spectacular as those of the first method, this approach clearly produces benefits for the program. In general, it seems to be useful to have a layered evaluation function where the faster layers can be used to feed moves to the slower more accurate layers.

This research suggests two critical areas for further work. The first is the evaluation function. Though minimum distance is useful and produces a reasonably strong program, evaluation functions need to be devised that more accurately take into account various aspects of a board position, such as potential versus secure territory, the mobility of the amazons, etc.

The second has to do with the opening. Since there is virtually no opening theory for the game, little progress has been made concerning making strong moves in the opening. Surely a separate evaluation geared towards the first 10 to 15 moves will prove extremely useful. Additional evidence of this importance is seen where we analyze the faster evaluation function. Most of the error added by that technique occurs in the opening, surely because minimum distance simply does not understand the opening game. A better opening evaluation will not only improve the quality of the moves being selected, it will probably also improve the performance of the layered evaluation technique.

# References

1. Hsu, F.: IBM's DEEP BLUE chess grandmaster chips. IEEE Micro **19** (1999) 70–81
2. Schaeffer, J.: One Jump Ahead: Challenging Human Supremacy in Checkers. Springer-Verlag (1997)
3. Buro, M.: The Othello match of the year: Takeshi Murakami vs. LOGISTELLO. International Computer Chess Association Journal **20** (1997) 189–193
4. Müller, M.: Review: Computer Go 1984-2000. In Marsland, T., Frank, I., eds.: Second International Conference on Computers and Games (CG 2000). Volume 2063 of Lecture Notes in Computer Science. Springer-Verlag (2001) 405–413
5. Matsubara, H., Iida, H., Grimbergen, R.: Natural developments in game research: From chess to shogi to Go. International Computer Chess Association Journal **19** (1996) 103–112
6. Takizawa, T., Grimbergen, R.: Review: Computer shogi through 2000. In Marsland, T., Frank, I., eds.: Second International Conference on Computers and Games (CG 2000). Volume 2063 of Lecture Notes in Computer Science. Springer-Verlag (2001) 433–442
7. Müller, M., Tegos, T.: Experiments in computer Amazons. In Nowakowski, R., ed.: More Games of No Chance. Cambridge University Press (2001) 243–260
8. Buro, M.: Simple Amazons endgames and their connection to Hamilton circuits in cubic subgrid graphs. In Marsland, T., Frank, I., eds.: Second International Conference on Computers and Games (CG 2000). Volume 2063 of Lecture Notes in Computer Science. Springer-Verlag (2001) 250–261
9. Hensgens, P.: A knowledge-based approach of the game of Amazons. Master's thesis, Department of Informatics, Universiteit Maastrict (2001) Technical report CS 01-09.
10. Reinfeld, A.: An improvement to the Scout tree search algorithm. International Computer Chess Association Journal **6** (1983) 4–14

# Playing Games with Multiple Choice Systems

Ingo Althöfer and Raymond Georg Snatzke

University of Jena, Germany
althofer@mathematik.uni-jena.de, rgsnatzke@aol.com

**Abstract.** Humans and computers have different strengths—which should be combined and not wasted. Multiple choice systems are an efficient way to achieve this goal. An instructive example is 3-Hirn, where two programs each make one proposal and a human boss has the final choice between them. In chess, Go, and several other brain games, many experiments with 3-Hirn and other settings proved the usefulness of multiple choice systems and the validity of the underlying ideas.

## 1 Introduction

Humans are able to think, to feel, and to sense. We can also compute, but not too well. In contrast, computers excel at computing, however they cannot do anything else. By combining the gifts and strengths of humans and machines in appropriate ways it is possible to achieve impressive results. One such way is a *multiple choice system*; one or more programs compute a clear handful of candidate solutions and a human chooses amongst these candidates. Brain games are ideal test-beds for decision support systems. Performance in play against other entities is a direct indicator for quality. So it is natural to test multiple choice systems in game playing.

In Sect. 2 we tell a story of success in chess. First experiments with multiple choice systems in Go are described in Sect. 3. In Sect. 4 we conclude with a short discussion.

## 2 3-Hirn and Other Multiple Choice Systems in Chess

In chess, performance is internationally measured by the Elo rating. The stronger a player, the higher his or her rating. If players $A$ and $B$ play against each other the difference $\text{Elo}(A) - \text{Elo}(B)$ allows one to make a statistical prediction of the outcome. For instance, players with the same rating should have equal chances. If the difference is 200 Elo points, then the stronger player should make about 75 percent of the points.

Starting in 1985, the first author made several experiments with multiple choice systems in chess. We first describe the constructions he used.

**3-Hirn:** ("Triple Brain" in English): Two independent chess programs propose one candidate move each. A human has the final choice amongst these candidates. If both moves coincide the human has to realize this single candidate. The human is not allowed to outvote the programs.

J. Schaeffer et al. (Eds.): CG 2002, LNCS 2883, pp. 142–153, 2003.

**Double-Fritz with Boss:** In 1995, FRITZ was the first chess program for the mass market which had a $k$-best mode. In $k$-best mode the $k$ best moves are computed, not just the best move. In Double-FRITZ with Boss, FRITZ was used in 2-best mode. The human is the boss and has the final choice amongst these two FRITZ candidates.

**List-3-Hirn:** This combines the concepts of 3-Hirn and Double-FRITZ with Boss. Two independent chess programs are used which both have $k$-best modes. Each program gives a list with its top $k$ candidate moves (typically $k = 3$ or something similar). Then the human has the final choice amongst the moves from the two lists of candidates.

In all three approaches the human gets the candidate moves, the corresponding principal variations, and the evaluations. Furthermore, the human is allowed to manage the time for the programs.

Table 1 gives an overview over the most important experiments with multiple choice systems in chess. All games had tournament type settings. The human boss was always Ingo Althöfer (Elo = 1900).

**Table 1.** Chronology of experiments with 3-Hirn [1].

| Year | Computer Elo Rating | Number of Games | 3-Hirn Performance |
|------|------|------|------|
| 1985 | 1500, 1500 | 20 | 1700 |
| 1987 | 1800, 1800 | 20 | 2050 |
| 1989 | 2090, 1950 | 8 | 2250 |
| 1992-1994 | 2260, 2230 | 40 | 2500 |
| 1995 | 2400, 2330 | 8 | 2550 |

In 1996, Double-FRITZ with Boss played 15 games and achieved a rating of 2520. FRITZ ran on a PC with PentiumPro processor at 200 MHz speed. The solo strength of FRITZ on this machine was about 2350.

In 1997, List-3-Hirn with computer programs of strength 2530 in solo play had a match of eight games against Germany's strongest grandmaster, Arthur Yusupov (Elo rating 2640 then), and won by a score of 5–3. This gave List-3-Hirn a provisional rating of above 2700.

A match of List-3-Hirn against a world top-10 human player was intended for 1998. However, after Yusupov's loss in 1997 the top humans were no longer willing to compete against 3-Hirn.

On average, 3-Hirn and its variants were about 200 Elo-points stronger than the computers in the team. Our explanation for this success is that the strengths of human and computers were well combined in 3-Hirn.

It is especially interesting to note that the increase in playing strength still occurred when the human boss was considerably weaker than the programs, as was the case from 1992 onwards. Even when the programs reached grandmaster level, the human boss (Ingo Althöfer, Elo = 1900) was able to add more than an $\epsilon$ to the system.

When one describes the two major factors of game playing skills as "search" and "knowledge", the programs were best at "search" and "technical knowledge". But the human boss contributed "soft knowledge" like pattern recognition, long-range planning, and perception of the opponent. The strengths and weaknesses are summarized in Table 2.

**Table 2.** Strengths and Weaknesses of Programs, Humans and 3-Hirn.

| Aspect/ Entity | Tactics | Memory | Long-range Planning | Learning | Psychology |
|---|---|---|---|---|---|
| Program | strong | strong | weak | weak | ?? |
| Human | weak | weak | strong | strong | ?? |
| 3-Hirn | strong | strong | strong | strong | strong |

The question marks in the psychology column indicate that a human's awareness of the (human) opponent has advantages and disadvantages. He realizes when the opponent is nervous, angry, or in time trouble. On the other hand, a tricky opponent may intentionally send false signals, or a "bold" opponent (like grandmaster Garry Kasparov) may scare the human.

In top-level correspondence chess nowadays (in the year 2002) all players without a single exception use chess programs for checking their analysis and improving their playing strength. Such help is explicitly permitted (in the rules of the main German correspondence chess federation BDF since summer 1996). For most players multiple choice analysis is part of this computer assistance.

## 3    Multiple Choice Systems in Go

As our recent research was focused on Go, we will go into more detail here, including offering explicit examples.

In Go playing strength is measured in the usual way for Eastern martial arts, in dan (master) and kyu (pupil) grades for amateurs. Professional players use their own scale of dan grades. The best amateurs in the world reach 6 or 7 dan, while 1 dan / 1 kyu is the level of a good club player. A beginner is approximately 35 kyu.

The average playing strength of a nominal "1 dan" is different in different parts of the world. All of the rankings mentioned in this paper refer to the central European scale of ratings.

Although interest in Go is widespread—especially in the Far East—and is said to be the most played board-game on Earth, Go-playing programs are extremely weak compared to chess programs. The best programs reach mediocre amateur-level at best, but it is very hard to assign a specific strength to them.

"Something around 10 kyu" is probably the only thing one can say. In addition Go programs are vulnerable to anti-computer strategies and all of them have special, easily exploitable weaknesses.

For multiple choice systems this means that the human boss of the team cannot rely on the tactical soundness of the proposals offered, which is one of the great advantages of the multiple choice approach in chess. The team also cannot be expected to play strongly in an objective sense, but only to play stronger than the participating programs themselves.

## 3.1   Experiments

*N*-Best Multiple Choice Systems: Since 2000 we conducted experiments with two different multiple choice systems in Go. In the experiment described first a special version of THE MANY FACES OF GO (MANY FACES), one of the leading Go programs of today, generated a list of its 20 best proposals for every move. Martin Müller, 5 dan from Edmonton, Canada, chose among these proposals. His opponent in a two-game match was Arnoud van der Loeff, 4 dan from the Netherlands. Via the Internet they played two even games with 6.5 komi on a board of size $13 \times 13$. ("Komi" are a bonus for White as compensation for going second. The ordinary board size in Go is $19 \times 19$.)

Both games were hard fought, but van der Loeff won them convincingly (see http://www.turbogo.com/games/Martin_ManyFaces_Ingo/). Although two games are not a lot of data, one may safely conclude that even 20 proposals on such a small board as a $13 \times 13$ are not enough to allow a strong player like Müller to play his game. He was still handicapped by being limited to the 20 candidate moves and only nearly, but not fully, reached his normal playing strength [2].

Classical 3-Hirn: Many experiments were undertaken in Go with the canonical 3-Hirn setting [3, 4, 5]. Most of the time we used the programs HANDTALK, multiple former world champion, and MANY FACES to produce one candidate move each. The human bosses of the team varied. With this configuration 3-Hirn played in five human Go-tournaments in Germany between Fall 2000 and January 2002. After 25 tournament games (with 11 wins), we can now rank 3-Hirn quite confidently as 8 kyu.

In these tournaments the strength of 3-Hirn seemed to be relatively independent of the playing strength of the human boss. In two tournaments the boss was a 15 kyu (Ingo Althöfer – Paderborn 2000, Jena 2001), in one tournament an 8 kyu (Stefan Mertin – Paderborn 2001) and in two tournaments a 1 dan (Georg Snatzke – Augsburg 2001, Erding 2002). Despite several problems with hardware and software 3-Hirn had no problems with the normal tournament time limits (between 45 minutes and 60 minutes plus overtime for the whole game).

Whereas 3-Hirn's playing strength is around 8 kyu, the Go-playing programs within 3-Hirn are probably not better than 10 kyu, but their exact strength is very hard to determine. If one plays against a program for the first time or doesn't even know that the opponent is a program, one might come to the premature conclusion that programs are several levels stronger.

The prime reason for this is that the playing strength of Go programs is based on other factors (tactics instead of strategy) than that of humans. Compared to equally strong humans, Go programs play better shape locally and on average have better tactics, spectacular tactical blunders notwithstanding. They are quite capable of utilizing tactical opportunities, if the situation is not too complex. They also don't suffer from oversights.

But Go programs are much weaker strategically. A typical behaviour of the programs is to make nice shape moves, where even a beginner can see by looking at the surrounding territory that the moves are either unnecessary or futile.

As long as one plays peacefully against a program, the program is usually able to make reasonable moves and stay in the game. As soon as one discovers the weaknesses of a given program and confronts it with unorthodox moves, aggressive play, and an intelligent whole-board strategy, all Go programs crumble. A set of experimental games demonstrated this impressively.

**Table 3.** Score of Guido Tautorat, 3 stone handicap, 13 × 13 board.

| LEARNING EFFECT AGAINST GO PROGRAMS | | |
|---|---|---|
| Opponent | Before Studying | After Studying |
| HANDTALK | 42, 27, 9, 38 | 171, 181, 40, 181 |
| MANY FACES | 26, 9, 50, 98 | 92, 175, 94, 171 |
| 3-Hirn (15 kyu Boss) | 54, 56, 77, 53 | 15, 53, 51, 23 |
| 3-Hirn (1 dan Boss) | 53, -5 | 18, 0, 15, 29 |

Table 3 gives the winning scores of Guido Tautorat, a 4 dan player from Jena, Germany, against several opponents. All games were on a 13 × 13 board with a 3 stone handicap against Tautorat. The human was instructed to try to win with as many points as possible, being rewarded with 5 Euro-Cents for each point in the score.

Between the games of the second and third column, Tautorat studied HANDTALK and MANY FACES intensively for several days. The table shows that Tautorat's play against the programs enhanced dramatically. After studying the programs he normally was able to annihilate the programs completely, which he did not manage a single time before. Annihilation amounts to a victory of about 170 points. Tautorat's concentration errors "saved" MANY FACES and HANDTALK three times in the second round.

On the other hand there is no evidence of any positive learning effect for Tautorat against 3-Hirn. There seemed to be a positive learning effect for the bosses of 3-Hirn, who employed special defensive strategies in the second round of matches. Against 3-Hirn with the 1 dan Boss Tautorat even suffered one 5 point defeat in the first match and a tie in the second match. 3-Hirn also never got annihilated. This clearly demonstrates that 3-Hirn offers much more resistance against anti-computer strategies and efforts to exploit program-specific weaknesses. See [6] for a more detailed report. ·

Surprisingly, in this series of 3-Hirn games the strength of the human boss seemed to play a role. However, as 3-Hirn's performance in the tournaments gives no hint in this direction it is probably premature to draw such a conclusion. Additionally, it seems that experience helps to make a good boss of 3-Hirn. But whatever the influence of the human boss really is, our experience with all of our experiments (with Tautorat, in five tournaments and many free sessions) strongly suggests that 3-Hirn outperforms its constituting programs by far.

## 3.2   Examples

In this section we want to show some specific examples of strengths and weaknesses of 3-Hirn that were typical. Unfortunately, if the reader is not accustomed to Go he has to trust our judgment in the following examples. Keep in mind that Go is a game about territory. Both players strive to surround more area with their own stones (and the edges of the board) than the opponent. Stones are not invulnerable; when they are completely surrounded by stones of the opponent, they are captured and taken off the board.

**Playing More Efficiently and Reducing Blunders:** Go programs love to waste moves. This may not be a spectacular problem, but it occurs very often and sums up to many lost points. These plays most often happen in the middle game and endgame, when defensive moves are made for groups that are either unconditionally alive or dead anyway. Another popular variation is to place new dead stones within the opponent's territory.

Most of the times at least one of the programs offers a useful move. Therefore the boss in 3-Hirn is able to select something constructive, not wasting time and points. This is especially important because both participating programs at various points in a game tend to insist on a certain move, which the boss does not want to play.

A typical example can be seen in Fig. 1. During the whole opening MANY FACES wanted to invade the lower right corner at nearly every move. Such a move is particularly uninteresting as it is quite likely that 3-Hirn will not be able to handle the following fight and lose its invasion stones. But with a second program to make proposals there always is another choice. While Black's opening moves are far from perfect, at least every move is doing something useful.

Every Go program has a few special weaknesses that are bound to produce a couple of spectacular blunders in nearly every game. Not defending a group of stones or a certain weak point at a critical moment are the most common examples.

Surprisingly HANDTALK and MANY FACES somehow practically never make the same errors at the same time. At least one of the programs noticed the need to defend, or at least offered some other move that was so threatening that the opponent first had to answer it, leaving the opportunity to defend 3-Hirn's

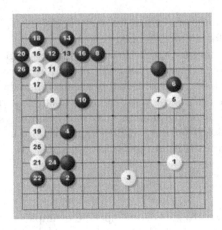

**Fig. 1.** Efficiency: Playing more useful moves (3-Hirn plays Black).

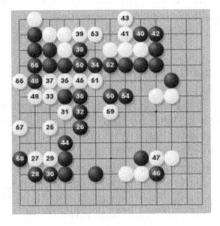

**Fig. 2.** Cooperation: Filling in the partner's defects (3-Hirn plays Black).

vulnerable spot later. Figure 2 shows an example. Stone 36 for Black is obvious and automatic for even weak human players. A failure to play at this point is clearly a spectacular blunder; the game stands or falls with playing that move. Yet only one of the programs wanted to play there. The other program would have thrown the game away.

Move 46 is another example. This move clearly belongs at 50. While one program would have played a much weaker move at 51, the other program at least offered a huge (but futile) threat elsewhere, 46, which had to be answered by White (and later found the move at 50). Using such external threats to hope for a better alternative later is a fundamental 3-Hirn tactic in Go.

**Cooperative Play:** Above we discussed examples where for one specific move one of the programs was able to remedy the defect of the other. But 3-Hirn is able to produce an even more elaborate sort of cooperation that extends over a whole sequence of moves. Neither program can complete the sequence itself, but together they are able to finish it. This can happen in certain Josekis, which are established move sequences in the corners. More impressive are life-and-death situations, where one program finds the one and only move to keep the opponent's group of stones dead, but the other program misses.

The already-mentioned Fig. 2 is an impressive example. With move 26 3-Hirn starts an impressive sequence of moves. This ends at move 60, when the White invasion group on the left side is safely captured. It can neither escape nor live on its own. MANY FACES alone as well as HANDTALK would never have been able to accomplish this task. MANY FACES would have made 5 and HANDTALK 4 critical errors during this sequence. However, they didn't fail on the same move.

Unfortunately they did fail simultaneously a couple of moves later, and the dead White group revived. One also has to admit that Black played some strange moves during this sequence, but up to move 60 everything worked out.

**Weaknesses:** Despite all the positive aspects of 3-Hirn Go shown previously, both programs have severe common deficiencies which are impossible to overcome even in a 3-Hirn setting. Both MANY FACES and HANDTALK don't understand Ko-Positions. A "Ko" is an unsettled position. The problem is that both programs tend to assume that they can settle the Ko in their favor and build their whole strategy on this assumption. When the Ko subsequently gets fought over, they normally are unable to handle the situation and their game collapses. This can be really depressing for the 3-Hirn boss, especially when he foresees this situation and is totally unable to do anything about it.

MANY FACES and HANDTALK are also very weak at invading and fail regularly when they try it. Sometimes invasions are necessary, for example when playing against a handicap. Figure 3 shows two typical examples in one game. Before move 45 Black has two large but hollow areas staked off, one around the lower right corner and the other one around the upper right corner. White invasions into these areas are difficult, but possible. At least one invasion has to succeed to give White a chance in the game. The programs recognize this and invade, first at 45 then at 51. Both invasion groups are abandoned immediately thereafter.

The inability to sustain invasions makes 3-Hirn much weaker when playing against handicaps, where invasions normally are a must. Therefore in even games without handicaps the strategic approach for 3-Hirn is to build up its position slowly and steadily, keeping its stones together, avoiding invasions. More often than not the programs force 3-Hirn into invasions, which they cannot handle later on.

Figure 4 gives an interesting example. 3-Hirn played very carefully up to that point, reducing the opponent's territory instead of invading it and keeping its stones together. After move 69 White has the lead and should be able to

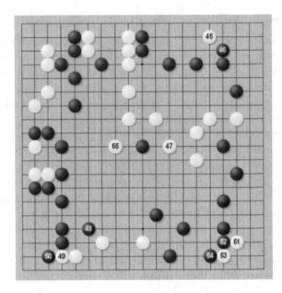

**Fig. 3.** Invasion: Abandoning the invasion stones (3-Hirn plays White).

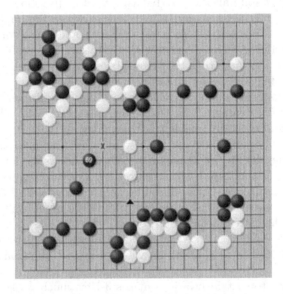

**Fig. 4.** Balance: Offering an unfavorable fight (3-Hirn plays White).

coast to an easy victory by connecting its two central stones to the left by playing on or around the point marked $X$. Instead both programs wanted to run deeper into Black's territory, offering the opportunity to start the fight that Black desperately needed. White played on the point marked ▲, Black cut the retreat line of the White stones and captured them after a long struggle and won the game.

This last example shows that MANY FACES and HANDTALK cannot assess the big picture. Instead of recognizing their strategic lead, they spot the tactical opportunity to invade, getting destroyed in the course of the fighting. Large, open areas, which are only loosely surrounded by stones of one color (so called "Moyos", potential territory) are especially critical. Both Go programs are unable to use a Moyo strategy themselves and they cannot deal with it when the opponent employs it.

While 3-Hirn eliminates many weaknesses of the single programs, several ideas for anti 3-Hirn strategies present themselves here. At the strategic level, build Moyos and force 3-Hirn to invade them. At the tactical level, create Ko-positions, leave them alone, and return to them later when they become critical. 3-Hirn's obvious counter-strategy is to avoid these situations wherever possible, but more often than not both programs steer right into the trouble, leaving their boss helpless.

## 4  Discussion

1. One of the most serious risks with multiple choice systems is that of getting **micro mutations instead of true alternatives**. "Simple" $k$-best algorithms tend to result in candidates which differ only in small details. The problem becomes more serious the more fine-grained the set of legal moves is. For instance, in Go "micro mutative candidates" occur more frequently on $19 \times 19$ boards than on those of dimension $9 \times 9$.

2. All of our experiments were performed with generally available chess and Go programs. The only feedback was that some programmers added $k$-best modes to their programs. Multiple choice systems are available in games other than chess and Go. For instance, the freeware Othello program W-ZEBRA has a multi-best move mode. For the solitaire card game FreeCell, André Grosse and Stefan Schwarz's freeware program BIGBLACKCELL gives an ordered list of all legal actions in each situation.

3. Multiple choice systems may be very helpful in many "serious" applications with strict real-time conditions. Some examples include:
   - optical pattern recognition,
   - vehicle routing and traffic control in general,
   - diagnoses in medicine (e.g., ECG-interpretation),
   - language translation, and
   - stock market actions and (financial) soundness checks.

   Multiple choice systems may be used to help improve the performance of applications that are not real-time critical, such as:
   - recognition of protein structures and DNA-alignments,
   - drug design, and
   - theorem proving in mathematics.

4. Why does it make sense to present the approach of multiple choice systems and 3-Hirn experiments in a scientific conference? From the viewpoint of hard sciences our investigations have two weaknesses. First, the 3-Hirn experiments are not fully reproducible. Second, we only have a limited number of data points. Point one is the case in all sciences where humans are involved, like medicine, economics, and humanities in general. Point two is really a weak spot, but the generation of a single data point (e.g., one game under tournament conditions) may take an average of two to six hours, where at least one human (namely the boss of 3-Hirn) has to be present.

Research on multiple choice systems is still in its infancy. One of the intentions behind this presentation in a games conference is to make the approach more public, to encourage other researchers also to work on the subject, and thus to get more data. Programmers of computer games should become motivated to make their products more easily usable in interactive settings like 3-Hirn. Here our wish list includes: easy take-back of single moves (and not only of move pairs like in many Go programs), clear presentation of candidate moves, and an implementation of a $k$-best mode.

5. Let us conclude with six theses.

   - **Combinatorial Games are nice test-beds for decision support systems.**
   - **Multiple choice systems are helpful tools not only in games, but also in "serious" applications with decision making.**
     This will hold especially for situations where neither human nor computers have a full understanding of the problem and thus may supplement each other.
   - **Have only one human in the multiple choice team.**
     A problem with more than one human in a team is that typically humans are touchy when their work or their solutions are not accepted. In contrast, a computer/program does not realize when you reject its proposal and press the reset button, or when you prefer other computers/programs most of the time.
   - **Multiple choice systems with more than two computers are problematic.**
     A natural generalization of 3-Hirn is to use more than two computer programs. The human collects the proposals of $N$ programs and makes the final choice amongst them. In principle, such an $(N+1)$-Hirn should at least be not weaker than a 3-Hirn which uses only the first two programs. Game-playing practice means decisions must be made under real-time constraints: a tournament chess player has an average of three minutes per move, while a Go player has to move every minute on average. Under such conditions it is a heavy burden for the controller to operate more than two programs simultaneously. The real-time strain for several hours without regular breaks is really a hard burden for the boss. Keep in mind that not only the move of the opponent has to be entered in the programs, but also not-realized move proposals have to be undone and substituted by the move selected.

- **In multiple choice systems the human is not allowed to outvote the programs. This is an advantage.**
  Of course this crude statement is exaggerated. It cannot be fully true when you look at the current experiences in Go. What we mean is that almost all humans overestimate the importance of human influence in decision support systems. It can be an advantage for the human boss when he only has to select and does not have to look for even better alternatives, because these tasks are structurally different. Concentrating just on the selection process is faster and easier for the human mind than to do both things at once. A selecting human may use his free capacities to consider other aspects like long-range planning and observation of the opponent. In [7] an overview is given of different grades of veto rights for multiple choice systems.
- **Human and computers together may be much stronger than humans alone or computers alone.**

## Acknowledgments

Thanks are due to Ken Chen (GO INTELLECT), David Fotland (THE MANY FACES OF GO), Anders Kierulf (SMARTGO), and Mick Reiss (GO4++) for providing multiple choice versions of their Go programs for our experiments. Thanks are also due to the Go players Arnoud van der Loeff, Stefan Mertin, Martin Müller, and Guido Tautorat for their participation in multiple choice experiments. Finally, thanks go to two anonymous referees for their constructive criticism.

## References

1. Althöfer, I.: 13 Jahre 3-Hirn—Meine Schach-Experimente mit Mensch-Maschinen-Kombinationen. Published by the author (1998)
2. Althöfer, I.: A 20-choice experiment in Go for human+computer. International Computer Games Association Journal **26** (2003) 108–114
3. Althöfer, I.: Go mit dem 3-Hirn. Deutsche Go-Zeitung (2000) July/August, 40–42
4. Gerlach, C.: Comments on the exhibition game 3-Hirn vs. Tautorat. Deutsche Go-Zeitung (2000) July/August, 42–45
5. Althöfer, I.: Das 3-Hirn beim Turnier in Paderborn. Deutsche Go-Zeitung (2000) November/December, 44–47
6. Althöfer, I., Snatzke, R., Tautorat, G.: Zum Abkochen von Go-Programmen. Deutsche Go-Zeitung (2002) July/August, 42–43
7. Althöfer, I.: Graded rights of veto in multiple choice systems. In: 8th IFAC/IFIP/IFORS/IEA Symposium on Analysis, Design, and Evaluation of Human-Machine Systems. (2001) 657–663

# The Neural MoveMap Heuristic in Chess

Levente Kocsis, Jos W.H.M. Uiterwijk, Eric Postma, and Jaap van den Herik

Department of Computer Science, Institute for Knowledge and Agent Technology,
Universiteit Maastricht, The Netherlands
{l.kocsis,uiterwijk,postma,herik}@cs.unimaas.nl

**Abstract.** The efficiency of alpha-beta search algorithms heavily depends on the order in which the moves are examined. This paper investigates a new move-ordering heuristic in chess, namely the Neural MoveMap (NMM) heuristic. The heuristic uses a neural network to estimate the likelihood of a move being the best in a certain position. The moves considered more likely to be the best are examined first. We develop an enhanced approach to apply the NMM heuristic during the search, by using a weighted combination of the neural-network scores and the history-heuristic scores. Moreover, we analyse the influence of existing game databases and opening theory on the design of the training patterns. The NMM heuristic is tested for middle-game chess positions by the program CRAFTY. The experimental results indicate that the NMM heuristic outperforms the existing move ordering, especially when a weighted-combination approach is chosen.

## 1 Introduction

Most game-playing programs for two-person zero-sum games use the alpha-beta algorithm. The efficiency of alpha-beta is closely related to the size of the expanded search tree and depends mainly on the order in which the moves are considered. Inspecting good moves first increases the likelihood of cut-offs that decrease the size of the search tree.

For move ordering, the existing programs frequently rely on information gained from previous phases of the search. Major examples of such information support are transposition tables [1] and the history heuristic [2]. Additionally, the original move-ordering techniques which are independent of the search process are still valid. In chess, for example, checking moves, captures and promotions are considered before 'silent' moves.

The search-independent move-ordering techniques are usually designed based on information provided by a domain expert. Recently two attempts were proposed to obtain a better move ordering by using learning methods. The chessmaps heuristic [3] employs a neural network to learn the relation between the control of the squares and the influence of the move. In a similar way, we trained a neural network to estimate the likelihood of a move being the best in a certain position [4]. We termed this technique the Neural MoveMap (NMM) heuristic. There are two main differences between the chessmaps heuristic and the NMM heuristic. First, in the chessmaps heuristic the neural network is trained to evaluate classes of moves, while in the NMM heuristic the neural network is tuned

J. Schaeffer et al. (Eds.): CG 2002, LNCS 2883, pp. 154–170, 2003.

to distinguish between individual moves. Second, in the chessmaps heuristic the target vector represents the influence of the move on the squares of the board, while in the NMM heuristic the target information for a neural network is the likelihood of the move being the best in a certain position.

The training information of the NMM heuristic shares common ideas with the comparison paradigm [5, 6] developed for learning evaluation functions. In the original version designed by Tesauro, a neural network is trained to compare two board positions. The positions are encoded in the input layer, and the output unit represents which of the two is better. The comparison paradigm was also employed to evaluate moves in Go [7]. Accordingly, a neural network was trained to rate the expert moves higher than random moves. The technique is not efficient for move ordering in a search-intensive game program due to speed limitations.

In [4], we presented the results of the NMM heuristic in the game of Lines of Action (LOA). The promising results were confirmed when the heuristic was tested in MIA [8], one of the top LOA programs. The reduction of the search tree was more than 20 percent, with an overhead of less than 8 percent [9]. Despite this success, it was unclear how the NMM heuristic performs when it is included in tournament programs for more complex and, especially, more studied games. In this article we investigate the performance of the NMM heuristic in chess by inserting the move ordering in the strong tournament program CRAFTY [10]. In LOA, we inserted the NMM heuristic in the search by replacing the move ordering of the history heuristic by that of the neural network. In this article we improve upon this approach. Chess, compared to LOA, has two extra features that can influence the training of the neural network: the existence of game databases and the advanced opening theory. In the experiments we analyse the choices resulting from these two features too.

The article is organized as follows. Section 2 describes the NMM heuristic. The experimental set-up for evaluating the efficiency of the NMM heuristic is described in Sect. 3. The experimental results are given in Sect. 4. Finally, Sect. 5 presents our conclusions.

## 2   The Neural MoveMap Heuristic

For the NMM heuristic, a neural network is trained to estimate the likelihood of a move being the best in a certain position. During the search, the moves considered more likely to be the best are examined first. The essence of the heuristic is rather straightforward. However, the details of the heuristic are complex since they are crucial for the heuristic to be effective, i.e., to be fast and to result in a small search tree. The details include: the architecture of the neural network (Sect. 2.1), the construction of the training data (Sect. 2.2) and the way the neural network is used for move ordering during the search (Sect. 2.3).

### 2.1   The Architecture of the Neural Network

For the game of LOA, we analysed several architectures for the neural network [4]. We found that the best architecture encodes the board position in the input

units of the neural network and uses one output unit for each possible move
of the game. A move is identified by its origin and destination square (i.e., the
current location and the new location of the piece to move). The activation value
of an output unit corresponding to a move represents the score of that move.
The other analysed architectures include a more compact representation of the
moves, either in the input or in the output of the neural network.

The same move encoding can be used for chess. In the case of LOA, we
assigned one input unit to each square of the board, with +1 for a black piece,
−1 for a white piece and 0 for an empty square. In chess there are more piece
types, and consequently we assign 6 input units (one for every piece type) to each
square. An additional unit is used to specify the side to move. Consequently, the
resulting network has 385 (6×64+1) input units and 4096 (64×64) output units.

Although the network is very large, the move scores can be computed quickly,
since we have to propagate only the activation for the pieces actually on the
board, and to compute only the scores for the legal moves. To increase the speed
further, in the case of chess we removed the hidden layer, present in the network
used for LOA. For LOA, we already noticed that in this architecture the hidden
layer is not increasing significantly the performance [4]. This way, the resulting
move ordering requires just a little extra computation during the search, namely
a summation over the pieces on the board.

## 2.2   The Construction of the Training Data

The neural network described in Sect. 2.1 performs a linear projection, and any
learning algorithm should thus be reasonably fast. Consequently, we can use any
of the existent learning algorithms for neural networks without influencing significantly the training. However, the choice of the training data is an important
issue. A training instance consists of a board position, the legal moves in the position and the move which is the best. From these three components, determining
the legal moves by an algorithm poses no problem. The choices on the other two
components are more difficult. There are two questions to be answered: "Where
are the best moves coming from?" and "Which positions should be included in
the training phase?"

In each training instance, one of the legal moves is labeled as the best. The
labeling can have two sources: (1) the move played in the game (as specified in
the database), (2) the move suggested by a game program (the one that will
use the neural network for move ordering). If we choose the second source, the
neural network might be able to incorporate the program's bias (e.g., preference
to play with bishops). The second source has, however, the disadvantage of the
extra computation needed for analysing all the positions, since, as observed for
LOA, a large number of positions have to be used to obtain good results. In the
experimental section, we examine both sources.

Chess games are usually divided into three phases: opening, middle game
and endgame. In each of these phases different strategies are used. We focus
on middle-game positions. There the original opening usually has a substantial influence on the strategies employed by chess players. In the experiments,

we investigate whether it is better to have opening-specific neural networks, or whether it is sufficient if only one network is used for all middle-game positions. The training positions originate from game databases either clustered by opening or collected in a mixed form.

## 2.3   Using the Neural Network during the Search

When the neural network is used during the search the moves are ordered according to the network's estimation of how likely a certain move is the best. The move ordering has to be placed in the context of the move orderings already existent in the game program. In the following, we describe three approaches to include the neural network in the search: the pure neural-network approach, the straightforward-combination approach, and the weighted-combination approach. These approaches deal only with the moves that would be ordered by the history heuristic.

**Pure Neural-Network Approach:** The first approach to use the neural network during the search is by replacing in every node of the search tree the move ordering of the history heuristic by that of the neural network. In the following, we refer to this approach as the pure neural-network approach. This approach was tested for LOA [4].

**Straightforward-Combination Approach:** The move ordering of the neural network and that of the history heuristic are not necessarily mutually exclusive; they can be combined. Hence, the second approach investigated in this article, the straightforward-combination, is as follows. We first consider the move rated best by the neural network followed by the moves as ordered by the history heuristic (of course, excluding the move picked by the neural network).

**Weighted-Combination Approach:** The history-heuristic score is built up from information collected during the search process. Therefore, it has a dynamic nature, but it has no specific connection with the position under investigation. In contrast, the scores suggested by the neural network are specific to the position; they are static, and do not gain from the information obtained in the current search process. In principle, a combination of the two scores can benefit both from information obtained in the search, and from information obtained off-line by analysing a large number of positions. Such a combination can be more suitable for the position in which the moves have to be ordered.

One way to combine the history-heuristic score and the neural-network score is by adding them. Since the two scores are not in the same range at least one of them has to be scaled.

The distribution of the neural-network scores is specific to a certain neural network and does not depend on the search depth or opening line of a position[1]. A sample distribution is given in Fig. 1, left. The neural-network scores

---

[1] Although the independence of the opening line might seem counter-intuitive, this is what we observed experimentally.

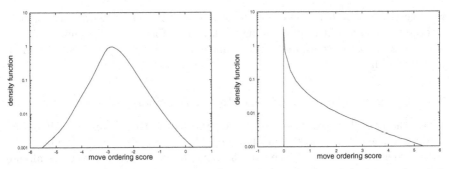

**Fig. 1.** The distribution of scores for the neural network (left) and the history heuristic (right).

are approximately normally distributed, and centered around a negative value, since most moves are not considered to be best. The history-heuristic scores are known to become larger as the search progresses. One way to normalize the history-heuristic scores is to divide them by the total number of history updates. After this normalization, the history-heuristic scores in the different parts of the search tree are approximately in the same range. However, they are still in a range different from the neural-network scores. Therefore, we divide the normalized history-heuristic scores by a coefficient that we term the history weight. The resulting distribution, using the value 500 for the history weight (see below and Sect. 4.2), is plotted in Fig. 1, right. The history-heuristic scores are only positive, and the peak is concentrated near 0, since most of the moves rarely produce a history update during the search. A second possibility for normalizing the history-heuristic scores is by dividing them by the standard deviation amongst the scores in the current position. Experimentally, we observed that the two choices to normalize the history-heuristic scores lead to similar performances. Since the second one is computationally more expensive than the first one (we have to compute the standard deviation in every position), we choose to normalize by the number of updates.

In conclusion, the score used to order the moves (*movescore*) is given by the following formula:

$$movescore = nnscore + \frac{hhscore}{hhupdate \times hhweight}$$

where *nnscore* is the score suggested by the neural network, *hhscore* is the original history-heuristic score, *hhupdate* represents the number of times the history table was updated during the search, and *hhweight* is the history weight.

To use the above formula for the move score we have to choose a value for the history weight. Choosing the proper value may have an important effect on the performance of the weighted-combination approach. If the weight is too large or too small, only one of the components of the score has influence. This effect might be beneficial in certain positions (or parts of the search tree), where one of the move-ordering methods performs poorly, but overall it is not desirable.

In essence, there are two ways to set the history weight. The first one is using off-line experiments with different values, and choosing the value leading to the best performance in these experiments. The second one adapts the weight during the search. Such an adaptive method is outlined below.

In the case of a good move ordering, the moves causing a cutoff are inspected early. Consequently, we can use the situation when the cutoff appears later as an error signal to modify the value of the history weight during the search. If the move causing a cutoff was considered later because both the neural network and the history heuristic scored it low, nothing could have been done. However, if, for instance, the neural network considered the move as promising, but it was scored overall lower than some other moves because the history heuristic scored it low, the history weight should be increased. The situation when the neural network scored a move producing a cutoff high, and the history heuristic low, we term it an *nn-miss*. The situation when the history heuristic scored the move high, but the neural network low, we term it an *hh-miss*. In the case of an nn-miss the history weight is increased by a small value $\beta_{nn}$, and in the case of an hh-miss the history weight is decreased by a small value $\beta_{hh}$. Depending on the update values, the history weight will converge to a value where the updates are balanced ($\Delta hhweight = 0$), or the ratio between the number of nn-miss ($n\_nnmiss$) and the number of hh-miss ($n\_hhmiss$) is inversely proportional to the ratio between the two update values:

$$n\_nnmiss \times \beta_{nn} = n\_hhmiss \times \beta_{hh}.$$

The exact values of $\beta_{nn}$ and $\beta_{hh}$ usually do not have a significant effect on the performance. They should have relatively small values (e.g., 0.1), in order to prevent large oscillations of the history weight, and should have similar magnitude (possibly equal) to give the same emphasis to both components.

## 3    Experimental Setup

In the experimental set-up we distinguish three phases: (1) the construction of the various pattern sets, (2) the training of the neural networks to predict the best move, and (3) the evaluation of the search performance of the move ordering.

In the first phase we construct the pattern sets. They consist of a board position, the set of legal moves, and the best move. To construct a pattern set we have to select the positions, to generate the legal moves in the selected positions, and to label one of the legal moves as best. The middle-game positions are selected from game databases, classified by opening. The legal moves are generated using a standard move-generation routine. A move is labeled best by one of two sources: by the choice of the player as is given in the game database, or by a game-playing program. For the latter, we employ CRAFTY with 2 seconds thinking time (approximately a 7-ply deep search). For the training in the second phase, we need three types of pattern sets: a *test set* to evaluate the move-ordering performance, a *validation set* for deciding when to stop the training,

**Table 1.** Opening lines of the learning sets.

| Code | Encyclopedia Code | Number of Positions | Name of Opening Line |
|------|-------------------|---------------------|----------------------|
| A30  | A30       | 122,578   | Hedgehog System of English Opening |
| A3x  | A30-A39   | 685,858   | Symmetrical Variation of English Opening |
| B84  | B84       | 92,285    | Classical Scheveningen Var. of Sicilian Defense |
| SI   | B80-B99   | 1,793,305 | Sicilian Defense (Scheveningen, Najdorf) |
| D85  | D85       | 124,051   | Exchange Variation of Grünfeld Indian |
| GI   | D70-D99   | 745,911   | Grünfeld Indian |
| E97  | E97       | 111,536   | Aronin-Taimanov Variation of King's Indian |
| E9x  | E90-E99   | 865,628   | Orthodox King's Indian |
| KI   | E60-E99   | 2,498,964 | King's Indian |
| ALL  | A00-E99   | 4,412,055 | all openings |

and a *learning set* to modify the weights of the neural networks. For our experiments, we construct four test sets, originating from the opening lines A30, B84, D85 and E97 [11]. Each of these sets consists of 3000 positions. To each test set corresponds a validation set taken from the database with the same opening line as the test set. These sets also consists of 3000 positions. The positions from the test sets and the validation sets are different. The learning sets are taken from the same opening line as the corresponding test set or from a more general one (see below). Each learning set, except ALL, includes all available middle-game positions from its opening line excluding the positions in the test and validation sets. The learning set ALL includes 10 percent of the positions from all opening lines. For instance, the test set A30 has three corresponding learning sets: A30 with 122578 positions, A3x with 685858 positions, and ALL with 4412055 positions (see Table 1). Analogously, test set E97 has four corresponding learning sets: E97, E9x, KI and ALL (see Table 1). Other details on the learning sets are given in Table 1.

During the second phase, the neural networks are trained to predict whether a certain move is the best. The learning algorithm used is RPROP [12], known for its good generalization. Like most of the supervised-learning algorithms for neural networks, RPROP also minimizes the mean square of the error (i.e., the difference between the target value and the output value). In RPROP, however, the update of the weights uses only the sign of the derivatives, and not their size. RPROP uses an individual adaptive learning rate for each weight, which makes the algorithm more robust with respect to initial learning rates.

The error measure for evaluation and for stopping purposes is the rate of not having the highest output activation value for the 'best' move. In each epoch the entire learning set is presented. The neural network updates the weights after each epoch (i.e., batch learning). The error on the validation set is tested after each update. The training is stopped if the error on the validation set did not decrease after a certain number of epochs. In this phase we observe how often the neural network predicts the best move. For a tournament program, however, it is more important how the neural network is behaving during the search.

The third phase evaluates the information gained about the performance of the neural network. In this phase we measure the size of the search tree investigated using the neural network for move ordering. For this purpose we insert the neural network in the move ordering of CRAFTY and collect the statistics for search with various depths. In an internal node, CRAFTY considers subsequently (1) the move from the transposition table, (2) the capture moves, (3) the killer moves and (4) the remaining moves sorted according to their history-heuristic scores. In the following, we refer to this move ordering as the reference move ordering. In this list we insert the predictions of the neural-network moves, using one of the three approaches described in Sect. 2.3. The performance of a neural network is measured by dividing the size of the investigated search tree by the size of the investigated search tree without neural-network move ordering. The search performance is measured using various sets of test positions. The set of positions employed in this phase is identical to the positions from the test sets used in the training phase.

# 4   Experimental Results

Section 4.1 deals with the choices on the construction of the pattern sets. In particular, the choices to be made are on the best move and on the positions. Section 4.2 compares the three approaches in which the neural networks are included in the search. The results of this subsection also evaluate the performance of our move-ordering technique in comparison with the existing techniques. Section 4.3 gives insight into the predictive quality of the neural networks. Although the predictive quality is not the main evaluation criterion (the size of the investigated search tree is the main criterion), it highlights some of the strengths and weaknesses of the neural networks trained for move ordering. Moreover, an example is given for illustration.

## 4.1   Choices on the Construction of the Pattern Sets

The open questions on the construction of the pattern sets posed in Sect. 2.2 are about the source of the best move and the selection of the positions. Below, we describe two preliminary experiments addressing the questions. Experiment 1, called Game vs. CRAFTY, compares two sources of the best move: by taking the move from the game database and by taking CRAFTY's move. Experiment 2, called Specific vs. General, compares more opening-specific learning sets (e.g., E97) with more general learning sets (e.g., E9x, KI or ALL).

The different choices lead to different neural networks. As a measure to compare the different neural networks we take the search performance (i.e., the size of the search tree investigated using the neural networks for move ordering). Out of the three approaches to include neural networks in the search (described in Sect. 2.3), we only use the pure neural-network approach and the straightforward-combination approach. The weighted-combination approach needs additional computation to obtain the history weight, and therefore we

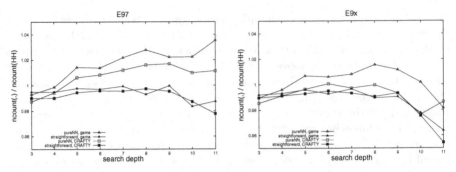

**Fig. 2.** Training the neural networks on game moves or on CRAFTY moves.

leave it out for the two preliminary experiments. In Figs. 2 and 3, the performance is plotted as a function of the search depth. The performance is defined as $ncount(.)/ncount(HH)$—the size of the search tree when the neural network is included divided by the size of the search tree investigated with the reference move ordering.

**Game vs. CRAFTY:** The first experiment is on the source of the best moves. It gives insight into the choice between the game database move and the move suggested by CRAFTY as training target. For this purpose we use the test set E97 with two learning sets: E97 and E9x. First we focus on the learning set E97. We label the moves of E97 either by the moves from the game database or by CRAFTY's suggestions. Doing so we obtain two learning sets, and accordingly, after the training process, two neural networks (one for each learning set). The search performances corresponding to the two neural networks are plotted in Fig. 2, left. The graph shows the search performance as a function of the search depth for the neural network trained with the learning set labeled either by CRAFTY or by the game database. Since we are using either the pure neural-network approach or the straightforward-combination approach, we have four curves. Similarly, Fig. 2 (right) shows the search performances corresponding to the learning set E9x. We observe that if the pure neural-network approach is used, the CRAFTY labeling is beneficial for both E97 and E9x, since it investigates a smaller search tree. If the straightforward-combination approach is used the advantage disappears. Since the CRAFTY labeling process is very time consuming[2] (especially for larger learning sets), we conclude that it is not useful to label the positions with a game program, and it is preferable to use the moves from the game database. The following experiments are using only the moves resulting from the original games (as given in the game database).

**Specific vs. General:** The second experiment is designed for giving an answer on how specific the learning set should be. Next to the two neural networks resulting from E97 and E9x, we trained two more neural networks with learning

---

[2] E.g., to label E9x we needed 20 days.

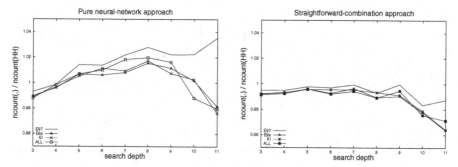

**Fig. 3.** Training sets varying from specific to general.

sets KI and ALL (using again E97 as test set). Out of the four sets the most specific learning set is E97, and then in this order E9x, KI, and ALL. The search performances for the four neural networks are plotted in Fig. 3. The performance of the neural networks corresponding to the learning sets E9x, KI and ALL are roughly equal. The neural network corresponding to the most specific learning set (E97) is performing worse than the other three for both the pure neural-network approach (Fig. 3, left) and the straightforward-combination approach (Fig. 3, right). The results suggest that using more specific learning sets does not improve the performance. For very specific sets, the performance may be decreased due to the lack of sufficient training data (in the case of E97: 111,536).

### 4.2   Comparison of the Three Neural-Network Approaches

In this subsection we compare experimentally the three approaches to include the neural network in the search. As a reference value, we also compare these approaches to the reference move ordering. To gain an even better picture on the performance of the new move-ordering techniques we use three more test sets (A30, B84 and D85), next to E97. For each of the four test sets (A30, B84, D85 and E97) a specific (A3x, SI, GI and E9x) and a general (ALL) learning set is used. With these learning sets, we obtain four neural networks specific to a certain test set and a neural network common to all test sets. The search performance is measured in the same way as in Sect. 4.1.

According to the pure neural-network approach and the straightforward-combination approach the neural networks can be included in the search directly. Thus the corresponding performance can be measured without any preparation. However, to measure the search performances for the weighted-combination approach, we first have to determine the history weight.

As described in Sect. 2.3, the history weight can be set according to two methods, either off-line or during the search. For the first method, we have to measure the search performance corresponding to different values of the history weight. Below we illustrate this mechanism for test set E97. Figure 4 plots the performance for different history-weight values, using the neural networks trained either with E9x (the left graph in Fig. 4) or with ALL (right). The

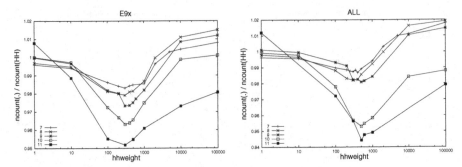

**Fig. 4.** Varying the history weight. The performance is plotted for the E97 test set using two neural networks, E9x and ALL, and five search depths, from 7 to 11.

five curves correspond to the performances obtained for search depths 7 to 11. The best value for the history weight is the one that corresponds to the best search performance (i.e., the lowest point for a certain curve). We observe that the best history-weight value is not changing drastically with search depth. It does, however, change with the neural network[3]. The reason for this is that the best history weight is inversely proportional to the standard deviation of the neural-network scores. As measured experimentally, the standard deviation for E9x is almost twice as large as the standard deviation for ALL, and thus the best history weight for E9x is roughly the half of the best value for ALL.

For the second method, i.e., when setting the value for the history weight during the search, using the update rules described in Sect. 2.3, we noticed that the history weight does indeed converge to a value where the equilibrium equation holds. The search performances are similar to the performances corresponding to the best static history weights from Fig. 4.

Using the value for the history weight obtained according to one of the above-mentioned methods, we measured the performance of the weighted-combination approach and compared it to the pure neural-network approach and the straight-forward-combination approach. The performances of these approaches using the neural networks mentioned earlier are plotted in Fig. 5 (for test sets A30, B84, D85 and E97 separately). In the graphs, the search performance of the reference move ordering has the value of 1. From the reference move ordering, the neural-network approaches are replacing the history heuristic. Henceforth their performance is compared to that of the history heuristic. We observe, that the pure neural-network approach has a slight advantage over the history heuristic for very shallow searches (depth 3 or 4), then it has a slight disadvantage for medium depths (5 to 9), and finally again better results for depths 10 or 11. Interestingly, a similar shape (although with different values) was observed for LOA in [4][4]. The straightforward-combination approach outperforms the his-

---

[3] An appropriate history weight value for E9x is 300, and for ALL is 500.

[4] The reason for this particular shape is unclear to us. It might be due to some properties of the history heuristic.

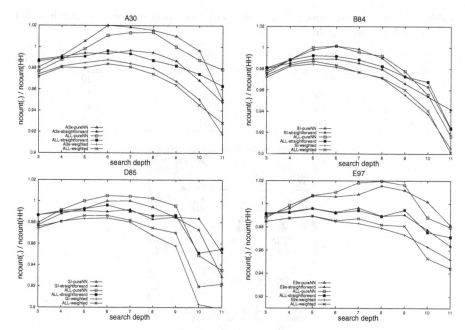

**Fig. 5.** Comparing the three move orderings using the four test sets (A30, B84, D85 and E97).

tory heuristic with a slight margin (approximately 1 percent) for search depths 3 to 9 and with an increasing amount for depths 10 and 11 (3 to 8 percent). The weighted-combination approach outperforms all the other ones on all test sets and search depths. The reduction compared to the history heuristic ranges between 5 and 10 percent for the largest search depth investigated.

The time overhead for the approaches using neural networks, compared to the reference move ordering, is approximately 5 to 8 percent. However, this value can be decreased by not using the neural network for move ordering close to the leaves. The performance loss in this case is negligible (less than 1 percent).

### 4.3   Predictive Quality of the Neural Networks

In this subsection, we provide some more insight into the predictive quality of the neural networks. In order to test the predictive quality, we measure how often the move made in the game is considered to be the best by the neural network. The neural networks used for testing are those from Sect. 4.2. The results are presented in Table 2. In the table, beside the success rate of the networks trained on specific learning sets and the ALL learning set, CRAFTY's predictive quality is also listed for comparison purposes. We observe that the networks are able to predict the game move in one out of three positions. This performance is almost as good as that of CRAFTY with a 7-ply deep search.

**Table 2.** Rate of correct predictions of the neural networks without any search and of CRAFTY using 7-ply searches.

| test set | specific learning set | learning set ALL | CRAFTY's prediction |
|---|---|---|---|
| A30 | 0.33 | 0.32 | 0.39 |
| B84 | 0.36 | 0.33 | 0.40 |
| D85 | 0.35 | 0.31 | 0.45 |
| E97 | 0.39 | 0.34 | 0.37 |

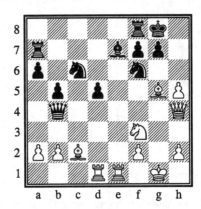

**Fig. 6.** A position from the Kramnik-Anand game played in Dortmund, 2001 (WTM). Kramnik's move was h6. The move ordering of the neural network: Qb4 Ne5 Bf6 a3 h6 Qh3 Bb1 Bh6 Nd4 b3 Qd4 Bb3 Rd3 Qg3 Re2 Rb1 Bc1 Bg6 Rd5 Bf5 Qf4 Re3 Rf1 Bf4 Qg4 Kh1 Kf1 Re7 h3 Bd3 Ra1 Bd2 Re5 Rc1 Re6 Bh7 Rd2 Re4 Ba4 Be3 a4 Kg2 Be4 Nd2 Qe4 Rd4 Qc4.

Below we illustrate this success by an example. Other example positions are given in the Appendix. The position in Fig. 6 is encountered between two of the best chess players, Kramnik and Anand. It is an open position, with some threats by White on the king side. The move made by Kramnik (h6), forces Black to a line of exchanges that leads in the end to an inferior ending. Without search, for the neural network the complications on the board are hidden, of course. Its main tool is to look for similarities between this position and the positions seen during the training, respectively between the legal moves in this position and the moves labeled best in the learning set. Most of the pieces in this position are in familiar places. However, the move made by White is not frequent at all in middle-game positions. Remarkably, the neural network still scores the move in the top five. Some other promising moves, like Ne5, Bf6 and Qh3 are also scored high.

## 5   Conclusions

This paper investigates the use of a new move-ordering heuristic, the Neural MoveMap heuristic, in chess. The heuristic uses a neural network to estimate the likelihood of a move being the best in a certain position. The moves considered more likely to be the best are examined first. The move ordering was tested for middle-game positions using the chess program CRAFTY. We explored the various details of the NMM heuristic, developing a new approach to include the neural network in the search, the weighted-combination approach.

**Construction of the Pattern Sets:** The preliminary experiments dealt with the details of the construction of the pattern sets. From the experiments we conclude that labeling with CRAFTY is not giving advantage over the choice of using the moves from the game database. This is especially true if a more advanced approach is used to include the neural network in the search. Because of the extra time required by the labeling process, using the game moves is to be preferred. A second conclusion is that, although it seems counter-intuitive, using learning sets specialized on opening lines is not improving the performance. Since training one neural network instead of many (one for each opening line) is faster, and using one neural network during the search is also more convenient than to use many, we conclude that using a single general neural network is the preferable solution. In summary, we have to train one neural network with a learning set resulting directly from the game databases. Thus, the training phase can be done in a relatively short time.

**Search Performance:** In the main experiments we investigated the approaches to use the neural network during the search. We designed three approaches: the pure neural-network approach, the straightforward-combination approach, and the weighted-combination approach. These approaches were also compared to the reference move ordering of CRAFTY. Although for shallower search depths the pure neural-network approach investigated slightly larger trees, for deeper searches it proved to be superior to the reference move ordering. The two approaches that combine the neural network and the history heuristic improve upon the reference move ordering for all investigated search depths. The node reduction increases with deeper searches. Out of the two combinations, the weighted-combination approach leads to better results.

If we compare the performance of the Neural MoveMap heuristic in chess to that in LOA, we conclude that the heuristic is more beneficial in domains where there is little human knowledge on how to order the moves, but even in the presence of this knowledge, it results in a notable improvement over the existent techniques.

Both for LOA and chess, we introduced the NMM heuristic as an alternative to the history heuristic. When a game programmer decides between these two heuristics, other issues than performance have to be considered. In the following we describe the differences between the two heuristics in terms of these issues. The history heuristic is trivial to implement and requires little tuning. The NMM heuristic is more difficult to implement and requires a training phase. Neither of the two heuristics require game-dependent knowledge. In the initial development cycle speed of implementation is very attractive. In this cycle, the program is not well-tuned yet, and the NMM heuristic is likely to have a more significant impact on the performance. Consequently, there is a trade-off between the two heuristics. In the later cycles, when the program is well developed, the performance is the most important issue for selecting the move-ordering heuristic. Therefore, the NMM heuristic should be preferred.

**Predictive Quality:** The predictive quality of the neural networks is surprisingly good, almost as good as a full search of CRAFTY. This suggests that the neural networks can be used for forward pruning too, in a similar way as in [13] for LOA. Such a forward-pruning algorithm for chess has to be developed in more detail. A promising line of research in this respect is the search control mechanism designed in [14].

# References

1. Marsland, T.: A review of game-tree pruning. International Computer Chess Association Journal **9** (1986) 3–19
2. Schaeffer, J.: The history heuristic. International Computer Chess Association Journal **6** (1983) 16–19
3. Greer, K.: Computer chess move-ordering schemes using move influence. Artificial Intelligence **120** (2000) 235–250
4. Kocsis, L., Uiterwijk, J., van den Herik, J.: Move ordering using neural networks. In Monostori, L., Váncza, J., Ali, M., eds.: Engineering of Intelligent Systems. Springer-Verlag (2001) 45–50 Lecture Notes in Artificial Intelligence, Vol. 2070.
5. Tesauro, G.: Connectionist learning of expert preferences by comparison training. In Touretzky, D., ed.: Advances in Neural Information Processing Systems 1. Morgan Kaufmann (1989) 99–106
6. Utgoff, P., Clouse, J.: Two kinds of training information for evaluation function learning. In: Ninth National Conference of the American Association for Artificial Intelligence (AAAI-91), AAAI Press (1991) 596–600
7. Enderton, H.: The GOLEM Go program. Technical Report CMU-CS-92-101, School of Computer Science, Carnegie-Mellon University (1991)
8. Winands, M., Uiterwijk, J., van den Herik, J.: The quad heuristic in Lines of Action. International Computer Games Association Journal **24** (2001) 3–15
9. Winands, M.: Personal communication (2002)
10. Hyatt, R., Newborn, M.: CRAFTY goes deep. International Computer Chess Association Journal **20** (1997) 79–86
11. Matanović, A., ed.: Encyclopaedia of Chess Openings. Volume A–E. Chess Informant (2003)
12. Riedmiller, M., Braun, H.: A direct adaptive method for faster backpropagation learning: The RPROP algorithm. In: IEEE International Conference on Neural Networks. (1993) 586–591
13. Kocsis, L., Uiterwijk, J., van den Herik, J.: Search-independent forward pruning. In: Belgium-Netherlands Conference on Artificial Intelligence. (2001) 159–166
14. Björnsson, Y., Marsland, T.: Learning search control in adversary games. In van den Herik, J., Monien, B., eds.: Advances in Computer Games 9. Universiteit Maastricht (2001) 157–174

**Fig. 7.** An early middle-game position from the E97 test set (WTM). The 'theoretical' moves are Nd2, Nd4, cd6, a4, and h3. The move ordering of the neural network: Nd2 cd6 Nb5 Nd4 e5 Qd2 a4 h3 Re1 a3 Qd4 Nh4 Qb3 b5 Qa4 c6 h4 Nb1 Qe1 Ne5 Ne1 Na4 Qd3 Ba6 Qc2 g4 Kh1 Nd3 Bb5 Ra1 Rb1 Rc2 Bc4 g3 Ng5.

**Fig. 8.** A position close to the endgame from the D85 test set (BTM). The move made in the game was Kf8. The move ordering of the neural network: Rb2 Rb1 Nb5 Rf8 Ne6 Rb3 Rb6 a5 Nc2 e6 Re8 Kg7 Rd8 g5 Ne2 e5 f5 Rb4 h6 f6 Nc6 h5 Kh8 a6 Nb3 Kf8 Nf3 Nf5 Rb5 Rc8 Rb7.

# Appendix: Examples of Move Ordering Using the Neural Network

In Figs. 7, 8 and 9, we illustrate some of the weaknesses and strengths of the move ordering performed with the neural networks. The positions in the figures were not included in the learning sets. The neural network employed to order the moves was the one trained with positions resulting from all opening lines (ALL). The first two positions are from the edges of the middle-game phase, illustrating some interesting properties of these parts of the game. The third position is a closed position from the 'core' of the middle game.

In the position in Fig. 7, an early middle-game position, the moves typical for this line of opening are easy to recognize. Consequently, although the position was never 'seen' by the neural network, the move ordering suggested is excellent, rating the 5 'theoretical' moves among the top 8 moves.

The second position (Fig. 8) is close to the endgame. Although, some of the first-rated moves make some sense, the good moves (including the one made in the game), are in the tail of the list. The shortcoming is resulting from the fact

**Fig. 9.** A position from a tournament game played by the first author (WTM). White's move Qg1 forces black to play a6, allowing Bb6. The move ordering of the neural network: h4 Bb6 Qg1 Qb3 Re1 Nf4 Nb5 a6 Qd2 gh5 g5 Ne5 Bf1 Ba7 Bg3 Rc1 Kh2 Qc1 Ra3 Bh4 Qb1 Rf1 Na4 Rg1 Kf1 Rh2 Ra2 Rb1 Qf1 Ra4 Bd4 Be1 Qe1 Qc2 Bg1 Na2 Ne1 Nb1 Nc5 Qa4 Be3 Nc1 Bc5 b5 Kg1 Nb2.

that in the position the middle-game principles are not true anymore, and it is completely natural to move the king closer to the center of the board. This suggests that if too many simplifications were made, and especially if the queens disappeared from the board, it is better not to use the neural networks for move ordering, even if under some criteria the position is considered to be the middle game.

The third position (Fig. 9) is a closed middle-game position (as opposed to the open position of Fig. 6). Although the move made in the game is again atypical, the neural network scores it relatively high again.

# Board Maps and Hill-Climbing for Opening and Middle Game Play in Shogi

Reijer Grimbergen[1] and Jeff Rollason[2]

[1] Department of Information Science, Saga University, Japan
grimbergen@fu.is.saga-u.ac.jp
[2] Oxford Softworks, England
100031.3537@compuserve.com

**Abstract.** Most strong game-playing programs use large, well tuned opening books to guide them through the early stages of the game. However, in *shogi* (Japanese chess) the classic approach of building a large opening book of known positions is infeasible. In this paper, we present a different approach for opening and middle game play in shogi. This method uses board maps that assign values to each square for each piece in a number of different formations. Hill-climbing is then used to guide pieces to optimal squares. We define board maps for defensive piece formations (castles), attacking formations (assaults) and for recognizing the type of opening position. Results show that using board maps in combination with hill-climbing significantly improves the playing strength of a shogi program. Furthermore, using maps for both castles and assaults is better than using only maps for castles.

## 1 Introduction

In general, two-player perfect information games have different stages. Depending on positional features and the number of moves from the starting position, these stages are usually called opening, middle game and endgame. For game programs, the challenge in the endgame is to find the game theoretical value of a position as early as possible (preferably this should be a win in all variations).

In the middle game searching as deep as possible has been the most successful method to find the best possible next move from a certain position. In this stage a game program is basically on its own and has to rely on the evaluation function for its actions.

Finally, in the opening the aim is to develop one's pieces in such a way that the chances of winning are at least equal to those of the opponent. Although search plays an important role here as well, a game program can make use of the fact that for most games the initial position is known. Furthermore, there is usually a fair amount of information about how to play the opening in the form of expert games and analysis. This information can be stored in a database called *opening book*. When the board position no longer matches any position in the database, the program is *out of book*.

J. Schaeffer et al. (Eds.): CG 2002, LNCS 2883, pp. 171–187, 2003.

## 1.1   Opening Books in Games

For most strong game programs, the opening book is an important contribution to the playing strength. For the chess program BELLE, Ken Thompson spent one hour a day for three years typing in the opening lines of the *Encyclopedia of Chess Openings*. This substantially improved the performance of the program [1]. The checkers program CHINOOK needed a book to avoid known book losses which might require a search of more than 40 ply. Initially the opening book had only 2600 entries, but especially for the famous 1994 match against World Champion Marion Tinsley, the large opening book of the program COLOSSUS was added to CHINOOK and COLOSSUS' programmer Martin Bryant to the CHINOOK team [2].

Both CHINOOK and the Othello program LOGISTELLO used opening book learning to find improvements of the opening theory. CHINOOK's opening book contained "cooked" moves that were not given in the literature, but a deep search had shown to improve the winning chances [2]. The opening book learning in LOGISTELLO was important for the 6-0 victory over World Champion Takeshi Murakami in 1997 [3].

Perhaps the most heavily tuned opening book was the one used by the chess program DEEP BLUE. The opening book of DEEP BLUE was built from 700,000 games. For the famous 1997 match against Kasparov the opening book was tuned by no less than four grandmasters (Benjamin, De Firmian, Fedorowicz and Illescas) [4]. As a result, in the match Kasparov did not dare to play the opening in his usual aggressive style, and this might have been one of the reasons for his historic defeat [5].

## 1.2   Shogi Compared to Chess

*Shogi* (Japanese chess) is a two-player perfect information game that is similar to chess. The goal of the game is the same as in chess, namely to capture the king of the opponent. However, there are some important differences between shogi and chess:

**Different Board Size.** A shogi board has 81 (9 × 9) squares instead of 64 (8 × 8) squares like in chess.

**Different Pieces.** In shogi each player has one king, one rook, one bishop, two golden generals, two silver generals, two knights, two lances and nine pawns. There is no queen like in chess, but instead there are the shogi-specific pieces golden general, silver general and lance. Also, shogi has more pieces than chess: 40 instead of 32.

**Re-use of Captured Pieces.** In shogi, captured pieces can be re-used. When it is a player's turn to move, a choice can be made between moving one of the pieces on the board or putting one of the pieces previously captured back on an empty square. This second type of move, where a captured piece is returned on the board, is called a *drop*.

**Less Mobility for Each Individual Piece.** In chess, the majority of pieces can move to multiple squares at a time. In shogi, each side has only one bishop and one rook with the same move capabilities as in chess. The other pieces can only move to the adjacent squares. Exceptions are the knight and lance. The knight can jump like in chess, but is limited in its movement compared to chess. A shogi knight can only make the two knight jumps to the front. The lance can move multiple squares vertically, but is not allowed to move backwards. Because of these limitations, neither the knight nor the lance can be considered mobile pieces in shogi.

**Different Castling Rules.** In shogi there are no standard castle moves. In chess there are only two different castle formations: castling on the king side or castling on the queen side of the board. Furthermore, castling in chess only takes a single move. To put the king in a safe position in shogi, a castle has to be built and the king has to be moved into this castle. There are a number of standard castle formations and there is opening theory about the move order in which this formation should be built based on the moves of the opponent. Although there are more than 50 standard castle formations, new ones are still being invented regularly. Castles can take a long time to complete. For example, building the *anaguma* castle, which is the strongest castle formation is shogi, takes more than 10 moves. Also, castle formations in shogi can have multiple stages. For example, the *mino* castle can be turned into the stronger *high mino*, which can be rebuilt into a *silver crown*.

**Different Promotion Rules.** In chess only pawns can promote and the pawn can promote to any piece (usually a queen). In shogi most pieces can promote, but the only choice is between promotion and non-promotion. Most pieces in shogi promote to gold (pawn, lance, knight and silver all promote to gold). However, the rook and bishop promote to pieces that add the ability to move one square vertically and horizontally (promoted bishop) or one square diagonally in each direction (promoted rook). Gold and king do not promote in shogi. Another difference between chess and shogi is that in shogi promotion is possible on the top three ranks instead of only on the top rank like in chess.

Recently, the strength of shogi programs has been steadily increasing and the top programs are considered to be about 3-dan, which is strong amateur level [6]. However, only in the mating stage of the game do shogi programs outperform human experts [7]. In all other stages of the game, there is still much room for improvement and one important research area is the opening, which according to top programmers is still one of the weakest points. Yamashita, author of former world champion program Yss, judges the opening play of his program about 1-kyu [8]. This means that the opening play of the program is between 300 and 400 ELO points weaker than its average level.

## 1.3   Using a Standard Opening Book in Shogi

The drop rule (i.e. the re-use of pieces) is generally considered to be the distinguishing feature of shogi. Most research efforts have been spent on dealing with the search explosion that is caused by this rule. Because of this rule, the average branching factor of the search tree in shogi is 80, while in chess the average branching factor is 35 [9]. However, in this paper we will focus on how some of the other differences between chess and shogi affect the opening play.

In the opening, the combination of a bigger board and pieces with less mobility make it less likely that the pieces come into contact early. Therefore, early tactical complications in shogi are quite rare. Furthermore, because multiple moves are needed to build a castle, in most games the priority in the opening is given to building a good defensive castle formation. Of course, these types of quiet, strategic openings also occur in chess, but in shogi they are the rule rather than the exception.

As an initial effort to improve the opening play of the shogi program SPEAR by the first author, a standard opening book was built containing more than a 1000 professional games and opening variations taken from more than 20 books on opening theory. The complete opening book built in this way has more than 110,000 positions. This is much smaller than for example the opening book of DEEP BLUE, but much larger than the ones used in most shogi programs. It is hard to build an opening book that is similar in size to the ones used in chess programs, as worldwide the number of expert shogi players is much smaller than the number of expert chess players. Therefore, in shogi the number of publicly available expert games and the number of books on opening play is much smaller than in chess.

The problem of using an opening book in shogi is illustrated with a test of the use of the opening book in games against strong shogi programs. 25 games each were played against the four strongest shogi programs on the market: AI SHOGI (winner of the Computer Shogi World Championship in 1997), KAKINOKI SHOGI (winner of the Computer Shogi Grand Prix in 1999), TODAI SHOGI (winner of the Computer Shogi World Championship in 1998, 2000 and 2001) and KANAZAWA SHOGI (winner of the Computer Shogi World Championship in 1995, 1996 and 1999). In these games, we recorded where SPEAR got out of book. The results of this test are given in Fig. 1.

The figure shows that despite the large number of positions in the opening book, the program gets out of book rather quickly. In 32 games the program is out of book within five moves and in 71 games the program is out of book in ten moves or less. On average, the program is out of book after 8.5 moves (note that this is the number of ply or half-moves, because in shogi a move by each side is counted separately).

To move into the simplest of castle formations called the *boat* castle, at least four moves are required. As said, it takes more than ten moves to move into the *anaguma*, the strongest castle in shogi. Therefore, it is unlikely that a shogi program can use an opening book to complete its opening formation[1]. A different approach is needed for strong opening play.

---

[1] A simpler test would have been to play SPEAR with and without the book against the other programs and compare the results. Unfortunately, SPEAR is not strong enough to win enough games (independent of opening book use) to draw any conclusions after such a test.

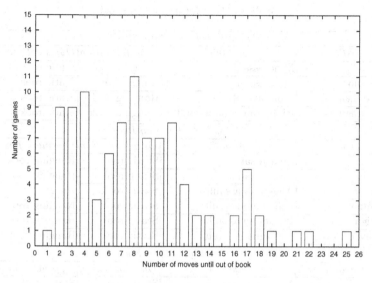

**Fig. 1.** Full board database matches in 100 games against AI SHOGI, KAKINOKI SHOGI, KANAZAWA SHOGI and TODAI SHOGI.

## 2  Board Maps for Opening and Middle Game Play in Shogi

In shogi, the rule of thumb for opening play is to build a castle with three generals. The fourth general, the bishop, rook and one knight are used to assault the castle of the opponent. The castle formation and assault formation depend on the position of the opponent's rook. There are basically two types of opening in shogi. One is the type where the rook stays on its original square. This type is called *Ibisha* or *Static rook*. The second type is the type where the rook moves sideways to a different file. This type is called *Furibisha* or *Ranging rook*. Therefore, we need a method where a shogi program can build formations for castles and assaults based on the opponent's strategy, without the help of a standard opening book. In this section we will show how board maps for pieces in combination with hill-climbing can be used to build piece formations in shogi. The definitions have three different levels of abstraction: piece formations, individual pieces and square values.

### 2.1  Piece Formations in Shogi

To guide the building of castle formations and assault formations, we have defined *castle maps* for a number of common castle formations in shogi and the *assault maps* for each castle. The list of castle maps and assault maps currently implemented is given in Table 1. The list includes the different castle formations (*Castle*), the best way to assault these castles (*Assault*) and for which type of

**Table 1.** Piece formations in shogi.

| Ranging rook vs. Static Rook | | | |
|---|---|---|---|
| Castle | Assault | Own formation | Opponent formation |
| Mino | Mino assault | Ranging rook | Static rook |
| High mino | Mino assault | Ranging rook | Static rook |
| Silver crown | Mino assault | Ranging rook | Static rook |
| Right anaguma | Right anaguma assault | Ranging rook | Static rook |
| Static rook vs. Ranging Rook | | | |
| Castle | Assault | Own formation | Opponent formation |
| Boat | Boat assault | Static rook | Ranging rook |
| Lozenge | Boat assault | Static rook | Ranging rook |
| Anaguma | Anaguma assault | Static rook | Ranging rook |
| Left mino | Left mino assault | Static rook | Ranging rook |
| Left high mino | Left mino assault | Static rook | Ranging rook |
| Left silver crown | Left mino assault | Static rook | Ranging rook |
| Static rook vs. Static Rook | | | |
| Castle | Assault | Own formation | Opponent formation |
| Yagura | Yagura assault | Static rook | Static rook |
| Yagura nakabisha | Yagura nakabisha assault | Static rook | Static rook |
| Snow roof | Climbing silver assault | Static rook | Static rook |
| Aigakari | Aigakari assault | Static rook | Static rook |
| Nakahara formation | Aigakari assault | Static rook | Static rook |
| Right side king | Mino assault | Static rook | Static rook |
| Ranging rook vs. Ranging Rook | | | |
| Castle | Assault | Own formation | Opponent formation |
| Minogakoi | Aifuribisha assault | Ranging rook | Ranging rook |
| Takamino | Aifuribisha assault | Ranging rook | Ranging rook |
| Ginkanmuri | Aifuribisha assault | Ranging rook | Ranging rook |
| Maikin | Aifuribisha assault | Ranging rook | Ranging rook |

position this castle is best suited (split into *Own formation* and *Opponent formation*). For example, the *mino* castle is best build when one's own rook has moved sideways, while the opponent has adopted a static rook strategy. On the other hand, the *boat castle* is best played when one's own rook is still in the original position, while the opponent has a ranging rook formation. The best assault against the mino castle is defined by the maps in *mino assault*, while the best assault against the boat castle is defined by the maps in *boat assault*.

## 2.2   Individual Board Maps

For each of the formations in Table 1 we need to define for each individual piece where it is best positioned. For the mino castle formation the board maps for each individual piece are given in Table 2. As mentioned before, gold and king do not promote, so there is no board map for these promoted pieces. Pieces that promote to gold (i.e. pawn, lance, knight and silver) have the same maps as a

**Table 2.** Individual board maps for a mino castle.

| Piece | Board maps | |
|---|---|---|
| | Unpromoted piece map | Promoted piece map |
| Pawn | Mino pawn | Mino gold |
| Lance | Mino lance | Mino gold |
| Knight | Mino knight | Mino gold |
| Silver | Mino silver | Mino gold |
| Gold | Mino gold | – |
| Bishop | Mino bishop | Mino bishop |
| Rook | Mino rook | Mino rook |
| King | Mino king | – |

| 9 | 8 | 7 | 6 | 5 | 4 | 3 | 2 | 1 | |
|---|---|---|---|---|---|---|---|---|---|
| −9 | −9 | −9 | −9 | −9 | −9 | −9 | −9 | −9 | a |
| −9 | −9 | −9 | −9 | −9 | −9 | −9 | −9 | −9 | b |
| −9 | −9 | −9 | −9 | −9 | −9 | −9 | −9 | −9 | c |
| −9 | −9 | −9 | −9 | −9 | −9 | −9 | −9 | −9 | d |
| −9 | −9 | −9 | −9 | −9 | −9 | −9 | −9 | −9 | e |
| −9 | −9 | −9 | −9 | −8 | −1 | −1 | −1 | −1 | f |
| −9 | −9 | −9 | −9 | −7 | 0 | 4 | 8 | 5 | g |
| −9 | −9 | −9 | −8 | −6 | 2 | 8 | 14 | 6 | h |
| −9 | −9 | −9 | −7 | −5 | −1 | 8 | 8 | 6 | i |

**Fig. 2.** *Mino king*: The board map for the king in the mino castle.

gold. Promoted rook and promoted bishop have the same board maps as the rook and bishop respectively. Promotion in shogi is only possible in the enemy camp (top three ranks of the board) and castles are built in one's own camp (bottom three ranks). Therefore, board maps for promoted pieces do not play an important role in the opening and middle game as castles are never built that far up the board and most pieces in shogi have very limited backward movement, making it highly unusual that a promoted piece can move all the way back to one's own camp. Exceptions are the rook and bishop, for which it is not unusual to move back in defence. Especially a promoted bishop is considered a strong defensive piece and the board map for the promoted bishop can be important for the castle formation.

## 2.3   Board Maps for Individual Pieces

Finally, for each individual piece we need to define where it is positioned best on the board. This is done by giving a value to each square on the 9 × 9 board for each individual piece. An example of the board map for the king in the mino castle formation (i.e. *Mino king* in Table 2) is given in Fig. 2.

From the values in this board map it can be inferred that the king in the mino castle is best positioned on square 2h which has the value 14. To move the king to this square, optimal paths can be constructed with a hill-climbing

| 9 | 8 | 7 | 6 | 5 | 4 | 3 | 2 | 1 | |
|---|---|---|---|---|---|---|---|---|---|
| -9 | -9 | -9 | -9 | -9 | -9 | -9 | -9 | -9 | a |
| -9 | -9 | -9 | -9 | -9 | -9 | -9 | -9 | -9 | b |
| -9 | -9 | -9 | -9 | -9 | -9 | -9 | -9 | -9 | c |
| -9 | -9 | -9 | -9 | -9 | -9 | -9 | -9 | -9 | d |
| -9 | -9 | -9 | -9 | -9 | -9 | -9 | -9 | -9 | e |
| -9 | -9 | -9 | -9 | -8 | -1 | -1 | -1 | -1 | f |
| -9 | -9 | -9 | -9 | -7 | 2 | 1 | 4 | 1 | g |
| -9 | -9 | -9 | -8 | -6 | 1 | 7 | -8 | 1 | h |
| -9 | -9 | -9 | -7 | -1 | 1 | 2 | -8 | 0 | i |

**Fig. 3.** *Mino silver:* The board map for the silver general in the mino castle.

| 9 | 8 | 7 | 6 | 5 | 4 | 3 | 2 | 1 | |
|---|---|---|---|---|---|---|---|---|---|
| -9 | -9 | -9 | -9 | -9 | -9 | -9 | -9 | -9 | a |
| -9 | -9 | -9 | -9 | -9 | -9 | -9 | -9 | -9 | b |
| -9 | -9 | -9 | -9 | -9 | -9 | -9 | -9 | -9 | c |
| -9 | -9 | -9 | -9 | -9 | -9 | -9 | -9 | -9 | d |
| -9 | -9 | -9 | -9 | -9 | -9 | -9 | -9 | -9 | e |
| -9 | -9 | -9 | 0 | 1 | 2 | 2 | 0 | 0 | f |
| -9 | -9 | -9 | 2 | 2 | 5 | 4 | 2 | 0 | g |
| -9 | -9 | -9 | 2 | 4 | 3 | -1 | 1 | -1 | h |
| -9 | -9 | -9 | 1 | 2 | 9 | -4 | 0 | -1 | i |

**Fig. 4.** *Mino gold:* The board map for the golden general in the mino castle.

approach: move the piece to a neighbouring square with a higher value. If more than one neighbouring square has a higher value, choose the square with the highest value.

In the example of Fig. 2, there are two natural paths to the optimal square from the starting position of the king on 5i (the square with value −5):

1. K5i ⇒ K4h ⇒ K3h ⇒ K2h
2. K5i ⇒ K4h ⇒ K3i ⇒ K2h

Both paths have square values {−5, 2, 8, 14}.

To obtain a sequence of moves to built a formation, the maximum improvement over all board maps for the pieces is obtained. In Fig. 3 and Fig. 4 the board maps for the silver general and golden general in the mino castle are given. If the king is again supposed to be on square 5i, the golden generals on squares 4i and 6i and the silver general on square 3i, the optimal sequence is as follows:

K5i $\overset{+7}{\Longrightarrow}$ K4h, K4h $\overset{+6}{\Longrightarrow}$ K3h, K3h $\overset{+6}{\Longrightarrow}$ K2h, S3i $\overset{+5}{\Longrightarrow}$ S3h, G6i $\overset{+3}{\Longrightarrow}$ G5h

Note that the second sequence for moving the king that was given earlier is not optimal here, as the silver has to move from square 3i first to allow the king to move to that square. The silver move from 3i to 3h only gives an improvement of +5, while the move K4h-3h gives an improvement of +6.

The assault maps are defined in exactly the same way. For each assault there is a table like Table 2 to define the maps of each piece. For each individual piece there is a map like Fig. 2, where the square evaluation of each piece for this particular assault formation is given.

# 3 Using the Maps to Guide the Opening and Middle Game Play

In our implementation, the maps are used to select the castle formation that can best be built from the current position. This is the castle formation that resembles the current formation most, i.e. the castle formation for which the total of the square values of all pieces is the highest. When the castles for both sides have been identified, the assaults for both castles are defined by Table 1. After this selection, the hill-climbing defined by the map values guides the search towards moves that improve the chosen castle formation and assault formation. This is done by adding the castle and assault values to the positional evaluation of each individual piece in the evaluation function.

Note that one of the side effects of the method is that formations which for some reason have gone astray can be fixed. Even if the optimal move order is no longer possible, the hill-climbing approach might still generate non-optimal moves to get the pieces into the right formation. For example, if there is a piece on 4h in the mino castle but not on 4i, the map of Fig. 2 can still generate a path for the king to move into the mino castle (K5i $\Rightarrow$ K4i $\Rightarrow$ K3h $\Rightarrow$ K2h).

There are several other ways in which the maps can be useful in a shogi program. The maps establish the correct order in which a castle or assault formation should be built, guide the use of the opening book and help in establishing the difference between opening and middle game. We will now discuss each of these in detail.

## 3.1 Playing Moves in the Correct Order

For most castle and assault formations in shogi, there is a standard order in which the formation should be built. Building a formation in the wrong order can give the opponent the time to anticipate on the formation and take counter-measures that make it impossible to reach the intended formation, resulting in an unfavourable position.

If castle values and assault values are only added in the evaluation function, this might result in the program playing moves in the wrong order. From the point of view of the evaluation function, playing king move $K$, silver move $S$ and gold move $G$ in the order $K \Rightarrow S \Rightarrow G$ will have the same value as when the order would have been $G \Rightarrow S \Rightarrow K$. As shown in the example above, the maps define the optimal move order for building a formation. To enforce this move order, we add bonus values to moves. These values are relative to the improvement of the castle value and assault value and are also relative to the search depth. Playing the move with the best improvement first will therefore have a higher bonus than playing this move later in the search.

## 3.2    Opening Book Guidance

The maps can help with a problem that often occurs in game programs with a large opening book. This is the problem that the program blindly follows the opening book and ends up in a position that human experts judge as better but the program does not understand because of the limitations of the evaluation function. Rather than trying to reach a position that is objectively better, the program should aim for a position it can understand. The maps provide a way to do this. If the opening book move decreases the map value of the moving piece, it should not be trusted as it might put the pieces in a position the program does not like. Therefore, such a move should not be played without search. Of course, it should not be discarded either, as it might have been the only way to avoid losing the game.

## 3.3    Establishing the Difference between Opening and Middle Game

Maps can help in establishing the game stage. In a shogi program, it is important to make a difference between the opening, middle game and endgame. The weights of the evaluation function features change dramatically based on the stage of the game. Especially the judgment of the transition from opening to middle game is a problem in shogi. The maps can provide extra information to make the decision whether the current position is still an opening position or a middle game position. Our current implementation uses a threshold on the castle map value combined with information about the optimal position of the king. If the king is in an optimal position and the castle value exceeds a certain minimal value, the position is judged to be a middle game position.

# 4    Extensions of the Basic Method

The method described so far is static. Once a castle formation has been selected at the start of the search, the program sticks with it, even if during search it turns out that a better castle could have been built. Also, the judgment of the initial castle can have errors, especially in the early stages of the game where several castles can be almost equally likely candidates.

Therefore, in addition to the basic method, some extensions are needed to make the method more dynamic and handle exceptions properly. We have implemented *swapping rules*, *scaling of formation values* and *multiple castles*.

## 4.1    Swapping Rules

There are situations in which the selection of the correct castles can not be made based on the maps alone. For example, in the very early stages of the opening, when it is still possible to build many different castle formations, the selection process needs extra support in the form of common shogi opening knowledge. This has been implemented with a number of swapping rules for castle formations and assault formations. An example of such a swapping rule is:

```
IF Castle = Yagura AND Bishop exchanged
THEN Castle = Kakugawari
```

The difference between a *yagura castle* and a *kakugawari castle* only depends on whether the bishops are exchanged (kakugawari) or not (yagura). This rule makes sure that this is handled correctly.

The swapping rules for assaults are especially important, since there can be more than one assault on a castle. However, from the data structure of Fig. 1 only a single assault can be derived for each castle formation. The option of changing this assault based on positional features is therefore vital.

## 4.2   Scaling of Formation Values

Related to the idea of swapping is the idea of scaling the total values for each formation. The importance of a castle formation and assault formation changes during the game. In the opening it is very important to quickly build the right castle formation, but in the endgame it is more important to mate the opponent king. Therefore, in the endgame keeping a castle formation intact does not have the highest priority anymore. A global scaling factor based on the stage of the game is attached to each map to account for this.

Scaling factors can also be used to force the program to make the right choice when there is only a subtle difference between two formations. An example of a scaling rule is:

```
IF Castle = Anaguma AND Corner lance on initial square
THEN Decrease castle value by 30%
```

This rule states that even though in general it is a good thing to aim for the strong anaguma castle, this is not necessarily the right castle if the lance in the corner is still on its initial square. In this case the castle value is decreased by 30%, making the selection of a different castle formation as the prime target more likely.

In the current implementation, swapping rules and scaling are hardcoded in the program. At some point it might be necessary to compile the rules into an external knowledge-base which can be loaded at run time.

Swapping rules and scaling are only used when the game is in the opening stage. As soon as a player has committed himself fully to a certain castle or assault formation, scaling and swapping do not play an important role as it is unlikely that there will be time to switch a completely different castle as soon as the fighting starts.

In the opening, swapping and scaling is performed at all nodes in the search tree. By doing this, it is possible to repair castle formations during the search. If the rules are used exclusively at the root node, only moves that improve the formations chosen at the root will get a positive score and opportunities to switch to a better castle might be missed.

## 4.3  Multiple Castles

A final extension of the method deals with the problem of only selecting a single castle formation and assault formation for each side. Especially in the early stages of the game, where neither player has committed to a certain formation, the values of different formations will be very similar and picking one based on a small difference in the total formation value is almost the same as picking one at random. However, a wrong choice can lead to a quick disaster. To avoid this problem, our implementation has the option of keeping a set of castle formations during the hill-climbing instead of a single formation. Only castle formations that are similar to the castle with the highest value should be kept, so there is a threshold on the value difference between the castle formations:

```
IF Castlevalue ≥ Threshold
THEN Add castle to list of alternatives
```

At each node of the search, only the formation value of the castles in the list of alternatives are calculated. This is usually a small set of only a few castles. The castle with the highest total value is then chosen as the target formation from that particular node in the search tree.

## 5  Results

The castle maps and assault maps were first implemented in the shogi program SHOTEST by the second author in the version that participated in the Computer Shogi World Championship in 1999. For the tournament in 2001, the method was implemented in the program SPEAR. The number of maps was increased considerably and extensively hand-tuned by playing hundreds of games against other commercial programs. The latest version has maps for 35 different castle formations and 20 assault formations[2]. Also implemented are the extensions of the basic method given in Sect. 4. As these have thus far only been implemented in SPEAR, all the results given below are obtained from tests with this program.

To show the importance of using maps for playing strength we performed a number of self-play experiments. We played three versions of the shogi program SPEAR against each other. One version was using both castle maps and assault maps (*AllMaps*), one version was using only castle maps (*CasMaps*) and one version was not using any maps (*NoMaps*). The basic shogi program has the following features:

- Iterative alpha-beta search.
- Principal variation search [10].
- Quiescence search [11].
- History heuristic and killer moves [12].
- Null-move pruning [13].

---

[2] Some of the swapping rules introduce castles not in the basic set of Table 1, for example the *kakugawari* castle mentioned earlier.

**Table 3.** Results of self play experiments between a version of SPEAR using both castle maps and assault maps (*AllMaps*), a version using only castle maps (*CasMaps*) and a version not using any maps (*NoMaps*). *Book* is the maximum number of moves the book was used in each game.

| Match | Book | | | | | Result |
|---|---|---|---|---|---|---|
| | 5 | 10 | 20 | 30 | 40 | |
| *AllMaps-NoMaps* | 31-19 | 38-12 | 30-20 | 27-23 | 25-25 | 151-99 |
| | 62% | 76% | 60% | 54% | 50% | 60% |
| *CasMaps-NoMaps* | 29-21 | 36-14 | 31-19 | 25-25 | 25-25 | 146-104 |
| | 58% | 72% | 62% | 50% | 50% | 58% |
| *AllMaps-CasMaps* | 28-22 | 27-23 | 29-21 | 29-21 | 29-21 | 142-108 |
| | 56% | 54% | 58% | 58% | 58% | 57% |
| Total | 88-62 | 101-49 | 90-60 | 81-69 | 79-71 | 439-311 |
| | 59% | 67% | 60% | 54% | 53% | 58% |

- Hash tables for transposition and domination [14].
- Specialized mating search [15].

In this experiment, we specifically wanted to know how the interaction between the opening book and the maps influenced playing strength. Therefore, we had the programs follow the opening book for 5, 10, 20, 30 and 40 moves (a maximum of 20 moves by black and 20 moves by white).

When the opening book is used for only 5 moves (three moves by black and two moves by white), it is impossible to use the opening book for building a castle and assault formation. In this case the program must build a proper formation by using only the board maps.

In Sect. 1.3 we showed that a standard opening book can be expected to be useful for 8.5 moves on average. The tests in which the opening book is used for 10, 20 and 30 moves are therefore performed to see how the playing strength is influenced after the program is out of book during different stages of the opening.

If the opening book is used for over 30 moves, the book can typically be used for building a full castle and assault formation, so it was expected that the maps will play a less important role. The test using an opening book for 40 moves was performed to confirm this hypothesis.

When the program gets out of book, the resulting position is played twice: once with each program version playing the black pieces. Each program version played the other versions fifty times, i.e. 25 times with black and 25 times with white. The time limit was 25 minutes per side per game on a 1GHz Pentium III. This is the same time limit that is used in the Computer Shogi World Championship. The results of this self-play experiment are given in Table 3.

From the results, the following conclusions can be drawn. First, when the program has almost no support from the opening book, it benefits significantly from having the castle maps and assault maps. If the database was used for 10 or 20 moves, the version with both castle maps and assault maps outperforms the version without maps with a score of 68-32. The version with only castle maps has almost the same result with a score of 67-33. The 68-32 result gives a 99.98% probability that the program with maps is stronger than the program without the maps. For the 68-32 result, this probability is 99.97%.

When the opening book is used for 30 or 40 moves, there is almost no difference in playing strength between the version without maps and the two other versions. This is consistent with our expectation that the opening book can be used long enough to build proper castle formations and assault formations.

If the opening book is used for only 5 moves, the results are worse for the two matches *AllMaps-NoMaps* and *CasMaps-NoMaps* compared to the results when the opening book is used for 10 moves. We explained in Sect. 2 that it is important for the choice of the castle formation to know whether the opponent has built a static rook formation or a ranging rook formation. However, when the opening book is used for only 5 moves, it frequently happens that there is not enough information about which formation the opponent will choose. Especially the *NoMaps* program will not commit itself and not play a known formation after that, confusing the versions using maps into building the wrong castle formation. Although this is in part a tuning problem (building reasonable formations against uncommitting opponents), it also suggests that using only the board maps is not sufficient. The best performance is achieved by using a standard opening book to set up the formations and use the board maps to complete them.

A final observation is that the two versions with maps have almost the same result against the version without maps: the match *AllMaps-NoMaps* ended in 151-99 and the match *CasMaps-NoMaps* ended in 146-104. However, the version with both castle maps and assault maps outperforms the version with only castle maps consistently throughout the use of the opening book, the version with both castle maps and assault maps winning three matches 29-21, one match 28-22 and one match 27-23. It seems that if the program has no clue about the castle of the opponent or makes a mistake in assessing the castle, the assault maps have no significant impact. However, if the opponent plays a known castle, the use of assault maps will lead to a significant improvement in playing strength which lasts for a long period of the game.

## 6    Related Work

We believe that our work is very similar to the so-called *otoshiana* method, which was first used in the shogi program Yss [16] made by Hiroshi Yamashita. The *otoshiana* method also uses a hill-climbing approach to build castle formations and it seems likely that the implementation is similar to ours. This is difficult to judge, as Yamashita's description of the *otoshiana* method is short and lacks implementation details. The original description mentions only 9 different castle formations and no assault formations. This could be an important difference, as our results indicate that the use of assault maps considerably improves the playing strength. Furthermore, it is unclear how the *otoshiana* method assesses the type of position that the opponent is building. Yamashita mentions that the position of the rook is taken into account, "among other things", but no details are given. Still, from private conversations with other programmers it can be concluded that most strong shogi programs use a method similar to the *otoshiana* method.

Another idea for shogi, proposed by Kotani, is to use a similarity measure of the current position with positions in the opening book. When the position is similar, the move that was played in an earlier (slightly different) position might also be a good candidate in the current position. A similarity measure could then be given as a bonus to the move during a normal search [17].

Kotani's idea has some similarities to DEEP BLUE's *extended opening book* [4], where information of a large database of games is used to guide the search in the direction of the expert consensus. Moves that were played more often are given a higher search bonus than moves that were played infrequently.

In an earlier version of the program SPEAR, a combination of the Kotani and DEEP BLUE ideas was tried. A similarity measure combined with move frequency was added as a search bonus. However, we were not able to produce stable opening play with this method and the idea was abandoned in favour of the method presented in this paper.

## 7   Conclusions

In this paper we have described a new method for constructing piece formations in the opening and middle game in shogi. This method uses board maps to decide the best formation to aim for and uses hill-climbing to improve the selected formation.

In shogi, three different sets of formations are needed: castle formations, assault formations and two special formations that determine the type of position. We have shown that if the program gets out of book in the first 20 moves, the playing strength can be improved considerably by using castle maps and assault maps. Furthermore, using both castle maps and assault maps gives significantly better results than using only castle maps.

We feel that further testing of the method is necessary to establish objectively how important this method is for strong opening play. We have not had the opportunity to separately test the importance of the extensions to the basic method. One problem is that the set of swapping rules and castle scales is not stable yet. Also, the experimental set-up can be a problem, as swapping rules and castle scales do not always influence the opening play. However, as soon as the set of swapping rules and castle scales have stabilized, we need to establish how important these are for the overall performance.

Another problem of our tests is that self-play experiments have the important limitation that only relative improvement is measured. Therefore, it is necessary to have some external measure to evaluate the performance of our idea. One test would be to play our programs against the strong commercial programs mentioned in Sect. 1.3. Another idea is to have a strong player (preferably a top professional player) evaluate a set of positions that the program produces at different points in the opening and middle game.

We also want to investigate automatic learning of board maps. To apply our method successfully to shogi, a lot of hand tuning was needed. Currently, we are looking into ways to automatically construct the maps from games played by expert players.

As a final note, it is interesting that the *otoshiana* method also includes a number of positional elements unrelated to castle formations and assault formations. For example, having a lance on the back rank is better than having a lance higher up the board and there are penalties for having a knight high up the board or a rook on the third rank. Our current research has focused only on castle and assault formations, but the method might have a more general use for positional evaluation of other piece formations as well. We are planning to investigate such extensions of the method in the near future.

# References

1. Baird, H., Thompson, K.: Reading chess. IEEE Transactions on Pattern Analysis and Machine Intelligence **12** (1990) 552–559
2. Schaeffer, J.: One Jump Ahead: Challenging Human Supremacy in Checkers. Springer-Verlag (1997)
3. Buro, M.: The Othello match of the year: Takeshi Murakami vs. LOGISTELLO. International Computer Chess Association Journal **20** (1997) 189–193
4. Campbell, M., Hoane Jr., A., Hsu, F.: DEEP BLUE. Artificial Intelligence **134** (2002) 57–83
5. Schaeffer, J., Plaat, A.: Kasparov versus DEEP BLUE: The rematch. International Computer Chess Association Journal **20** (1997) 95–101
6. Takizawa, T., Grimbergen., R.: Review: Computer shogi through 2000. In Marsland, T., Frank, I., eds.: Second International Conference on Computers and Games (CG 2000). Volume 2063 of Lecture Notes in Computer Science. Springer-Verlag (2001) 433–442
7. Grimbergen, R.: A survey of tsume-shogi programs using variable-depth search. In van den Herik, J., Iida, H., eds.: First International Conference on Computers and Games (CG 1998). Volume 1558 of Lecture Notes in Computer Science. Springer-Verlag (1999) 300–317
8. Yamashita, H.: YSS. `http://plaza15.mbn.or.jp/~yss/index_j.html` (2003) In Japanese.
9. Matsubara, H., Iida, H., Grimbergen, R.: Natural developments in game research: from chess to shogi to Go. International Computer Chess Association Journal **19** (1996) 103–112
10. Pearl, J.: Heuristics: Intelligent Search Strategies for Computer Problem Solving. Addison Wesley Publishing Company (1984)
11. Beal, D.: A generalised quiescence search algorithm. Artificial Intelligence **43** (1990) 85–98
12. Schaeffer, J.: The history heuristic and alpha-beta search enhancements in practice. IEEE Transactions on Pattern Analysis and Machine Intelligence **11** (1989) 1203–1212
13. Beal, D.: Experiments with the null move. In Beal, D., ed.: Advances in Computer Chess 5, Elsevier Science Publishers B.V. (1989) 65–79

14. Seo, M.: On effective utilization of dominance relations in tsume-shogi solving algorithms. In: Game Programming Workshop in Japan '99. (1999) 129–136 In Japanese.
15. Seo, M.: The C* algorithm for AND/OR tree search and its application to a tsume-shogi program. Master's thesis, University of Tokyo, Faculty of Science (1995)
16. Yamashita, H.: Yss: About its datastructures and algorithm. In Matsubara, H., ed.: Computer Shogi Progress 2, Kyoritsu Suppan Co. (1998) 112–142 In Japanese.
17. Kotani, Y.: Example-based piece formation by partial matching in shogi. In van den Herik, J., Monien, B., eds.: Advances in Computer Games 9, Universiteit Maastricht, Department of Computer Science (2001) 223–232

# Solitaire Clobber

Erik D. Demaine[1], Martin L. Demaine[1], and Rudolf Fleischer[2]

[1] MIT Laboratory for Computer Science, Cambridge, MA, USA
{edemaine,mdemaine}@mit.edu
[2] HKUST Department of Computer Science, Hong Kong
rudolf@cs.ust.hk

**Abstract.** Clobber is a new two-player board game. In this paper, we introduce the 1-player variant Solitaire Clobber where the goal is to remove as many stones as possible from the board by alternating white and black moves. We show that a $n$ stone checkerboard configuration on a single row (or single column) can be reduced to about $n/4$ stones. For boards with at least two rows and columns, we show that a checkerboard configuration can be reduced to a single stone if and only if the number of stones is not a multiple of three, and otherwise it can be reduced to two stones. But in general it is NP-complete to decide whether an arbitrary Clobber configuration can be reduced to a single stone.

## 1 Introduction

Clobber is a new two-player combinatorial board game with complete information, recently introduced by Albert, Grossman, and Nowakowski (see [1]). It is played with black and white stones occupying some subset of the squares of an $n \times m$ checkerboard. The two players, White and Black, move alternately by picking up one of their own stones and *clobbering* an opponent's stone on a horizontally or vertically adjacent square. The clobbered stone is removed from the board and replaced by the stone that was moved. The game ends when one player, on their turn, is unable to move, and then that player loses.

We say a stone is *matching* if it has the same color as the square it occupies on the underlying checkerboard; otherwise it is *clashing*. In a *checkerboard configuration*, all stones are matching, i.e., the white stones occupy white squares and the black stones occupy black squares. And in a *rectangular* configuration, the stones occupy exactly the squares of some rectangular region on the board. Usually, Clobber starts from a rectangular checkerboard configuration, and White moves first (if the total number of stones is odd we assume that it is White who has one stone less than Black).

At the recent Dagstuhl Seminar on Algorithmic Combinatorial Game Theory [2], the game was first introduced to a broader audience. Tomáš Tichý from Prague won the first Clobber tournament, played on a $5 \times 6$ board, beating his supervisor Jiří Sgall in the finals. Not much is known about Clobber strategies, even for small boards, and the computation of combinatorial game values is also only in its preliminary stages [1].

J. Schaeffer et al. (Eds.): CG 2002, LNCS 2883, pp. 188–200, 2003.
© Springer-Verlag Berlin Heidelberg 2003

In this paper we introduce *Solitaire Clobber*, where a single player (or two cooperative players) tries to remove as many stones as possible from the board by alternating white and black moves. If the configuration ends up with $k$ immovable stones, we say that the initial board configuration is *reduced* to $k$ stones, or $k$-*reduced*. Obviously, 1-reducibility can only be possible if half of the stones are white (rounded down), and half of the stones are black (rounded up). But even then it might not be possible.

We prove the following necessary condition for a Clobber position to be 1-reducible: The number of stones plus the number of clashing stones cannot be a multiple of three. Surprisingly, this condition is also sufficient for truly two-dimensional rectangular checkerboard configurations (i.e., with at least two rows and two columns). And if the condition is not true, then the board is 2-reducible (with the last two stones separated by a single empty square), which is the next-best possible. A similar 2-coloring argument can be used to solve Question 3 of the 34th International Mathematical Olympiad 1993 [3] which asked to prove that the peg solitaire game (a peg can jump over an adjacent peg onto an empty square, and the jumped over peg is removed) on an $n \times n$ grid can be reduced to a single peg when $n \equiv 1 \bmod 3$. However, in general, we show that it is NP-complete to decide whether an arbitrary non-rectangular non-checkerboard configuration is 1-reducible.

If we play one-dimensional Solitaire Clobber (i.e., the board consists of a single row of stones) reducibility is more difficult. We show that the checkerboard configuration can be reduced to $\lceil n/4 \rceil + \{1$ if $n \equiv 3 \pmod 4\}$ stones, no matter who moves first, and that this bound is best possible even if we do not have to alternate between white and black moves. This result was obtained independently by Grossman [4].

This paper is organized as follows. In Sect. 2, we analyze the reducibility of checkerboard configurations on a line. In Sect. 3, we study reducibility of two-dimensional rectangular checkerboard configurations, and in Sect. 4 we show that deciding 1-reducibility is NP-complete in general. We conclude with some open problems in Sect. 5.

## 2    One-Dimensional Solitaire Clobber

In this section we study Solitaire Clobber played on a board consisting of a single row of stones. Let $A_n$ denote the checkerboard configuration, i.e., an alternating sequence of white and black stones. By symmetry, we can assume throughout this section that $A_n$ always starts with a black stone, so we have $A_n = \bullet \circ \bullet \circ \cdots$. We first show an upper bound on the $k$-reducibility of checkerboard configurations.

**Theorem 1.** *For* $n \geq 1$, *the configuration* $A_n$ *can be reduced to* $\lceil n/4 \rceil + \{1$ *if* $n \equiv 3 \pmod 4\}$ *stones by an alternating sequence of moves, no matter who is to move first.*

*Proof.* Split the configuration $A_n$ into $\lceil n/4 \rceil$ substrings, all but possibly one of length four. Each substring of length one, two, or four can be reduced to one

**Table 1.** Reducibility of one-dimensional checkerboard Clobber configurations.

| Configuration | Reducibility |
|---|---|
| $A_1$  ● | 1 |
| $A_2$  ●○ | 1 |
| $A_3$  ●○● | 2 |
| $A_4$  ●○●○ | 1 |
| $A_5$  ●○●○● | 2 |
| $A_6$  ●○●○●○ | 2 |
| $A_7$  ●○●○●○● | 3 |
| $A_8$  ●○●○●○●○ | 2 |
| $A_9$  ●○●○●○●○● | 3 |
| $A_{10}$  ●○●○●○●○●○ | 3 |
| $A_{11}$  ●○●○●○●○●○● | 4 |
| $A_{12}$  ●○●○●○●○●○●○ | 3 |

stone by alternating moves, no matter which color moves first. And a substring of size three can be reduced to two stones by one move, no matter which color moves first. □

In this move sequence, we end up with one isolated stone somewhere in the middle of each block of four consecutive stones. One might wonder whether a more clever strategy could end up with one stone at the end of each subblock, and then we could clobber one more stone in each pair of adjacent stones from the subblocks. Unfortunately, this is not possible, as shown by the following matching lower bound. The lower bound holds even if we are not forced to alternate between white and black moves. We give a simple proof for the theorem due to Grossman [4].

**Theorem 2.** *Let $n \geq 1$. Even if we are not restricted to alternating white and black moves, the configuration $A_n$ cannot be reduced to fewer than $\lceil n/4 \rceil + \{1$ if $n \equiv 3 \pmod 4\}$ stones.*

*Proof.* First, it is not possible to reduce $A_3$ or $A_5$ to a single stone. Second, each stone in the final configuration comes from some contiguous substring of stones in the initial configuration. But each of these substrings can have only one, two, or four stones. Thus, there are at least $\lceil n/4 \rceil$ stones left at the end, and even one more if $n \equiv 3 \pmod 4$. □

Somewhat surprisingly, the tight bound of Theorems 1 and 2 is not monotone in $n$, the number of stones in the initial configuration. See Table 1.

## 3    Rectangular Solitaire Clobber

In this section we study reducibility of rectangular checkerboard configurations with at least two rows and two columns. We first show a general lower bound on the reducibility that holds for arbitrary Clobber configurations. For a configuration $C$, we denote the quantity "number of stones plus number of clashing stones" by $\delta(C)$.

As it turns out, $\delta(C)$ (mod 3) actually divides all clobber configurations into three equivalence classes. Any configuration will stay in the same equivalence class, after any number of moves. Because one of the three equivalence classes (with $\delta(C) \equiv 0$ (mod 3)) does not contain configurations with a single stone, all configurations in this equivalence class are not 1-reducible. As in the 1-dimensional lower bound of Theorem 2, this is true even if we allow arbitrary non-alternating move sequences.

**Theorem 3.** *For a configuration $C$, $\delta(C)$ (mod 3) does not change after an arbitrary move sequence.*

*Proof.* If we move a matching stone in $C$ then $\delta$ drops by one because we clobber another matching stone, and $\delta$ rises by one because our stone becomes clashing, so $\delta$ actually does not change in this move. If we move a clashing stone then $\delta$ drops by two because we clobber another clashing stone, and $\delta$ drops by another one because our stone becomes matching, resulting in a total drop of three for the move. □

**Corollary 1.** *A configuration $C$ with $\delta(C) \equiv 0$ (mod 3) is not 1-reducible.*

*Proof.* A single stone can only have $\delta$ equal to one or two (depending on whether it is a matching or clashing stone). Thus, by the previous theorem, configurations $C$ with $\delta(C) \equiv 0$ (mod 3) are not 1-reducible. □

The rest of this section is devoted to a proof that this bound is actually tight for rectangular checkerboard configurations:

**Theorem 4.** *For $n, m \geq 2$, a rectangular checkerboard configuration with $n$ rows and $m$ columns is 2-reducible if $nm \equiv 0$ (mod 3), and 1-reducible otherwise.*

We present an algorithm that computes a sequence of moves that reduces the given checkerboard configuration to one or two stones as appropriate.

We distinguish cases in a somewhat complicated way. There are finitely many cases with $2 \leq n, m \leq 6$; these cases can be verified trivially, as shown in Appendix 5. The remaining cases have at least one dimension with at least seven stones; by symmetry, we ensure that the configuration has at least seven columns. These cases we distinguish based on the parities of $n$ and $m$:

- **Case EE:** Even number of rows and columns [Sect. 3.2]

- **Case OE:** Odd number of rows, even number of columns [Sect. 3.3]

- **Case EO:** Even number of rows, odd number of columns [Sect. 3.4]

- **Case OO:** Odd number of rows and columns [Sect. 3.4]

Cases OE and EO are symmetric for configurations with at least seven rows and at least seven columns. By convention, we handle such situations in Case EO. But when one dimension is smaller than seven, we require that dimension to be rows, forcing us into Case OE or Case EO and breaking the symmetry. In fact, we solve these instances of Case OE by rotating the board and solving the simpler cases E3 and E5 (even number of rows, and three or five columns, respectively).

Section 3.1 gives an overview of our general approach. Section 3.2 considers Case EE, which serves as a representative example of the procedure. Section 3.3 extends this reduction to Case OE (when the number of rows is less than seven), which is also straightforward. Finally, Sect. 3.4 considers the remaining more tedious cases in which the number of columns is odd.

## 3.1 General Approach

In each case, we follow the same basic strategy. We eliminate the stones on the board from top to bottom, two rows at a time. More precisely, each *step* reduces the topmost two rows down to $O(1)$ stones (usually one or two) arranged in a fixed pattern that touches the rest of the configuration through the bottom row.

There are usually four types of steps, repeated in the order

$$(1), \underbrace{(2), (3), (4),}_{} \underbrace{(2), (3), (4),}_{} \underbrace{(2), (3), (4),}_{} \ldots.$$

Step (1) leaves a small remainder of stones from the top two rows in a fixed pattern. Step (2) absorbs this remainder and the next two rows, in total reducing the top four rows down to a different pattern of remainder stones. Step (3) leaves yet another pattern of remainder stones from the top six rows. Finally, step (4) leaves the same pattern of remainder stones from step (1), so the sequence can repeat (2), (3), (4), (2), (3), (4), ....

In some simple cases, steps (1) and (2) leave the same pattern of remainder stones. Then just two types of steps suffice, repeating in the order (1), (2), (2), (2), .... In other cases, three steps suffice.

In any case, the step sequence may terminate with any type of step. Thus, we must also show how to reduce each pattern of remainder stones down to one or two stones as appropriate; when needed, these final reductions are enclosed by parentheses because they are only used at the very end. In addition, if the total number of rows is odd, the final step involves three rows instead of two rows, and must be treated specially.

In the description below, a single move is denoted by $\rightarrow$. But we often do not show long move sequences completely. Instead, we usually 'jump' several moves at a time, denoted by $\overset{a}{\rightarrow}$ or $\underset{a}{\rightarrow}$, depending on whether White or Black moves first, where $a$ denotes the number of moves we jump.

## 3.2   Case EE: Even Number of Rows and Columns

We begin with the case in which both $n$ and $m$ are even. This case is easier than the other cases: the details are fairly clean. It serves as a representative example of the general approach.

Because the number of columns is even and at least seven, it must be at least eight. Every step begins by reducing the two involved rows down to a small number of columns. First, we clobber a few stones to create the following configuration in which the lower row has two more stones than the upper row, one on each side:

$$\bullet\circ\cdots\bullet\circ \xrightarrow{3} \bullet\circ\cdots\circ\cdot \xrightarrow{3} \cdot\bullet\cdots\circ\cdot$$

Then we repeatedly apply the following reduction, in each step removing six columns, three on each side:

$$\cdot\bullet\circ\bullet\cdots\circ\bullet\circ\cdot \xrightarrow{4} \cdot\circ\cdot\bullet\cdots\circ\cdot\bullet \xrightarrow{4} \cdots\bullet\cdots\circ\cdots \xrightarrow{4} \cdots\cdots\circ\cdots\bullet\cdots$$

We stop applying this reduction when the bottom row has just six, eight, or ten columns left, and the top row has four, six, or eight columns, depending on whether $m \equiv 2$, 1, or 0 (mod 3), respectively.

The resulting two-row configuration has either a black stone in the lower-left and a white stone in the lower-right $\left(\begin{smallmatrix}\cdot\bullet\\\bullet\circ\end{smallmatrix}\cdots\begin{smallmatrix}\circ\cdot\\\bullet\circ\end{smallmatrix}\right)$, or vice versa $\left(\begin{smallmatrix}\cdot\circ\\\circ\bullet\end{smallmatrix}\cdots\begin{smallmatrix}\bullet\cdot\\\circ\bullet\end{smallmatrix}\right)$. We show reductions for the former case; the latter case is symmetric.

*Case 1: $m \equiv 2$ (mod 3)*

(1)  [stone diagram] $\xrightarrow{3}$ [stone diagram] $\xrightarrow{3}$ [stone diagram] $\xrightarrow{2}$ [stone diagram] $\left(\to \text{[stone diagram]}\right)^{1}$

(2)  [stone diagram] $\xrightarrow{2}$ [stone diagram] $\xrightarrow{2}$ [stone diagram] $\xrightarrow{2}$ [stone diagram] $\xrightarrow{3}$ [stone diagram]

(3)  [stone diagram] $\xrightarrow{2}$ [stone diagram] $\xrightarrow{2}$ [stone diagram] $\xrightarrow{2}$ [stone diagram] $\xrightarrow{2}$ [stone diagram]

(4)  [stone diagram] $\xrightarrow{2}$ [stone diagram] $\xrightarrow{2}$ [stone diagram] $\xrightarrow{2}$ [stone diagram] $\xrightarrow{2}$ [stone diagram] $\left(\to \text{[stone diagram]}\right)$

*Case 2: $m \equiv 1$ (mod 3)*

(1)  First we clear another six columns and obtain

[stone diagram] $\xrightarrow{12}$ [stone diagram] $\to$ [stone diagram]

(2)  [stone diagram] $\to$ [stone diagram] $\xrightarrow{3}$ [stone diagram] $\to$ [stone diagram] $\xrightarrow{2}$ [stone diagram] $\xrightarrow{4}$ [stone diagram]

(3)  [stone diagram] $\to$ [stone diagram] $\xrightarrow{3}$ [stone diagram] $\xrightarrow{2}$ [stone diagram] $\xrightarrow{2}$ [stone diagram] $\xrightarrow{4}$ [stone diagram]

(4)  [stone diagram] $\xrightarrow{2}$ [stone diagram] $\xrightarrow{4}$ [stone diagram] $\xrightarrow{2}$ [stone diagram] $\xrightarrow{4}$ [stone diagram] $\xrightarrow{3}$ [stone diagram]

---

[1] Parenthetical moves are made only if this is the final step.

*Case 3:* $m \equiv 0 \pmod 3$

(1) First we clear another six columns and obtain

```
     ·●○●○●○·    12  ····○●····    4  ··········
     ●○●○●○●○○●  →   ···○●○●···    →   ···○·●····
```
(2)
```
     ···○·●···   6  ···○·●···    4  ···○····     4  ··········   4  ··········
     ·●○●○●○○·   →  ···●○●○···   →  ··●●·○·   →   ···●●·····   →   ··········
     ●○●○●○●○●○      ·○●○●○●●·      ··○○●·●○●        ·○●●·○○●·        ···○·●···
```
(3)
```
     ···○·●···   6  ···○·●···    4  ····○●···    4  ··········   4  ··········
     ·●○○●○●○·   →  ···○○●○···   →  ···○●●···   →   ··○●·○○●··   →   ···○·●····
     ●○●○●○●○●○      ·○●○●○●●·      ··○○●○●●·        ··○●·○●●·        ···○·●····
```

## 3.3  Case OE: Odd Number of Rows, Even Number of Columns

To extend Case EE from the previous section to handle an odd number of rows, we could provide extra termination cases with three instead of two rows for any step. Because these steps are always final, they may produce an arbitrary result configuration with one or two stones.

However, as observed before, we only need to consider configurations with three or five rows in Case OE (any other configuration can be rotated into a Case EO). It turns out that we can describe their reduction more easily (and conform with all other cases) by first rotating them. Thus, the following reductions use the general approach from Sect. 3.1 to reduce configurations with three or five columns and an even number of rows.

*Three Columns:*

(1)
```
     ●○●   2  ··●   2  ···
     ○●●   →  ●○○   →  ○·●
```
(2)
```
     ○·●   2  ○··   2  ···   2  ···
     ●○●   →  ●·●   →  ○·●   →  ···
     ○●○      ○●○      ●·○      ○·●
```

*Five Columns:*

(1)
```
     ●○●○●   2  ·●●○●   2  ·●●●·   2  ·○●··   3  ·····
     ○●○●○   →  ·○○●○   →  ·○○○·   →  ··○●·   →  ··○··
```
(2)
```
     ··○··       ··○··                      ·····
     ●○●○●       ·●●●·       ·●○··   3  ·····
     ○●○●○   →   ·○○○·   →   ·○●··   →  ·○···
              4          3
```
(3)
```
     ·○···       ·○···                 ·····
     ●○●○●       ·●●●·       ·○·●·   2  ·····
     ○●○●○   →   ·○○○·   →   ·●·○·   →  ·○·●·
              4          3
```
(4)
```
     ·○·●·   4  ·○···   2  ·····   2  ·····   3  ·····
     ●○●○●   →  ·●●●○   →  ·●·●○   →  ···○·   →  ······
     ○●○●○      ·○○○·      ·○○●·      ·●○●·      ··○··
```

## 3.4  Cases EO and OO: Odd Number of Columns

Finally we consider the case of an even or odd number of rows and an odd number of columns. For each step, we give two variants, one reduction from two rows and one reduction from three rows. The latter case is applied only at the end of the reduction, so it does not need to end with the same pattern of remainder stones. Also, for an odd number of rows, the initial symmetrical removal of columns from both ends of the rows in a step is done first for the final three-row step, before any other reduction; this order is necessary because the three-row symmetrical removal can start only with a White move.

The number of columns is at least seven. Every step begins by reducing the two or three involved rows down to a small number of columns.

*Two Rows.* First, we clobber a few stones to create the following configuration in which the upper row has one more stone on the left side than the lower row, and the lower row has one more stone on the right side than the upper row:

$$\begin{smallmatrix}●○\\○●\end{smallmatrix}\cdots\begin{smallmatrix}○●\\●○\end{smallmatrix} \xrightarrow{3} \begin{smallmatrix}·○·\\·\end{smallmatrix}\cdots\begin{smallmatrix}○●\\●○\end{smallmatrix} \xrightarrow[3]{} \begin{smallmatrix}·○·\\·●·\end{smallmatrix}\cdots$$

Similar to Case EE, we repeatedly apply the following reduction, in each step removing six columns, three on each side:

$$\begin{smallmatrix}○●○●\\·○●○\end{smallmatrix}\cdots\begin{smallmatrix}●○●·\\○●○●\end{smallmatrix} \xrightarrow[3]{3} \begin{smallmatrix}○●○●\\·○●○\end{smallmatrix}\cdots\begin{smallmatrix}●···\\○●○·\end{smallmatrix} \to \begin{smallmatrix}·●○●\\··○·\end{smallmatrix}\cdots\begin{smallmatrix}●···\\○●○·\end{smallmatrix} \xrightarrow{2} \begin{smallmatrix}··●●\\···○\end{smallmatrix}\cdots\begin{smallmatrix}●···\\○○··\end{smallmatrix} \xrightarrow{4} \begin{smallmatrix}···●\\····\end{smallmatrix}\cdots\begin{smallmatrix}●···\\·○··\end{smallmatrix}$$

We stop applying this reduction when the total number of columns is just five, seven, or nine, so each row has four, six, or eight occupied columns, depending on whether $m \equiv 1, 0,$ or $2 \pmod 3$, respectively.

The resulting two-row configuration has either (a) a black stone in the upper-left and a white stone in the lower-right, $\begin{smallmatrix}●○○\\·○●\end{smallmatrix}\cdots\begin{smallmatrix}●○·\\○●○\end{smallmatrix}$, or (b) vice versa, $\begin{smallmatrix}○●○\\·○●\end{smallmatrix}\cdots\begin{smallmatrix}○●·\\●○●\end{smallmatrix}$. We will show reductions from both configurations. It turns out that configuration (a) is more difficult to handle because it is not always possible to end up with a single stone (or pair of stones) on the bottom row. In that case, we will make the last move parenthetical, omitting it whenever this step is not the last.

Sometimes we also need to start from the configuration (a') $\begin{smallmatrix}·○●\\○●○\end{smallmatrix}\cdots\begin{smallmatrix}●○●\\·○·\end{smallmatrix}$ or (b') $\begin{smallmatrix}·●○\\●○●\end{smallmatrix}\cdots\begin{smallmatrix}○●○\\·●·\end{smallmatrix}$ which are the mirror images of the configurations (a) and (b). These starting points can be achieved by applying the reductions above upside-down.

*Three rows.* First, we clobber a few stones to create the following configuration:

$$\begin{smallmatrix}●○\\○●\\●○\end{smallmatrix}\cdots\begin{smallmatrix}○●\\●○\\○●\end{smallmatrix} \xrightarrow{8} \begin{smallmatrix}·●·\\····\\·○·\end{smallmatrix}\cdots\begin{smallmatrix}○·\\··\\●·\end{smallmatrix}$$

Then, we reduce long rows by four columns at a time (not six as in the two-row reductions):

$$\begin{smallmatrix}●●○\\·○●\\○●○\end{smallmatrix}\cdots\begin{smallmatrix}○●○·\\·●○·\\○●●\end{smallmatrix} \xrightarrow[3]{3} \begin{smallmatrix}·●○\\··●\\·○○\end{smallmatrix}\cdots\begin{smallmatrix}○●○·\\·●○·\\○●●\end{smallmatrix} \to \begin{smallmatrix}··●\\·····\\··○\end{smallmatrix}\cdots\begin{smallmatrix}○●○·\\··●○·\\○●●\end{smallmatrix} \xrightarrow[3]{3} \begin{smallmatrix}··●\\····\\··○\end{smallmatrix}\cdots\begin{smallmatrix}○··\\○●·\\○●·\end{smallmatrix} \to \begin{smallmatrix}··●\\·····\\··○\end{smallmatrix}\cdots\begin{smallmatrix}○··\\····\\●··\end{smallmatrix}$$

Note that we can also obtain the symmetric configuration $\begin{smallmatrix}···○\\·········\\··●\end{smallmatrix}\begin{smallmatrix}●··\\·····\\○···\end{smallmatrix}$. Because we cannot perform this reduction with Black starting, we must perform this reduction at the very beginning of the entire algorithm, before any other steps.

We stop this reduction when we have reached one of the three configurations $\begin{smallmatrix}●●○\\·○·\\○●●\end{smallmatrix}$ or $\begin{smallmatrix}●●○●○\\·○●○·\\○●○●●\end{smallmatrix}$ or $\begin{smallmatrix}●●○●○●○\\·○●○●○·\\○●○●○●●\end{smallmatrix}$. We are not able to reduce the last configuration further because in some cases it isolates the remaining stones from the rows above.

*Reductions.* Now we show how to reduce the configurations described above following the general approach from Sect. 3.1. For the case of three rows, we only need to consider the following two reductions in step (1):

(1) $\quad \begin{smallmatrix}●●○\\·○·\\○●●\end{smallmatrix} \xrightarrow{2} \begin{smallmatrix}●●○\\·○·\\·●·\end{smallmatrix} \xrightarrow{3} \begin{smallmatrix}·●·\\···\\·○·\end{smallmatrix}$

(1') $\quad \begin{smallmatrix}●●○●○\\·○●○·\\○●○●●\end{smallmatrix} \xrightarrow{2} \begin{smallmatrix}·●·●○\\·○●○·\\○●○●●\end{smallmatrix} \xrightarrow{2} \begin{smallmatrix}·●·○·\\·○·●·\\○●○●●\end{smallmatrix} \xrightarrow{2} \begin{smallmatrix}·····\\·●·○·\\○●○●●\end{smallmatrix} \xrightarrow{2} \begin{smallmatrix}·····\\···○·\\·●○●●\end{smallmatrix} \xrightarrow{3} \begin{smallmatrix}·····\\·····\\·●·○·\end{smallmatrix}$

*Case 1:* $m \equiv 1 \pmod 3$

The initial configuration is of type (a) for $m = 13 + 12k$ columns and of type (b) for $m = 7 + 12k$ columns, for $k \geq 0$.

(1a)  $\begin{matrix}\bullet\circ\bullet\circ\cdot\\\cdot\bullet\circ\bullet\circ\end{matrix} \xrightarrow{2} \begin{matrix}\bullet\circ\bullet\circ\cdot\\\cdot\cdot\bullet\circ\cdot\end{matrix} \xrightarrow{2} \begin{matrix}\cdot\bullet\circ\cdot\cdot\\\cdot\cdot\bullet\circ\cdot\end{matrix} \xrightarrow{2} \begin{matrix}\cdot\cdot\bullet\cdot\cdot\\\cdot\cdot\circ\cdot\cdot\end{matrix} \left(\rightarrow \begin{matrix}\cdot\cdot\circ\cdot\cdot\\\cdot\cdot\cdot\cdot\cdot\end{matrix}\right)$

(1b)  $\begin{matrix}\circ\bullet\circ\bullet\cdot\\\cdot\circ\bullet\circ\bullet\end{matrix} \xrightarrow{2} \begin{matrix}\cdot\circ\bullet\cdot\cdot\\\cdot\circ\bullet\circ\bullet\end{matrix} \xrightarrow{2} \begin{matrix}\cdot\cdot\circ\cdot\cdot\\\cdot\circ\bullet\bullet\cdot\end{matrix} \xrightarrow{3} \begin{matrix}\cdot\cdot\cdot\cdot\cdot\\\cdot\cdot\circ\cdot\cdot\end{matrix}$

(2a)  [configuration diagram]

(2b)  [configuration diagram]

(3a)  [configuration diagram]

(3b)  [configuration diagram]

(4a)  [configuration diagram]

[configuration diagram]

(4b′) We must reduce these rows starting with the mirrored standard initial configuration.

[configuration diagram]

*Case 2:* $m \equiv 0 \pmod 3$

The initial configuration is of type (a) for $m = 15 + 12k$ columns and of type (b) for $m = 9 + 12k$ columns, for $k \geq 0$.

(1a)  [configuration diagram]

(1b)  [configuration diagram]

(2a)
```
 · · ○ · ● · ·   2   · · ○ · ● · ·   4   · · · · ● · ·   2   · · · · · · · ·   2   · · · · · · · ·   2   · · · · · · · ·
●○●○●○· →  ·●●○○·· →  ··○·○·· →  ·····●·· →  ·····●·· →  ········
 ·●○●○●○     ·●○●○●○      ··●●○●○      ··○●○●○      ··○·●○·      ··○·●··
```
For three rows, this case is identical to Case 1(4b).

(2b)
```
 · · ● · ○ · ·   2   · · · · ○ · ·   2   · · · · ○ · ·   2   · · · · · ○ · ·   6   · · · · · · ·
○●○●○● →  ·○●●○● →  ·○●·●● →  ·●··○● →  ····· ·
 ·○●○●○●      ·○●○●○●      ·○○·●○●      ·○○·●○●      ··●·○··
```
For three rows, this case is symmetric to (4a) in Case 1 (with the mirrored initial configuration).

*Case 3:* $m \equiv 2 \pmod 3$

The initial configuration is of type (a) for $m = 17 + 12k$ columns and of type (b) for $m = 11 + 12k$ columns, for $k \geq 0$.

(1a)
```
●○●○●○●○·  4   ·●●○●○·●·  4   ··○●●○···  4   ···○●·····  3   ·········
 ·●○●○●○●○ →  ·○·●○●○○· →  ····○●○●· →  ····○●··· →  ·····○····
```

(1b)
```
○●○●○●○●·  4   ·○○●○·○·  4   ··●●○●···  4   ···●·····  3   ·········
 ·○●○●○●○● →  ·●·○●○●● →  ···○●○○· →  ···○●○··· →  ····○····
```

(2a)
```
 · · · · ○ · · · ·        · · · · · · · ·        · · · · · · · ·        · · · · · · · ·         · · · · · · · ·
●○●○●○●○· →  ●○●○○○··· →  ●○●○○○··· →  ··○○○●··· ( →  ····○···· )
 ·●○●○●○●○  4   ·●○●○●○●  4   ·○·●·●···  4   ···●·····  4   ········
```
```
 · · ○ · · · ·        · · · · · · ·        · · · ○ · · · ·        · · · ○ · · · ·         · · · · · ·
●●○●○●○ →  ·●·○●○● →  ···●○●○ →  ···●○○· →  ····○·· →  ········
 ·○●○●○·  4   :○●○●○·  4   ·○●○●○· 4   ·○●·● ·  4   ··●●·  3   ···· ● ·
○●○●○●●      ○●●··○●      ··●·● ·
```

(2b)
```
 · · ○ · · · · ·        · · · ○ · · · ·        · · · ○ · · · ·   4   · · · · · · · ·   3   · · · · · · · ·
○●○●○●○●· →  ·●○●○●··· →  ··●●○····· →  ····●···· →  ·········
 ·○●○●○●○●  6   ···○●○●○·  3   ···○●●○··      ····○●○·      ····○····
```
For three rows, this case is identical to Case 1(3b).

(3a)
```
 · · ○○○● · · · ·        · · ○○● · · · ·        · · · ○○● · · · ·        · · · · · · · · · ·        · · · · · · · · ·
 · · · ● · · · · ·        · · · ● · · · · ·        · · · ● · · · · ·        · · · ○ · · · · · ·        · · · · · · · · · ·
●○●○●○●○ →  ●○●○●○●○· →  ··●○●○○○··· →  ···●●○○○··· →  ·········
 ·●○●○●○●○  4   ··○··●○●○  4   ·····●·●○  4   ·····●·●○  6   ·····●··○
( →  · · · · · · · · ·
     · · · · · · · · ·
     · · · · · · · ● · ○ )
```
```
○○○●      · ○ · ·      · · · ·      · · · ·
 · ● · ·      · ● · ·      · ○ · ·      · · · ·
·●●○ →  ·●●○ →  ·●○· →  ·●··
 · · ○ ·  4   · · ○ ·  4   · · ● ·  3   · · · ·
 · ○●●      · · ○●
```

(3b) We must reduce these rows starting with the mirrored standard initial configuration. (We could also solve the standard configuration, but then we could not continue with step (4b).)
```
 · · · · ○ · · · ·        · · · · ○ · · · ·        · · · · ○ · · · ·   4   · · · · · · · · ·   2   · · · · · · · · ·
 ·○●○●○●○ →  ···●○●○● →  ····●●●·· →  ·······●··· →  ·········
●○●○●○●○·  6   ·○●○●○···  3   ··○○●○···      ··○●·○···      ···○·●···
```
For three rows, this case is identical to Case 1(2b).

(4a) We must reduce these rows starting with the mirrored standard initial configuration, because of the white single stone left over at the right end of the row above.
```
 · · · · · ○ · · ·        · · · · · ○ · · ·        · · · · · · · · ·        · · · · · · · · ·        · · · · · · · ·
 · · · ● · · ○        · · · · · ● · · ·        · · · · · ○ · · ·        · · · · · · · · ·        · · · · · · · ·
 ·○●○●○●● →  ·○●○●○●○· →  ·●·○●○●·· →  ···○●○··· →  ····○····
○●○●○●○●  4   ○●○●○·●··  4   ○●○●○····  4   ·●○○○····  4   ···●○····
                                                                  · · · · · · · · ·
                                                              →  · · · · · · · · ·
                                                              2   · · · · ○ · · · ·
```
For three rows, we must reduce the number of columns a little bit asymmetrically (remove four additional columns on the left side) and then do the following reduction.

```
· · ○ · · · ·      · · ○ · · · ·      · · · · · · · ·      · · · · · · · ·      · · · · · · · ·      · · · · · · · ·
· · ● · · ○ ·  →   · · ● · · · · ·  →  · · ○ · · · ·  →   · · ○ · · · ·  →   · · ○ · · · ·  →   · · · · · · · ·
● ● ○ ● ○ ● ○     ● ● ○ ● ● ○ ○      ● ● ○ ● ○ ○ ·      ● ● ○ ● ○ · ·      ● ● ○ ○ · · ·      ● · ○ · · · ·
· ○ ● ● ○ ● ·  4  · ○ ● ○ · ○ ·  4   · ○ ● ○ · · ·  4   · ○ ● ○ · · ·  4    · · ● · · · · ·  4
○ ● ○ ● ○ ● ●     ○ ● ○ ● · ● ·      ○ ● ○ ● · · ·      · ● · · · · · ·      · · · · · · · ·
```
```
        · · ○ · ● · · · ·     · · · ○ · ● · · · ·    · · · · ○ · · · · ·   · · · · ○ · · · · ·        · · · · · · · · · ·
(4b)    ○ ● ○ ● ○ ● ○ ●   4  · ● ○ ● · ○ ○ ●   4  · ● ○ ● · ● · · ·  4  · · ● ● · · · ·  5    · · · · · · · · · ·
        · ○ ● ○ ● ○ ● ○ ● → · · · ○ ● ○ ● ○ ● → · · · ○ ● ○ ● ○ · → · · · ○ ● ○ · · · →   · · · · ○ · · · · ·
```

For three rows, this case is identical to Case 1(4b).

# 4  NP-Completeness of 1-Reducibility

In this section we consider arbitrary initial Clobber positions that do not need to have a rectangular shape or the alternating checkerboard placement of the stones. We show that then the following problem is NP-complete.

---

**Problem** SOLITAIRE-CLOBBER:
Given an arbitrary initial Clobber configuration, decide whether we can reduce it to a single stone.

---

The proof is by reduction from the Hamiltonian circuit problem in grid graphs. A *grid graph* is a finite graph embedded in the Euclidean plane such that the vertices have integer coordinates and two vertices are connected by an edge if and only if their Euclidean distance is equal to one.

---

**Problem** GRID-HAMILTONICITY:
Decide whether a given grid graph has a Hamiltonian circuit.

---

Itai *et al.* proved that GRID-HAMILTONICITY is NP-complete [5, Theorem 2.1].

**Theorem 5.** SOLITAIRE-CLOBBER *is NP-complete.*

*Proof.* We first observe that SOLITAIRE-CLOBBER is indeed in NP, because we can easily check in polynomial time whether a proposed solution (which must have only $n - 1$ moves) reduces the given initial configuration to a single stone.

We prove the NP-completeness by reduction from GRID-HAMILTONICITY. Let $G$ be an arbitrary grid graph with $n$ nodes, embedded in the Euclidean plane. Let $v$ be a node of $G$ with maximum $y$-coordinate, and among all such nodes the node with maximum $x$-coordinate. If $v$ does not have a neighbor to the left then $G$ cannot have a Hamiltonian circuit. So assume there is a left neighbor $w$ of $v$. Note that $v$ has degree two and therefore any Hamiltonian circuit in $G$ must use the edge $(v, w)$.

Then we construct the following Clobber configuration (see Fig. 1). We put a black stone on each node of $G$. We place a single white stone just above $w$, the *bomb*. We place a vertical chain of $n$ white stones above $v$, the *fuse*, and another single black stone, the *fire*, on top of the fuse. Altogether we have placed $n + 1$ white and $n + 1$ black stones, so this is a legal Clobber configuration.

If $G$ has a Hamiltonian circuit $C$ then the bomb can clobber all black nodes of $G$, following $C$ starting in $w$ and ending in $v$ after $n$ rounds. At the same

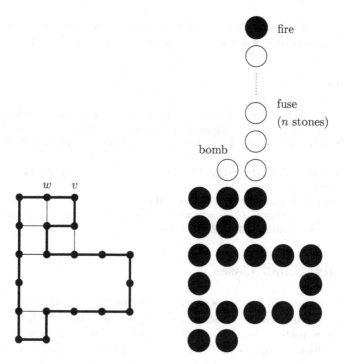

**Fig. 1.** An $n$-node grid graph with Hamiltonian circuit and the corresponding Clobber configuration that can be reduced to a single stone.

time, the black fire can clobber the $n$ stones of the fuse and end up just above $v$ after $n$ rounds. But then in a last step the bomb can clobber the fire, leaving a single stone on the board.

On the other hand, if the initial configuration can be reduced to a single stone then White cannot move any stone on the fuse (because that would disconnect the black fire from the stones on $G$), so it must move the bomb until Black has clobbered the fuse. But that takes $n$ steps, so White must in the meanwhile clobber all $n$ black stones of $G$, that is, it must walk along a Hamiltonian circuit in $G$.                                                                                      □

## 5   Conclusions and Open Problems

We have seen that reducing to the minimum number of stones is polynomially solvable for checkerboard rectangular configurations, and is NP-hard for general configurations. What about checkerboard non-rectangular configurations and rectangular non-checkerboard configurations?

We have also seen a lower bound on the number of stones to which a configuration can be reduced that is based on the number of stones plus the number of stones on squares of different color. It would be interesting to identify other structural parameters of a configuration that influence reducibility.

## References

1. Wolfe, D.: Clobber research. http://www.gac.edu/~wolfe/games/clobber (2002)
2. Demaine, E., Fleischer, R., Fraenkel, A., Nowakowski, R.: Seminar no. 02081. Technical Report 334, Dagstuhl Seminar on Algorithmic Combinatorial Game Theory (2002)
3. 34th International Mathematical Olympiad: Question 3. http://www.kalva.demon.co.uk/imo/imo93.html (1993)
4. Grossman, J.: Private communication (2002)
5. Itai, A., Papadimitriou, C., Szwarcfiter, J.: Hamilton paths in grid graphs. SIAM Journal on Computing **11** (1982) 676–686

## Appendix: Small Cases

Our proof of Theorem 4 requires us to verify reducibility for all instances with $2 \leq n, m \leq 6$. This fact can be checked easily by a computer, but for completeness we give the reductions here. By symmetry, we only need to show the cases with $n \leq m$. Reductions of $2 \times 3$, $2 \times 5$, $3 \times 4$, $3 \times 6$, $4 \times 5$, and $5 \times 6$ boards are already given in Sect. 3.3. Eight more small boards remain. We assume White moves first.

**2×2:** diagram sequence

**2×4:** diagram sequence

**2×6:** diagram sequence

**3×3:** diagram sequence

**3×5:** diagram sequence

**4×4:** We reduce the upper two rows as in the $2 \times 4$ board, then Black moves next: diagram sequence

**4×6:** We reduce the upper two rows as in the $2 \times 6$ board, then White moves next: diagram sequence

**6×6:** We reduce the upper four rows as in the $4 \times 6$ board, then White moves next: diagram sequence

# Complexity of Error-Correcting Codes Derived from Combinatorial Games

Aviezri S. Fraenkel and Ofer Rahat

Department of Computer Science and Applied Mathematics
The Weizmann Institute of Science, Rehovot, Israel
{fraenkel,ofer}@wisdom.weizmann.ac.il

**Abstract.** The main result of the present paper is to establish an algorithm for computing a linear error-correcting code called *lexicode* in $O(n^{d-1})$ steps, where $n$ and $d$ are the size and distance of the code, respectively. This is done by using the theory of combinatorial games, specifically, two-player cellular automata games. Previous algorithms were exponential in $n$.

## 1  Introduction

For a linear error-correcting code, denote by $n$ the common size of the codewords, by $k$ its dimension, and by $d$ the minimum distance of any two codewords. The main purpose of this paper is to present an $O(n^{d-1})$ algorithm for computing a basis for such a code, called *lexicode*, using the theory of combinatorial games. Previous methods [1, 2] required $O(2^{2n})$ steps. In [3, 4], an iterative algorithm for constructing the lexicode with time complexity $O(2^{n-k+1})$ is given. In [5, Theorem 3.5], the asymptotic lower bound $k \geq (2n)/d$ is given. In [3] the slightly tighter asymptotic lower bound $k \geq (3n)/d$ is given.

When comparing the dimension of the binary lexicode calculated in [2, Table VII] with those of the best linear codes listed in [6] for $n \leq 44$ and $k \leq 10$, it turns out that the lexicode has always either the parameters $(n, k, d)$ of the best code known, or its dimension $k$ is within one or two of the best ones. In particular, the lexicode contains both Hamming, Golay and other codes. For details see e.g., [2] and [7, Sect. 2]. The Golay code was called "... probably the most important of all codes, for both practical and theoretical reasons" in [8, Ch. 2].

The present results stem from a program to compute codes efficiently, using the theory of combinatorial games. See [9, end of Introduction, and Remark 4.2 in the penultimate paragraph]. It partly motivated the invention of the *Cellata* games [10, 11, 12], a special case of which will feature below, where they are called *Celata* games.

In Sect. 2 some necessary background concepts are collected, Sect. 3 contains a definition of a special case of a Celata game and of a *Celopt* game, where some of their properties are explored. It also contains a definition of the lexicode. An important connection between a Celopt game and a lexicode is stated in Theorem 1. The key results of the present paper are presented in Sect. 4, based

J. Schaeffer et al. (Eds.): CG 2002, LNCS 2883, pp. 201–212, 2003.

on further properties of Celopt games, derived from properties of Celata games. The latter results stem from previous results on *annihilation* games, which are a special case of Celata games [13, 14, 15, 16, 17]. For annihilation games with the last player losing, see [18]. Another key result is Theorem 6 in Sect. 5, which leads to the algorithm for computing the basis, each of whose members has weight $\geq d$. The second part of Theorem 6 points out that it can be arranged that all basis elements have weight precisely $d$. A brief wrap-up is presented in the final Sect. 6.

## 2    Preliminaries

We are concerned with games played on a digraph $G = (V, E)$. Tokens are initially distributed on a subset of $V$. A move consists of selecting a token and moving it to a neighboring vertex, unoccupied or occupied. The player making the last move wins, the opponent loses. If there is no last move, the outcome is a draw.

The corresponding *game-graph* is a digraph $\boldsymbol{G} = (\boldsymbol{V}, \boldsymbol{E})$ where $\boldsymbol{V}$ is the set of all collections of token distributions on $G$, and $(\boldsymbol{u}, \boldsymbol{v}) \in \boldsymbol{E}$ if there is a move from $\boldsymbol{u}$ to $\boldsymbol{v}$. Note that for the case of a single token on $V$ we have $\boldsymbol{G} = G$.

For this and the following notations and basic facts, see e.g., [19].

A vertex $\boldsymbol{u} \in \boldsymbol{V}$ (game-position) is labeled $N$, if the *N*ext player (the player moving from $\boldsymbol{u}$) has a winning move; by $P$ if the *P*revious player (who landed on $\boldsymbol{u}$) can win.

Denote by $\mathcal{P}$ and by $\mathcal{N}$ the set of all *P*-positions and all *N*-positions respectively.

For any digraph $G = (V, E)$, the set $F(u)$ of *followers* of $u \in V$ is defined by $F(u) = \{v \in V : (u, v) \in E\}$.

It can then be shown that

$$\boldsymbol{u} \in \mathcal{P} \text{ if and only if } F(\boldsymbol{u}) \subseteq \mathcal{N}, \tag{1}$$

$$\boldsymbol{u} \in \mathcal{N} \text{ if and only if } F(\boldsymbol{u}) \cap \mathcal{P} \neq \emptyset. \tag{2}$$

Further, if $G$ is acyclic, then the sets $\mathcal{P}$, $\mathcal{N}$ partition $V$: $\mathcal{P} \cup \mathcal{N} = V$, $\mathcal{P} \cap \mathcal{N} = \emptyset$.

Let $S \subset \mathbb{Z}_{\geq 0}$, $S \neq \mathbb{Z}_{\geq 0}$, and $\overline{S} = \mathbb{Z}_{\geq 0} \setminus S$. The *minimum excluded value* of $S$ is

$$\operatorname{mex} S = \min \overline{S} = \text{least nonnegative integer not in } S.$$

A (classical) *Sprague-Grundy function*, or simply *g-function* $g: V \to \mathbb{Z}_{\geq 0}$ on a digraph $G = (V, E)$ is defined recursively by

$$g(u) = \operatorname{mex} g\big(F(u)\big),$$

where for any set $T$ and any function $h$ on $T$,

$$h(T) = \{h(t) : t \in T\}.$$

We then have

$$\mathcal{P} = \{\boldsymbol{u} : g(\boldsymbol{u}) = 0\}, \quad \mathcal{N} = \{\boldsymbol{u} : g(\boldsymbol{u}) \neq 0\}.$$

We denote the sum of the integers $a, b \in \mathbb{Z}_{\geq 0}$ over GF(2) by $a \oplus b$, also known as XOR, or *Nim-sum*. The XOR of $a_1, \ldots, a_m$ may also be denoted by $a_1 \oplus \ldots \oplus a_m = \sum'^{m}_{i=1} a_i$, where the apostrophe denotes XOR.

We shall also be concerned with vectors over GF$(2)^n$ ($n$-dimensional binary vectors). The *numerical value* of a vector $\boldsymbol{u} = (u^0, \ldots, u^{n-1}) \in$ GF$(2)^n$ is $|\boldsymbol{u}| := \sum_{i=0}^{n-1} u^i 2^i$. The *weight* of $\boldsymbol{u}$ is $w(\boldsymbol{u}) = \sum_{i=0}^{n-1} u^i =$ number of 1-bits of $\boldsymbol{u}$. The *parity weight* of $\boldsymbol{u}$ is $w'(\boldsymbol{u}) = \sum'^{n-1}_{i=0} u^i$.

The Hamming distance between two vectors $\boldsymbol{u}, \boldsymbol{v}$ over GF(2) is defined by $H(\boldsymbol{u}, \boldsymbol{v}) = w(\boldsymbol{u} \oplus \boldsymbol{v})$. It is the number of positions in which $\boldsymbol{u}$ and $\boldsymbol{v}$ differ.

Following [20], a vector $\boldsymbol{u}$ with $w(\boldsymbol{u})$ even is said to be an *evil* number, and one with $w(\boldsymbol{u})$ odd is an *odious* number.

The *lexicode* on binary vectors (words) with $n$ bits and distance $d \geq 2$ is defined by applying the greedy algorithm to the sequence of lexicographically ordered vectors over GF(2): adjoin $\Phi = (0, \ldots, 0)$, and at each step adjoin the vector $\boldsymbol{u}$ with smallest numerical value which satisfies $H(\boldsymbol{u}, \boldsymbol{v}) \geq d$ for every previously adjoined $\boldsymbol{v}$. The result is a linear code. See [1, 2, 9, 7].

The basic fact used in this paper is that the set $\mathcal{P}$ of any game $\Gamma$ forms a linear code: The $P$-positions constitute a (graph-theoretic) independent set of the game-graph of $\Gamma$. If its $P$-positions are represented as vectors, there is a certain minimum Hamming distance between them.

# 3    Celata and Celopt Games

*Two-player cellular automata games*, or *Celata* games for short, were defined in [12]. See also [10], [11]. Below is a brief description of a class of Celata games.

Given a finite digraph $G = (V, E)$, also called *groundgraph*, order $V$ ($|V| = n$) in some way, say

$$V = \{z_0, \ldots, z_{n-1}\}.$$

This ordering is assumed throughout.

The special case of Celata games on $G$ of interest here depends on a parameter $s \in \mathbb{Z}_{>0}$. A move from an occupied vertex $z_k$ consists of *firing* (see below) $z_k$ and some neighborhood of $z_k$ of size $q_k$, where

$$q = q_k = \min(s, |F(z_k)|). \tag{3}$$

It will be called *Celata game with parameter s*, or simply *Celata* game.

*Firing* $z_k$ means to *complement* $z_k$ and a $q_k$-neighborhood of $z_k$: the token on $z_k$ is removed, and the token-occupancy of $q$ of its neighbors is complemented.

Let $\boldsymbol{G} = (\boldsymbol{V}, \boldsymbol{E})$ denote the following game-graph of the Celata game played on $G = (V, E)$. The digraph $\boldsymbol{G}$ is also called the *Celata graph* of $G$. Any vertex in $\boldsymbol{G}$ can be described in the form $\boldsymbol{u} = (u^0, \ldots, u^{n-1})$ over the field GF(2), where

$u^k = 1$ if $z_k$ is occupied by a token, $u^k = 0$ if $z_k$ is unoccupied. In particular, $\Phi = (0, \ldots, 0)$ is a leaf of $V$, and $|V| = 2^n$.

Note that $V$ is an abelian group under the addition $\oplus$ of GF(2), also called Nim-addition, with identity $\Phi$. Every nonzero element has order 2. Moreover, $V$ is a vector space over GF(2) satisfying $1u = u$ for all $u \in V$. For $i \in \{0, \ldots, n-1\}$, define unit vectors $z_i = (z_i^0, \ldots, z_i^{n-1})$ with $z_i^j = 1$ if $i = j$; $z_i^j = 0$ otherwise. They span the vector space. In particular, for any $u = (u^0, \ldots, u^{n-1}) \in V$, we can write $u = \sum_{i=0}^{n-1} u^i z_i = \sum_{i=0}^{\prime n-1} u^i z_i$.

For defining $E$, let $u \in V$ and let $0 \leq k \leq n-1$. Let $F^q(z_k) \subseteq F(z_k)$ be any subset of $F(z_k)$ satisfying

$$|F^q(z_k)| = q. \tag{4}$$

Define

$$(u, v) \in E \text{ if } u^k = 1,\ q > 0, \text{ and } v = u \oplus z_k \oplus \sideset{}{'}\sum_{z_\ell \in F^q(z_k)} z_\ell, \tag{5}$$

for every $F^q(z_k)$ satisfying (4).

Informally, an edge $(u, v)$ reflects the firing of $u^k$ in $u$ (with $u^k = 1$), i.e., the complementing of the tokens on $z_k$ and $F^q(z_k)$. Such an edge exists for every $F^q(z_k)$ satisfying (4). Note that if $z_k \in G$ is a leaf, then there is no move from $z_k$, since then $q = 0$.

If (5) holds, we also write $v = F_k^q(u)$. The set of all followers of $u$ is

$$F(u) = \bigcup_{u^k = 1} \ \bigcup_{F^q(z_k) \subseteq F(z_k)} F_k^q(u).$$

A $Celopt_s = Celopt$ game with parameter $s \in \mathbb{Z}_{>0}$ and game-graph called $Celopt\text{-}graph$, is defined the same way as a Celata game, except that we $opt$ for $p \in \{0, \ldots, s\}$ followers, i.e., (5) is replaced by

$$(u, v) \in E \text{ if } u^k = 1,\ q > 0, \text{ and } v = u \oplus z_k \oplus \sideset{}{'}\sum_{z_\ell \in F^p(z_k)} z_\ell, \tag{6}$$

for all $p \in \{0, \ldots, s\}$, where $F^p(z_k)$ — the only difference between (5) and (6) — is any subset of $p$ followers of $z_k$.

In other words, the edges (moves) describe the firing of $z_k$ and $p$ followers in $F^q(z_k)$ for all $p \in \{0, \ldots, s\}$. A single move consists of choosing one of the values $p \in \{0, \ldots, s\}$ and firing $z_k$ and $p$ followers. Note that $p = 0$ means to complement $z_k$ only. This is a legal move unless $z_k$ is a leaf, in which case $q = 0$, and then there is no move from $z_k$, by convention.

**Lemma 1.** *Every $Celopt_s$ game on a digraph $G = (V, E)$ reduces to a Celata game on $G' = (V', E')$, where $|V'| = |V| + s$, $|E| \leq |E'| \leq |E| + s|V|$, and the complexity of the strategy for both games is the same.*

**Proof.** Put $V' = V \cup \{\ell_1 \ldots, \ell_s\}$, where the $\ell_i$ are leaves. For $k \in \{0, \ldots, n-1\}$, denote by $L_k$ the number of edges leading from $z_k$ to leaves in $G$. If $L_k < q_k$, adjoin an edge $(z_k, \ell_i)$ for every $i \in \{1 \ldots, q_k - L_k\}$. Thus every non-leaf vertex $z_k$ of $G'$ satisfying $L_k < q_k$, has $q_k$ edges to leaves. It is now clear that Celopt on $G$ is the same as Celata on $G'$: firing a neighborhood of size $p$ of $z_k$ on $G$ corresponds to firing the same neighborhood on $G'$ as well as $q - p$ leaves. ∎

**Definition 1.** The *Nim-graph* of size $n$ is a digraph $G = (V, E)$ with $V = \{z_1, \ldots, z_n\}$, $E = \{(z_j, z_i) : 1 \leq i < j\}$ for $j = 1, \ldots, n$. A Celopt$_s$ game (of size $n$) on a Nim-graph (of size $n$) will be called a *Nim-Celopt$_s$* game.

*Remark 1.* (i) Note that the indices are in the integer interval $[1, n]$, rather than $[0, n - 1]$ as before. This happens to be more convenient for the present case. Therefore we also use here the notation $\boldsymbol{u} = (u^1, \ldots, u^n)$. When $n = 0$, the Nim-graph is empty.
(ii) The vertex $z_1$ is not a leaf. When $p = 0$, a token on $z_i$, for any $i \in \{1, \ldots, n\}$ can be removed; in particular from $z_1$.
(iii) From any vertex $z_i$, every follower is accessible. It follows that $g(z_i)$ is monotonically increasing: $1 = g(z_1) < 2 = g(z_2) < g(z_3) < \ldots$.

**Theorem 1.** *Let $\mathcal{P}$ be the set of P-positions of Nim-Celopt$_s$. Let $\mathcal{L}$ be the set of codewords of the lexicode with Hamming distance $d \geq s + 2$. Then $\mathcal{P} = \mathcal{L}$.*

**Proof.** Induction on the numerical value $|\boldsymbol{u}|$ of the vector $\boldsymbol{u}$. Clearly, $\Phi \in \mathcal{P}$ and $\Phi \in \mathcal{L}$. Let $|\boldsymbol{u}| \geq 1$. The induction hypothesis implies that for every $\boldsymbol{v}$ with $|\boldsymbol{v}| < |\boldsymbol{u}|$ we have $\boldsymbol{u} \in \mathcal{P}$ if and only if $\boldsymbol{u} \in \mathcal{L}$. We show $\boldsymbol{u} \in \mathcal{P}$ if and only if $u \in \mathcal{L}$.

Let $\boldsymbol{u} \in \mathcal{P}$. We have to show that for every $|\boldsymbol{v}| < |\boldsymbol{u}|$, if $\boldsymbol{v} \in \mathcal{L}$, then $H(\boldsymbol{u}, \boldsymbol{v}) \geq d$. Equivalently, for every $|\boldsymbol{v}| < |\boldsymbol{u}|$, if $H(\boldsymbol{u}, \boldsymbol{v}) \leq d - 1 = s + 1$, then $\boldsymbol{v} \notin \mathcal{L}$. Indeed, the last two conditions imply that $\boldsymbol{v} \in F(\boldsymbol{u})$ in Celopt. Moreover, $\boldsymbol{v} \in \mathcal{N}$ since $\boldsymbol{u} \in \mathcal{P}$, so $\boldsymbol{v} \notin \mathcal{L}$ by the induction hypothesis.

Conversely, suppose that $\boldsymbol{u} \notin \mathcal{P}$. Then $\boldsymbol{u} \in \mathcal{N}$. Thus there exists $\boldsymbol{v} \in F(\boldsymbol{u}) \cap \mathcal{P}$. The rules of Celopt imply $|\boldsymbol{v}| < |\boldsymbol{u}|$, and $H(\boldsymbol{u}, \boldsymbol{v}) \leq s + 1 < d$. By the induction hypothesis, $\boldsymbol{v} \in \mathcal{L}$. Hence $\boldsymbol{u} \notin \mathcal{L}$. ∎

In the next section we shall deal with Nim-Celopt games, for which, by Lemma 1, the results of [12] apply. There, the generalized Sprague-Grundy function $\gamma$ had to be used because the groundgraph contained cycles. Here we deal only with acyclic digraphs, so the classical Sprague-Grundy function $g$ suffices.

## 4   The Sprague-Grundy Function of Nim-Celopt Games

Calculation of the $\gamma$ values of a digraph similar to Nim-Celopt appears in [12, Example 1, Sect. 3]. Further examples appear in [10, 11].

We shall compute the $g$-values of the game-graph $\boldsymbol{G} = (\boldsymbol{V}, \boldsymbol{E})$ of Nim-Celopt in order to get its $P$-positions. We make use of the following result.

**Theorem 2.** *Let $\boldsymbol{G} = (\boldsymbol{V}, \boldsymbol{E})$ be the game-graph of any Celopt game. For $\boldsymbol{u}, \boldsymbol{v} \in \boldsymbol{G}$ we have, $g(\boldsymbol{u} \oplus \boldsymbol{v}) = g(\boldsymbol{u}) \oplus g(\boldsymbol{v})$.*

**Proof.** Theorem 1 of [12] implies that $g(\boldsymbol{u} \oplus \boldsymbol{v}) = g(\boldsymbol{u}) \oplus g(\boldsymbol{v})$ for any Celata game. The same holds for any Celopt game by Lemma 1.                    ∎

**Lemma 2.** *For the game-graph of Nim-Celopt$_s$,* $g(\boldsymbol{z}_k) = g_s(\boldsymbol{z}_k) = \text{mex}\{g(\boldsymbol{z}_{i_1}) \oplus \ldots \oplus g(\boldsymbol{z}_{i_j}) : 1 \le i_1 < \ldots < i_j < k, \quad j \le s\}$.

**Proof.** By definition, $g(\boldsymbol{z}_k) = \text{mex}\{g(\boldsymbol{z}_{i_1} \oplus \ldots \oplus \boldsymbol{z}_{i_j}) : 0 \le i_1 < \ldots < i_j < k, \quad j \le s\}$. The result now follows from Theorem 2.                    ∎

*Example 1.* We have,
$$g_1(\boldsymbol{z}_i)_{i \ge 1} = \{1, 2, 3, 4, 5, 6, 7, 8, 9, 10, \ldots\},$$
$$g_2(\boldsymbol{z}_i)_{i \ge 1} = \{1, 2, 4, 7, 8, 11, 13, 14, 16, \ldots\},$$
$$g_3(\boldsymbol{z}_i)_{i \ge 1} = \{1, 2, 4, 8, 15, 16, 32, 51, 64, \ldots\}.$$

**Definition 2.** The *left-shift* of $\boldsymbol{u} = (u^1, \ldots, u^n)$ is $L(\boldsymbol{u}) = (u^2, \ldots, u^n)$. The *right-shift* with *parity bit* $w'(\boldsymbol{u})$ is $R(\boldsymbol{u}) = (w'(\boldsymbol{u}), u^1 \ldots, u^n)$.

*Remark 2.* Note that $R(\boldsymbol{u})$ is always evil. In terms of numerical values, $|L(\boldsymbol{u})| = \lfloor |\boldsymbol{u}|/2 \rfloor$, $|\boldsymbol{u}| = \lfloor |R(\boldsymbol{u})|/2 \rfloor$, $L(R(\boldsymbol{u})) = \boldsymbol{u}$; and $R(L(\boldsymbol{u})) = \boldsymbol{u}$ if and only if $\boldsymbol{u}$ is evil.

Denote by $\mathcal{P}_s$, $\mathcal{N}_s$, $F_s(\boldsymbol{u})$ the set of P-positions, N-positions, followers of $\boldsymbol{u}$ in Nim-Celopt$_s$ respectively.

**Theorem 3.** *Let $s \in \mathbb{Z}_{>0}$ be odd, $s + 1 \le n$. Then $\mathcal{P}_{s+1} = \{R(\boldsymbol{u}) : \boldsymbol{u} \in \mathcal{P}_s\}$.*

**Notation 1** It is sometimes convenient to denote vectors $\boldsymbol{u} = (u^1, \ldots, u^n)$ by the superscripts $i$ for which $u^i = 1$. Thus $\boldsymbol{u} = (3, 7)$ denotes a vector for which $u^3 = u^7 = 1$, and $u^i = 0$ for all $i \notin \{3, 7\}$. This notation is used, inter alia, in Example 2 and in the proof of Theorem 6.

*Example 2.*

| n | 1 | 2 | 3 | 4 | 5 | 6 | 7 | 8 | 9 | 10 | 11 |
|---|---|---|---|---|---|---|---|---|---|----|----|
| $g_1(z_n)$ | 1 | 2 | 3 | 4 | 5 | 6 | 7 | 8 | 9 | 10 | 11 |
| $g_2(z_n)$ | 1 | 2 | 4 | 7 | 8 | 11 | 13 | 14 | 16 | 19 | 21 |

We have

$$\mathcal{P}_1 = \{(2, 4, 6), (2, 3, 4, 5), \ldots\}, \quad \mathcal{P}_2 = \{(1, 3, 5, 7), (3, 4, 5, 6), \ldots\},$$

where $R(2, 4, 6) = (1, 3, 5, 7)$, with $g_2(1) \oplus g_2(3) \oplus g_2(5) \oplus g_2(7) = 1 \oplus 4 \oplus 8 \oplus 13 = 0$; and $R(2, 3, 4, 5) = (3, 4, 5, 6)$, with $g_2(3) \oplus g_2(4) \oplus g_2(5) \oplus g_2(6) = 4 \oplus 7 \oplus 8 \oplus 11 = 0$.

**Proof of Theorem 3.** Let $\mathcal{P}'_{s+1} = \{R(\boldsymbol{u}) : \boldsymbol{u} \in \mathcal{P}_s\}$, $\mathcal{N}'_{s+1} = V \backslash \mathcal{P}'_{s+1}$. Since the groundgraph of Nim-Celopt is acyclic, its P- and N-positions are unique. By (1), (2) it therefore suffices to show the following: if $\boldsymbol{u} \in \mathcal{P}'_{s+1}$, then $F_{s+1}(\boldsymbol{u}) \subseteq \mathcal{N}'_{s+1}$, and if $\boldsymbol{u} \in \mathcal{N}'_{s+1}$, then $F_{s+1}(\boldsymbol{u}) \cap \mathcal{P}'_{s+1} \ne \emptyset$.

Let $u \in \mathcal{P}'_{s+1}$, $v \in F_{s+1}(u)$. The move from $u$ to $v$ is done by selecting an occupied $z_k$ and firing it together with $p \leq s+1$ followers of $z_k$. Remark 2 implies that $\mathcal{P}'_{s+1}$ has only evil vectors. Assume $v \in \mathcal{P}'_{s+1}$. Since $s$ is odd, we have $p \leq s$. Thus there exists of a move from $L(u) \in \mathcal{P}_s$ to $L(v) \in \mathcal{P}_s$, a contradiction.

Now let $u \in \mathcal{N}'_{s+1}$. Put $w = L(u)$. We consider two cases.

(i) $w \in \mathcal{N}_s$. Then there exists $z \in F_s(w) \cap \mathcal{P}_s$. Let $v = R(z)$. Then $v \in \mathcal{P}'_{s+1}$. The move from $w$ to $z$ involves complementing some occupied vertex $z_k$ and a $p$-neighborhood of $z_k$, for some $p \leq s$. Note that the occupied vertex $z_k$ in $w$, is the occupied vertex $z_{k+1}$ in $u$. Hence it is possible to move from $u$ to $v$ directly, by complementing $z_{k+1}$ and a suitable $p$-neighborhood of $z_{k+1}$ for some $p \leq s$, and by controlling the occupancy of $z_0$. See Fig. 1. Therefore $v \in F_{s+1}(u) \cap \mathcal{P}'_{s+1}$.

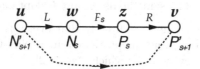

**Fig. 1.** Case (i) of the proof.

(ii) $w \in \mathcal{P}_s$. If $w = \Phi$ we are done, since then $w \in F_{s+1}(u) \cap \mathcal{P}'_{s+1}$. Hence we may assume that there is a token on $w$. Without loss of generality, assume that $w^0 = 1$, where $w = (w^0, \ldots, w^{n-1})$. Put $w' = (1 - w^0, \ldots, w^{n-1})$. Then $w' \in F_s(w) \cap \mathcal{N}_s$. Thus there exists $z \in F_s(w') \cap \mathcal{P}_s$. Put $v = R(z)$. Then $v \in \mathcal{P}'_{s+1}$. Note that we can move from $u$ to $v$ directly (see Fig. 2): it involves complementing at most $s + 3$ vertices $z_i$. Now if $u$ were evil, then Remark 2 would imply that $u = R(w) \in \mathcal{P}'_{s+1}$. Hence $u$ is odious. Also $v$ is evil. Therefore at most $s + 2$ vertices are complemented in moving from $u$ to $v$. Thus $v \in F_{s+1}(u) \cap \mathcal{P}'_{s+1}$. ■

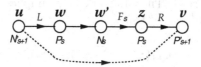

**Fig. 2.** Case (ii) of the proof.

**Corollary 1.** *If $u \in \mathcal{P}$ in a Nim-Celopt$_s$ game with $s$ even, then $u$ is evil.*

**Proof.** Follows directly from the statement of Theorem 3 and from Remark 2. ■

**Definition 3.** (i) For any Nim-Celopt game, if $g(z_k)$ ($\in \mathbb{Z}_{\geq 0}$) is not a power of 2, then $g(z_k)$ is called a *seed*.
(ii) Denote by $t = t(G) \in \mathbb{Z}_{\geq 0}$ the smallest number satisfying $g(u) \leq 2^t - 1$ for every $u$ in a Celopt game.
(iii) In a Celopt graph $G = (V, E)$, put $V_i = \{u \in V : g(u) = i\}$, for $0 \leq i < 2^t$.

Note that, for any $k \in \mathbb{Z}_{\geq 0}$, $g(z_k)$ is a Nim-sum of lower powers of two if and only if $g(z_k)$ is a seed.

**Theorem 4.** *Let $G = (V, E)$ be the Celopt graph of the finite acyclic digraph $G = (V, E)$. Then $g$ is a homomorphism from $V$ onto $GF(2)^t$ for $t$ given in Definition 3, with kernel $V_0$ and quotient space $V/V_0 = \{V_i : 0 \leq i < 2^t\}$, $\dim(V) = m + t$, where $m = \dim(V_0)$.*

**Proof.** This is a special case of Theorem 2 of [12] for the case of Celata played on an acyclic digraph. The same holds for any Celopt game by Lemma 1.    ∎

*Remark 3.* (i) Theorem 4 implies directly that every $g$-value in $\{0, \ldots, 2^t - 1\}$ is assumed in a Celopt game, in particular in a Nim-Celopt game. Further, $t = \lfloor \log_2 g(u) \rfloor + 1$ for every $u$ for which $g(u) > 2^{t-1}$.
(ii) Note that $\mathcal{P} = V_0$, so $\mathcal{P}$ is a linear subspace of $V$.
(iii) Every seed $g(u)$ produces a new $P$-position in a Nim-Celopt game: it is the (Nim)-sum $\sigma$ of smaller powers of 2, so $g(u) \oplus \sigma = 0$.
(iv) Theorem 4 and Lemma 2 imply that for every $k \in \{0, \ldots, t-1\}$, $2^k$ is assumed by some $g(z_i)$.
(v) In stating in Theorem 4 that $g$ is a homomorphism, we also identified the numerical value of $g$ with the binary vector representation of that value. We shall do so also below.

**Theorem 5.** *For all $n \in \mathbb{Z}_{>1}$, if $s$ is odd, then $g_s(z_{n-1}) = \lfloor g_{s+1}(z_n)/2 \rfloor$, and $g_{s+1}(z_n)$ is an odious number.*

**Proof.** The first part is proved by induction on $n$. For $n = 2$, $g_s(z_1) = 1$ for all $s \in \mathbb{Z}_{>0}$, and $\lfloor g_{s+1}(z_2)/2 \rfloor = \lfloor 2/2 \rfloor = 1$. So we may assume $n \geq 3$. We consider two cases.

(i) $g_{s+1}(z_n)$ is a seed. Then Remark 3(iv) implies that $g_{s+1}(z_n)$ can be written in the form $(u^1, \ldots, u^n)$ such that

$$g_{s+1}(z_n) = \sum_{i=1}^{n} u^i g_{s+1}(z_i), \tag{7}$$

where $u^n = 0$, and $u^i \neq 0 \implies g_{s+1}(z_i)$ is a nonnegative power of 2 for all $i \in \{1, \ldots, n\}$. Since $g_{s+1}(z_1) = 1$ we thus have,

$$\left\lfloor \frac{g_{s+1}(z_n)}{2} \right\rfloor = \left\lfloor \frac{\sum_{i=1}^{n} u^i g_{s+1}(z_i)}{2} \right\rfloor = \sum_{i=2}^{n} u^i \frac{g_{s+1}(z_i)}{2} =$$

$$\sideset{}{'}\sum_{i \in \{2, \ldots, n-1\}} u^i \frac{g_{s+1}(z_i)}{2}.$$

Note that (7) implies $(u^1, \ldots, u^{n-1}, 1) \in \mathcal{P}_{s+1}$. By Corollary 1, $(u^1, \ldots, u^{n-1}, 1)$ is evil, hence $g_{s+1}(z_n) = (u^1, \ldots, u^{n-1}, u^n)$ is odious ($u^n = 0$). Further, Theorem 3 implies $L(u^1, \ldots, u^{n-1}, 1) = (u^2, \ldots, u^{n-1}, 1) \in \mathcal{P}_s$, where the rightmost 1-bit is at position $n - 1$ of the vector. This and the induction hypothesis then imply,

$$g_s(z_{n-1}) = {\sum_{i \in \{2, \ldots, n-1\}}}' u^i g_s(z_{i-1}) = {\sum_{i \in \{2, \ldots, n-1\}}}' u^i \frac{g_{s+1}(z_i)}{2}.$$

Thus $g_s(z_{n-1}) = \lfloor g_{s+1}(z_n)/2 \rfloor$.

(ii) $g_{s+1}(z_n)$ is a power of 2, i.e., $g_{s+1}(z_n) = 2^t$, where $t \in \mathbb{Z}_{\geq 0}$ is the smallest number satisfying $g_{s+1}(z_{n-1}) \leq 2^t - 1$. In particular, $g_{s+1}(z_n)$ is odious. By monotonicity and Remark 3(iv), $g_{s+1}(z_{n-1}) \geq 2^{t-1}$. Since a power of 2 is not the Nim-sum of lower powers of 2, there is no binary vector $(u^1, \ldots, u^{n-1}, 1) \in \mathcal{P}_{s+1}$. Hence by Theorem 3, there is no binary vector $(u^2, \ldots, u^{n-1}, 1) \in \mathcal{P}_s$, so also $g_s(z_{n-1}) = 2^r$, where $r$ is the smallest nonnegative integer satisfying $g_s(z_{n-2}) \leq 2^r - 1$. By the induction hypothesis, $2^{t-2} \leq \lfloor g_{s+1}(z_{n-1})/2 \rfloor = g_s(z_{n-2}) < 2^{t-1}$. It follows that $r = t - 1$, so $g_s(z_{n-1}) = 2^{t-1} = g_{s+1}(z_n)/2$. ∎

## 5   An Efficient Computation of the Lexicode

As was remarked at the end of Sect. 2, the $P$-positions of a game constitute a code. In this section we present an algorithm for computing a basis for the $P$-positions of Nim-Celopt$_s$, played on the Nim-graph $G = (V, E)$, where $V = (z_0, \ldots, z_n)$. By Theorem 1, it is the lexicode with distance $d = s + 2$. We let $t \in \mathbb{Z}_{\geq 0}$ be the smallest number satisfying $g_s(z_n) \leq 2^t - 1$.

**Theorem 6.** *For any Nim-Celopt$_s$ game,*
(a) *dim $\mathcal{P}_s$ is the number of seeds $\leq 2^t - 1$.*
(b) *If $g_s(z_m)$ is a seed, then there is $u \in \mathcal{P}_s$ with $u^m = 1$; and $w(u) = s + 2$.*

*Example 3.* Illustration of Theorems 5 and 6.

| n | 1 | 2 | 3 | 4 | 5 | 6 | 7 | 8 | 9 | 10 | 11 |
|---|---|---|---|---|---|---|---|---|---|---|---|
| $g_1(z_n)$ | 1 | 2 | 3 | 4 | 5 | 6 | 7 | 8 | 9 | 10 | 11 |
| $g_2(z_n)$ | 1 | 2 | 4 | 7 | 8 | 11 | 13 | 14 | 16 | 19 | 21 |
| $t = \lfloor \log_2 g_2(z_n) \rfloor + 1$ | 1 | 2 | 3 | 3 | 4 | 4 | 4 | 4 | 5 | 5 | 5 |
| dim $\mathcal{P}_2 = n - t$ | 0 | 0 | 0 | 1 | 1 | 2 | 3 | 4 | 4 | 5 | 6 |

**Proof of Theorem 6.** (a) Any $P$-position can be written in the form

$$u = ((i_1, \ldots, i_l, m) : i_1 < \ldots < i_l < m, \; g(z_{i_1}) \oplus \ldots \oplus g(z_{i_l}) \oplus g(z_m) = 0)$$

for suitable $m \leq n$ (we used Notation 1 for $u = (u^0, \ldots, u^n)$). By Remark 1(ii), $g(z_{i_1}) < \ldots < g(z_{i_l}) < g(z_m)$. If $g(z_m) = 2^k$ is a power of 2, then $u$ cannot be a $P$-position, since $2^k$ is not the sum of distinct nonnegative integers $< 2^k$. Let $T := \{g(z_j) : 0 < j < m, \ g(z_j) \text{ is a power of } 2\}$. Whenever $g(z_m)$ is a seed, there clearly are powers of 2 such that $g(z_m) \oplus \sum'_{g(z_j) \in R} g(z_j) = 0$, where $R$ is a suitable subset of $T$. Thus, $v := g(z_m) \oplus \sum'_{g(z_j) \in R} \in P$. Denote by $S$ the set of all such $v$. All $v \in S$ are linearly independent, since every $z_m$ such that $g(z_m)$ is a seed appears in $S$ only once, and then $u^k = 0$ for all $k > m$. Thus $S$ constitutes a basis for $P$.

(b) The first part was already proved in (a). For $u \in S$, Lemma 2 implies that $g(z_m)$ is the *smallest* nonnegative integer which is not the Nim-sum of $p$ of its followers, for all $p \in \{0, \ldots, q\}$. Further, $g(z_m) = g(z_{i_l}) \oplus x$, where $x := \sum'^{l-1}_{j=1} g(z_{i_j})$. Clearly, $x < g(z_{i_l}) < g(z_m)$. Hence there are $0 < k_1 < \ldots < k_p$ with $p \le s$, such that $x = \sum'^{p}_{j=1} g(z_{k_j})$. Therefore $g(z_m) = g(z_{i_l}) \oplus \sum'^{p}_{j=1} g(z_{k_j})$ is the Nim-sum of at most $s + 1$ summands, so for $u = z_m \oplus z_{i_l} \oplus \sum'^{p}_{j=1} g(z_{k_j})$, we have $u \in P$ and $w(u) \le s + 2$. Since Theorem 1 implies that $P$ is a set of codewords of Hamming distance $d \ge s + 2$, so in particular $w(u) \ge s + 2$, we have $w(u) = s + 2$. ∎

The following is an algorithm for computing $P_s \cap \{0, 1\}^n$, i.e., the basis for a linear code (LinCod).

### Algorithm LinCod

1. For $m \in \{1, \ldots, n\}$, compute $g_s(z_m) = \mathrm{mex}\{g(z_{i_1}) \oplus \ldots \oplus g(z_{i_j}) : 1 \le i_1 < \ldots < i_j < m, \ j \le s\}$.

2. Compute a basis member of $P$ for each seed: it is the seed and all the powers of two appearing in it. (Thus, if the seed is 13, then the basis member induced by 13 is $\{13, 8, 4, 1\}$.) End.

### Verification and Complexity of Algorithm LinCod

Step 1 takes time $n \sum_{i=1}^{s} \binom{n}{i} = O\left(n(n + n^2 + \ldots + n^s)\right) = O(n^{s+1})$. This dominates the complexity of step 2. Then Theorem 6(b) and its proof show how to compute the basis members. By Theorem 1, the time complexity is $O(n^{d-1})$. The space complexity is $O(n^{d-2})$. It is possible to formulate a different algorithm, which takes time $O(n^{2d})$ and space $O(n)$, by doing an exhaustive search for a vector $u$ with the properties of Theorem 6(b).

*Remark 4.* (i) If we are only interested in computing the dimension of $P$, it suffices, by Theorem 6(a), to count the number of seeds, deleting step 2.
(ii) For $s$ even, Theorem 5 implies that the complexity is $O(n^s) = O(n^{d-2})$. For the special case $s = 2$, the algorithm is linear in $n$, since for $s = 1$ (ordinary Nim — see also Example 3), we have $g(n) = n$.

*Example 4.* Let $n = 8$, $s = 3$. Then $d = 5$ by Theorem 1. We get $(g(z_i)_{i=1}^{8}) = (1, 2, 4, 8, 15, 16, 32, 51)$. The seeds are $15, 51$, hence a basis is given by

$$\{11111000, 11000111\}.$$

The set of all $P$-positions is the linear span of the basis vectors, so $P = \{\Phi, 11111000, 11000111, 00111111\}$.

# 6    Epilogue

We note that the *Turning Turtles* and *Mock Turtles* games [20] Ch. 14, have been mentioned in connection with lexicodes in [2], [7]. It turns out that the *P*-positions of these games also constitute the lexicode. This can be proved the same way as Theorem 1 above. Initial segments of the $g$-values for various values of $s$ (in a different notation) of Turning Turtles are given in Table 3 in [20], Ch. 14. It was observed there that for $s = 2$ the $g$-values are odious; but it was not mentioned there that the $g$-values are in fact odious for all even $s$ (Theorem 5 above).

In [21], two classes of games were defined:

(i) Games People Play (*PlayGames*): games that are challenging to the point that people will purchase them and play them.

(ii) Games Mathematicians Play (*MathGames*): games that are challenging to a mathematician or other scientist to play with and ponder about, but not necessarily to "the man in the street".

Examples of PlayGames are chess, Go, Hex, Reversi; of MathGames: Nim-type games, Wythoff games, annihilation games, octal games.

Some "rule of thumb" properties, which seem to hold for the majority of PlayGames and MathGames were listed there. One of the referees of this paper asked us to state that Celata and Celopt games belong to the class of Math-Games.

Theorem 3 shows how to get a code with parameters $(n + 1, k, d + 1)$ from a code with parameters $(n, k, d)$, for $d$ odd, by adding the "parity check bit" $w'(u)$. This phenomenon is well-known in coding theory. It is equivalent to the Mock Turtle Theorem in [20], Ch. 14. See also [7].

To summarize, the central result of this paper is an $O(n^{d-1})$ algorithm for computing the basis elements of a lexicode. The previously known algorithms have complexity exponential in $n$: $O(2^{2n})$ ([2]), $O(2^{n(1-3/d)+1})$ ([3]). All the basis elements which the algorithm produces have weight $\geq d$. The results were obtained from the theory of combinatorial games; specifically, from two-player games on cellular automata. There aren't many meaningful applications of the theory of impartial combinatorial games to areas outside of game theory. This is one of them.

# Acknowledgment

We thank Michael Langberg for inspiring conversations, which led to an improvement in the complexity of the algorithm.

# References

1. Conway, J.: Integral lexicographic codes. Discrete Mathematics **83** (1990) 219–235
2. Conway, J., Sloane, N.: Lexicographic codes: error-correcting codes from game theory. IEEE Transactions on Information Theory **IT-32** (1986) 337–348

3. Trachtenberg, A.: Designing lexicographic codes with a given Trellis complexity. IEEE Transactions on Information Theory **48** (2002) 89–100
4. Trachtenberg, A., Vardy, A.: Lexicographic codes: Constructions, bounds, and Trellis complexity. In: 31st Annual Conference on Information Sciences and Systems, http://citeseer.nj.nec.com/448945.html (1997)
5. Brualdi, R., Pless, V.: Greedy codes. Journal of Combinatorial Theory *(Ser. A)* **64** (1993) 10–30
6. Brouwer, A.: Bounds on the minimum distance of linear codes. http://www.win.tue.nl/~aeb/voorlincod.html (2003)
7. Pless, V.: Games and codes. In Guy, R., ed.: Combinatorial Games, *Symposium in Applied Mathematics*. Volume 43. Amererican Mathematical Society (1991) 101–110
8. MacWilliams, F., Sloane, N.: The Theory of Error-Correcting Codes. North-Holland, Elsevier, Amsterdam (1978)
9. Fraenkel, A.: Error-correcting codes derived from combinatorial games. In Nowakowski, R., ed.: Games of No Chance. Volume 29. Cambridge University Press (1996) 417–431
10. Fraenkel, A.: Virus versus mankind. In Marsland, T., Frank, I., eds.: Second International Conference on Computers and Games (CG 2000). Volume 2063 of Lecture Notes in Computer Science. Springer-Verlag (2001) 204–213
11. Fraenkel, A.: Mathematical chats between two physicists. In Wolfe, D., Rodgers, T., eds.: Puzzler's Tribute: A Feast for the Mind. A K Peters (2002) 383–386
12. Fraenkel, A.: Two-player games on cellular automata. In Nowakowski, R., ed.: More Games of No Chance. Volume 42. Cambridge University Press (2002) 279–306
13. Fraenkel, A.: Combinatorial games with an annihilation rule. In LaSalle, J., ed.: The Influence of Computing on Mathematical Research and Education, *Symposium in Applied Mathematics*. Volume 20. American Mathematical Society (1974) 87–91
14. Fraenkel, A., Tassa, U., Yesha, Y.: Three annihilation games. Mathematics Magazine **51** (1978) 13–17 special issue on Recreational Mathematics.
15. Fraenkel, A., Yesha, Y.: Theory of annihilation games. Bulletin of the American Mathematical Society **82** (1976) 775–777
16. Fraenkel, A., Yesha, Y.: Theory of annihilation games — I. Journal of Combinatorial Theory *(Ser. B)* **33** (1982) 60–86
17. Yesha, Y.: Theory of annihilation games. PhD thesis, Weizmann Institute of Science, Rehovot, Israel (1978)
18. Ferguson, T.: Misère annihilation games. Journal of Combinatorial Theory *(Ser. A)* **37** (1984) 205–230
19. Fraenkel, A.: Scenic trails ascending from sea-level nim to alpine chess. In Nowakowski, R., ed.: Games of No Chance. Volume 29. Cambridge University Press (1996) 13–42
20. Berlekamp, E., Conway, J., Guy, R.: Winning Ways for Your Mathematical Plays. Academic Press (1982)
21. Fraenkel, A.: Complexity, appeal and challenges of combinatorial games (2003) Expanded version of a keynote address at Dagstuhl Seminar "Algorithmic Combinatorial Game Theory", Feb. 17–22, 2002. To appear in *Theoretical Computer Science*, special issue on Algorithmic Combinatorial Game Theory. Preprint at http://www.wisdom.weizmann.ac.il/~fraenkel.

# Analysis of Composite Corridors

Teigo Nakamura[1] and Elwyn Berlekamp[2]

[1] International Computer Science Institute, Berkeley, CA
and Kyushu Institute of Technology, Fukuoka, Japan
[2] University of California, Berkeley

**Abstract.** This work began as an attempt to find and catalog the mean values and temperatures of a well-defined set of relatively simple common Go positions, extending a similar but smaller catalog in Table E.10, Appendix E of the book *Mathematical Go* [1].
The major surprises of our present work include the following

- A position of chilled value *2 (previously unknown in Mathematical Go), shown at the end of Sect. 3.1.
- A surprisingly "warm" position, whose temperature is routinely underestimated even by very strong Go players, shown in Sect. 4.
- More insights into decompositions.

It is hoped that these results may someday provide the basis for further new insights and generalizations.

## 1 Introduction

Appendix E of the book *Mathematical Go* [1] contains a collection of small, relatively cool Go positions and their chilled values. Among the highlights of this Appendix is Fig. E.10 and its surroundings (pages 196–201 in the English edition; pages 199–204 in the Japanese translation), which is an extensive tabulation of simple corridors located along an edge of the board. This paper extends those results. Our preliminary long-range goal was to evaluate all positions in which:

- All points on the first line are initially empty except for the endpoints, each of which may be occupied by an immortal stone of either color.
- All points on the second line are either empty or occupied by immortal stones.
- Every empty point on the second line (a "gap") lies between second-line points occupied by stones of different colors.

The present paper investigates all such positions with only one gap, as well as many (but not all) cases with two gaps. For each such position, we obtain the mean value and a temperature. For games of temperature one, we also obtain the value of the infinitesimal to which the game chills. Our analysis includes not only these initial positions, but all of their orthodox descendants as well as some plausible unorthodox descendants.

J. Schaeffer et al. (Eds.): CG 2002, LNCS 2883, pp. 213–229, 2003.

## 1.1    Definition

We define the following terms for Go positions used in this article.

**corridor** All the stones on the second line are alive. There are two types of corridors. A *blocked corridor* is closed at one side of the first line with the stone of the same color as the corridor. The opponent may invade from the other side. An *unblocked corridor* is open to invasion on both sides. Figure 1(a) is the *blocked corridor of length i* and Fig. 1(b) is the *unblocked corridor of length i.* The short lines coming out of the stones on the second line denote these stones are alive.

(a) Blocked corridor                         (b) Unblocked corridor

**Fig. 1.** Example of corridors.

**gap** An empty point between the stones of different color on the second line is a gap. Since we focus on analysing the corridors just below the second line, we assume that the gap-closing move has no effect on the rest of the board. For example, the corridors with a gap in Fig. 2(a) should be regarded as the position like Fig. 2(b) or 2(c).

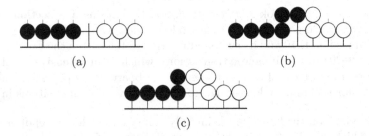

**Fig. 2.** Corridors with a gap.

**socket** We define two types of sockets. We call the point "a" of Fig. 3(a) a *full socket* (or just *socket*) and the point "b" of Fig. 3(b) a *half socket.*

(a)                                         (b)

**Fig. 3.** Two types of sockets.

# 2   Two Adjacent Corridors without Gaps

## 2.1   Blocked on Both Sides

The following two tables show the mean value, temperature, ish-type, Black's play and White's play for each $i$, $j$ and $k$ of two adjacent blocked corridors without gaps [2]. The rows of the tables are sorted in lexicographic order of $(j, i, k)$. Table 2 is in the case of $k = 0$ and we assume $i \leq j$ without loss of generality.

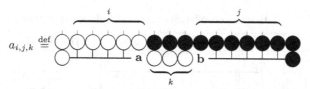

$$a_{i,j,k} \stackrel{\text{def}}{=}$$

**Table 1.** $a_{i,j,k}$: Blocked on both sides.

| i | j | k | Mean | Temp | ish | Black | White |
|---|---|---|------|------|-----|-------|-------|
| 0 | $\geq 0$ | $\geq 1$ | $j - 2 + 2^{1-j}$ | $1 - 2^{1-j}$ | | b | b |
| 1 | 0 | 1 | $\frac{1}{3}$ | $\frac{1}{3}$ | | a | a |
| 1 | 0 | 2 | 1 | 1 | $\downarrow$ | a | a |
| 1 | 0 | $\geq 3$ | $k - 1\frac{1}{12}$ | $k - 1\frac{1}{12}$ | | a | a |
| 2 | 0 | $\geq 1$ | $k - 1$ | $k$ | $*$ | a | a |
| $\geq 3$ | 0 | 1 | $-i + 2\frac{1}{6}$ | $1\frac{1}{6}$ | | a | a |
| $\geq 3$ | 0 | $\geq 2$ | $k - i + 1$ | $k$ | | a | a |
| $\geq 1$ | 1 | $\geq 1$ | $-i + 1$ | 0 | | | |
| 1 | $\geq 2$ | 1 | $j - 1\frac{1}{3}$ | $\frac{2}{3}$ | $*$ | b | b |
| 1 | $\geq 2$ | 2 | $j - 1$ | 1 | $\Downarrow *$ | b | b |
| 1 | $\geq 2$ | $\geq 3$ | $j - 1$ | 1 | $\rule[0.5ex]{1em}{0.4pt}$ | b | b |
| $\geq 2$ | $\geq 2$ | $\geq 1$ | $j - i$ | 1 | $\rule[0.5ex]{1em}{0.4pt}$ | b | b |

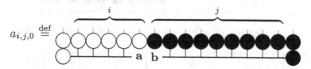

$$a_{i,j,0} \stackrel{\text{def}}{=}$$

**Table 2.** $a_{i,j,0}$: Blocked on both sides.

| i | j | k | Mean | Temp | ish | Black | White |
|---|---|---|------|------|-----|-------|-------|
| 0 | $\geq 0$ | 0 | $j - 2 + 2^{1-j}$ | $1 - 2^{1-j}$ | | b | b |
| 1 | 1 | 0 | 0 | 0 | | | |
| 1 | 2 | 0 | $\frac{1}{2}$ | $\frac{1}{2}$ | $*$ | b | b |
| $\geq 2$ | $\geq 2$ | 0 | $j - i$ | 1 | $*$ | a | b |
| 1 | $\geq 3$ | 0 | $j - 1\frac{2}{3}$ | $\frac{2}{3}$ | $\uparrow$ | b | b |

## 2.2   Blocked on Offensive Side and Unblocked on Defensive Side

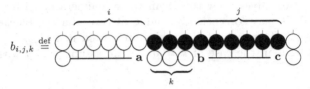

**Table 3.** $b_{i,j,k}$: Blocked on offensive side and unblocked on defensive side.

| i | j | k | Mean | Temp | ish | Black | White |
|---|---|---|------|------|-----|-------|-------|
| ≥0 | 0 | ≥1 | $-i$ | 0 | | | |
| 0 | ≥1 | ≥1 | $j-4+2^{3-j}$ | $1-2^{3-j}$ | | b,c | b,c |
| 1 | 1 | 1 | $-\frac{1}{3}$ | $\frac{2}{3}$ | * | b(=c) | b(=c) |
| 1 | 1 | 2 | 0 | 1 | ⇓ * | b(=c) | b(=c) |
| 1 | 1 | ≥3 | 0 | 1 | — | b(=c) | b(=c) |
| 2 | 1 | 1 | $-1$ | 1 | ↓ | b(=c) | b(=c) |
| ≥2 | 1 | ≥2 | $-i+1$ | 1 | — | b(=c) | b(=c) |
| ≥3 | 1 | 1 | $-i+1$ | 1 | — | b(=c) | b(=c) |
| 1 | 2 | ≥1 | 0 | 0 | | | |
| ≥2 | ≥2 | ≥1 | $j-i-2+2^{2-j}$ | $1-2^{2-j}$ | | b | c |
| 1 | 3 | 1 | $\frac{1}{3}$ | $\frac{1}{3}$ | | b | c |
| 1 | ≥3 | ≥2 | $j-3+2^{2-j}$ | $1-2^{2-j}$ | | b | c |
| 1 | ≥4 | 1 | $j-3\frac{1}{3}+\frac{2^{4-j}}{3}$ | $1-\frac{2^{4-j}}{3}$ | | c | c |

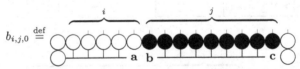

**Table 4.** $b_{i,j,0}$: Blocked on offensive side and unblocked on defensive side.

| i | j | k | Mean | Temp | ish | Black | White |
|---|---|---|------|------|-----|-------|-------|
| ≥0 | 0 | 0 | $-i$ | 0 | | | |
| 0 | ≥1 | 0 | $j-4+2^{3-j}$ | $1-2^{3-j}$ | | b,c | b,c |
| ≥0 | 1 | 0 | $-i+1-2^{-i}$ | $1-2^{-i}$ | | b(=c) | b(=c) |
| 1 | 2 | 0 | 0 | 0 | | | |
| ≥2 | ≥2 | 0 | $j-i-2+3\cdot2^{1-j}$ | $1-2^{1-j}$ | | a | a,b |
| 1 | 3 | 0 | 0 | 0 | | | |
| 1 | ≥4 | 0 | $j-3\frac{2}{3}+\frac{2^{4-j}}{3}$ | $1-\frac{2^{4-j}}{3}$ | | c | c |

## 2.3  Unblocked on Offensive Side and Blocked on Defensive Side

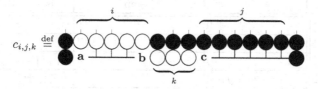

**Table 5.** $c_{i,j,k}$: Unblocked on offensive side and blocked on defensive side.

| i | j | k | Mean | Temp | ish | Black | White |
|---|---|---|------|------|-----|-------|-------|
| 0 | $\geq 0$ | $\geq 1$ | $j+2k$ | $0$ | | | |
| 1 | $\geq 0$ | $\geq 1$ | $k+j-1+2^{-j}$ | $k+1-2^{-j}$ | | a(=b) | a(=b) |
| 2 | 0 | 1 | $\frac{2}{3}$ | $\frac{2}{3}$ | ↓ | b | b |
| $\geq 2$ | 0 | $\geq 2$ | $k-i+3-3\cdot 2^{1-i}$ | $k-2^{1-i}$ | | b | b |
| 3 | 0 | 1 | $\frac{1}{4}$ | $\frac{3}{4}$ | * | b | b |
| $\geq 4$ | 0 | 1 | $-i+4\frac{1}{6}-\frac{11}{3}2^{1-i}$ | $\frac{7}{6}-\frac{5}{3}2^{1-i}$ | | b | b |
| 2 | 1 | $\geq 1$ | $0$ | $0$ | | | |
| $\geq 3$ | 1 | $\geq 1$ | $-i+3-2^{2-i}$ | $1-2^{2-i}$ | | a | a |
| 2 | $\geq 2$ | 1 | $j-1\frac{1}{6}$ | $\frac{5}{6}$ | * | c | c |
| $\geq 2$ | $\geq 2$ | $\geq 2$ | $j-i+2-2^{2-i}$ | $1$ | — | c | c |
| 3 | $\geq 2$ | 1 | $j-1\frac{5}{8}$ | $\frac{7}{8}$ | | c | c |
| 4 | $\geq 2$ | 1 | $j-2\frac{13}{48}$ | $\frac{47}{48}$ | | c | c |
| $\geq 5$ | $\geq 2$ | 1 | $j-i+2-2^{2-i}$ | $1$ | — | c | c |

We omit the table of $c_{i,j,0}$, since $c_{i,j,0}$ is $-b_{j,i,0}$ obvious from the definition.

## 2.4  Unblocked on Both Sides

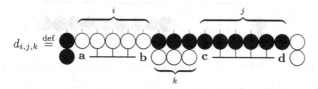

**Table 6.** $d_{i,j,k}$ : Unblocked on both sides.

| i | j | k | Mean | Temp | ish | Black | White |
|---|---|---|------|------|-----|-------|-------|
| $\geq 0$ | $i$ | $\geq 1$ | $0$ | $0$ | | | |
| $\geq 1$ | 0 | $\geq 1$ | $-i+2-2^{1-i}$ | $1-2^{1-i}$ | | a | a |
| $\geq 3$ | $>i$ | $\geq 2$ | $j-i+2-2^{2-j}-2^{2-i}$ | $1-2^{2-j}$ | | c | d |
| 0 | 1 | $\geq 1$ | $k$ | $k$ | | c(=d) | c(=d) |

## Table 6. (continued)

| i | j | k | Mean | Temp | ish | Black | White |
|---|---|---|------|------|-----|-------|-------|
| 2 | 1 | $\geq 1$ | 0 | 0 | | | |
| $\geq 3$ | 1 | $\geq 1$ | $-i+3-2^{2-i}$ | $1-2^{2-i}$ | | c(=d) | c(=d)s |
| 0 | $\geq 2$ | 1 | $j$ | 1 | $\downarrow +j.(\uparrow *)$ | d | d |
| 0 | $\geq 2$ | $\geq 2$ | $2k+j-2$ | 1 | $0^{j-2}\mid$ ✚ | d | d |
| 1 | 2 | $\geq 1$ | $k$ | $k$ | | a(=b) | a(=b) |
| $\geq 3$ | 2 | $\geq 1$ | $-i+3-2^{2-i}$ | $1-2^{2-i}$ | | a | a |
| $>j$ | $\geq 3$ | $\geq 2$ | $j-i+2^{2-j}-2^{2-i}$ | $1-2^{2-i}$ | | c | a |
| 1 | $\geq 3$ | $\geq 1$ | $k+j-3+2^{2-j}$ | $k+1-2^{2-j}$ | | a(=b) | a(=b) |
| 2 | 3 | $\geq 1$ | $\frac{1}{2}$ | $\frac{1}{2}$ | | c | b(,c,d) |
| 2 | $\geq 3$ | $\geq 2$ | $j-3+2^{2-j}$ | $1-2^{2-j}$ | | c | d |
| 4 | 3 | 1 | $-\frac{3}{4}$ | $\frac{3}{4}$ | | a,c | a |
| $\geq 5$ | $\geq 3$ | 1 | $j-i+2^{2-j}-2^{2-i}$ | $max\{1-2^{2-j},1-2^{2-i}\}$ | | c | a or d |
| 2 | 4 | 1 | $1\frac{1}{6}$ | $\frac{2}{3}$ | | c | d |
| 3 | 4 | 1 | $\frac{3}{4}$ | $\frac{3}{4}$ | | c | b,d |
| 2 | $\geq 5$ | 1 | $j-3\frac{1}{6}+\frac{2^{4-j}}{3}$ | $1-\frac{2^{4-j}}{3}$ | | d | d |
| 3 | 5 | 1 | $1\frac{1}{2}$ | $\frac{1}{2}$ | | d | b,d |
| 4 | 5 | 1 | $\frac{7}{8}$ | $\frac{7}{8}$ | | c | b,d |
| 3 | $\geq 6$ | 1 | $j-3\frac{5}{8}+2^{2-j}$ | $1-2^{2-j}$ | | d | d |
| 4 | 6 | 1 | $1\frac{13}{16}$ | $\frac{15}{16}$ | | c | d |
| 4 | 7 | 1 | $2\frac{37}{48}$ | $\frac{23}{24}$ | | c | d |
| 4 | $\geq 8$ | 1 | $j-4\frac{7}{24}+2^{3-j}$ | $1-2^{3-j}$ | | c | d |

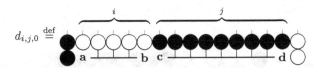

**Table 7.** $d_{i,j,0}$: Unblocked on both sides. $i \leq j$ without loss of generality.

| i | j | k | Mean | Temp | ish | Black | White |
|---|---|---|------|------|-----|-------|-------|
| $\geq 0$ | $i$ | 0 | 0 | 0 | | | |
| $\geq 2$ | $>i$ | 0 | $j-i-3\cdot 2^{1-i}+2^{2-j}$ | $1-2^{2-j}$ | | c | d |
| 0 | $\geq 1$ | 0 | $j-2+2^{1-j}$ | $1-2^{1-j}$ | | d | d |
| 1 | $\geq 2$ | 0 | $j-3+2^{2-j}$ | $1-2^{2-j}$ | | a(=b),d | a(=b),d |
| 2 | 3 | 0 | 0 | 0 | | | |
| 2 | $\geq 4$ | 0 | $j-3\frac{1}{2}+2^{2-j}$ | $1-2^{2-j}$ | | d | d |

# 3  Two Adjacent Corridors with Gaps

## 3.1  Blocked on Both Sides

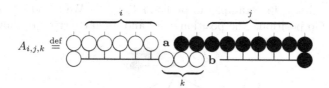

**Table 8.** $A_{i,j,k}$: Blocked on both sides.

| i | j | k | Mean | Temp | ish | Black | White |
|---|---|---|---|---|---|---|---|
| 0 | $\geq 0$ | $\geq 1$ | $j - 2 + 2^{1-j}$ | $1 - 2^{1-j}$ | | b | b |
| 1 | 0 | 1 | $-\frac{1}{3}$ | $\frac{2}{3}$ | | a | a |
| 1 | 0 | 2 | $1 - i$ | 1 | $\Downarrow *$ | a | a |
| 1 | 0 | $\geq 3$ | $1 - i$ | 1 | $\rightharpoondown$ | a | a |
| 2 | 0 | 1 | $1 - i$ | 1 | $\downarrow$ | a | a |
| 2 | 0 | $\geq 2$ | $1 - i$ | 1 | $\rightharpoondown$ | a | a |
| $\geq 3$ | 0 | $\geq 1$ | $1 - i$ | 1 | $\rightharpoondown$ | a | a |
| $\geq 1$ | 1 | $\geq 1$ | $-i + \frac{1}{2}$ | $\frac{1}{2}$ | | a | a |
| $\geq 1$ | 2 | $\geq 1$ | $1 - i$ | 0 | | a,b | b |
| 1 | $\geq 3$ | 1 | $j - 2\frac{1}{6}$ | $\frac{5}{6}$ | $*$ | a,b | b |
| 1 | $\geq 3$ | 2 | $j - 2$ | 1 | $\Downarrow$ | a,b | b |
| 1 | $\geq 3$ | $\geq 3$ | $j - 2$ | 1 | $\rightharpoondown |0$ | a,b | b |
| 2 | $\geq 3$ | 1 | $j - 3$ | 1 | $\Downarrow *$ | a,b | b |
| 2 | $\geq 3$ | $\geq 2$ | $j - 3$ | 1 | $\rightharpoondown |0$ | a,b | b |
| $\geq 3$ | $\geq 3$ | $\geq 1$ | $j - i - 1$ | 1 | $\rightharpoondown |0$ | a,b | b |

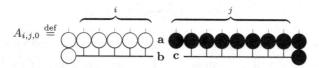

**Table 9.** $A_{i,j,0}$: Blocked on both sides.

| i | j | k | Mean | Temp | ish | Black | White |
|---|---|---|---|---|---|---|---|
| $\leq 1$ | $\leq 1$ | 0 | 0 | 0 | | | |
| 0 | $\geq 2$ | 0 | $j - 2 + \frac{1}{2}$ | $\frac{1}{2}$ | $*$ | c | c |
| 1 | $\geq 2$ | 0 | $j - 1\frac{1}{3}$ | $\frac{5}{6}$ | $\uparrow$ | b | b |
| 2 | 2 | 0 | 0 | 1 | $*2$ | a,b | a,b |
| $\geq 3$ | 2 | 0 | $2 - i$ | 1 | $\Uparrow *|0, *$ | b | a,b |
| 2 | $\geq 3$ | 0 | $j - 2$ | 1 | $0, *| \Downarrow *$ | a,b | b |
| $\geq 3$ | $\geq 3$ | 0 | $j - i$ | 1 | $0| \dashv\vdash || \rightharpoondown |0$ | b | b |

In Table 9, we omit some unimportant entries which follow immediately from symmetry. We found several new positions whose ish-types are previously unknown as Go positions. They are $*2$, $\{0, *| \Downarrow *\}$, $\{\Uparrow *|0, *\}$ and $\{0| + || - |0\}$ listed at the bottom four rows of Table 9. The existence of $*2$ was an open problem which is listed, for example, in problem 45 in [3]. We found $*2$ in the form of a chilled Go position. The position of $*2$ and its game tree are shown below.

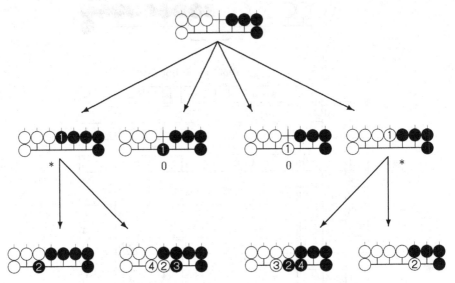

## 3.2   Blocked on Offensive Side and Unblocked on Defensive Side

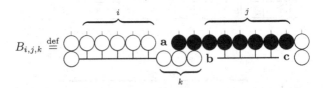

**Table 10.** $B_{i,j,k}$: Blocked on offensive side and unblocked on defensive side.

| i | j | k | Mean | Temp | ish | Black | White |
|---|---|---|------|------|-----|-------|-------|
| $\geq 1$ | $\leq 1$ | $\geq 1$ | $-i$ | $0$ | | | |
| $\geq 1$ | $2$ | $\geq 1$ | $-i + \frac{1}{2}$ | $\frac{1}{2}$ | | a | a |
| $0$ | $\leq 3$ | $\geq 1$ | $0$ | $0$ | | | |
| $1$ | $3$ | $1$ | $-\frac{1}{3}$ | $\frac{2}{3}$ | | a | a |
| $1$ | $\geq 3$ | $\geq 2$ | $j - 4 + 3 \cdot 2^{1-j}$ | $1 - 2^{1-j}$ | | a | a |
| $\geq 2$ | $\geq 3$ | $\geq 1$ | $j - i - 3 + 3 \cdot 2^{1-j}$ | $1 - 2^{1-j}$ | | a | a |
| $0$ | $\geq 4$ | $\geq 1$ | $j - 4 + 2^{3-j}$ | $1 - 2^{3-j}$ | | b,c | b,c |
| $1$ | $4$ | $1$ | $\frac{1}{4}$ | $\frac{3}{4}$ | | a | a |
| $1$ | $5$ | $1$ | $1$ | $\frac{2}{3}$ | | c | c |
| $1$ | $\geq 6$ | $1$ | $j - 4\frac{1}{6} + \frac{1}{3} \cdot 2^{4-j}$ | $1 - \frac{1}{3} \cdot 2^{4-j}$ | | c | c |

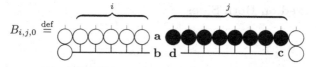

$$B_{i,j,0} \stackrel{\text{def}}{=}$$

**Table 11.** $B_{i,j,0}$: Blocked on offensive side and unblocked on defensive side.

| i | j | k | Mean | Temp | ish | Black | White |
|---|---|---|---|---|---|---|---|
| $\geq 0$ | $0$ | $0$ | $-i - 2^{-1-i}$ | $1 - 2^{-1-i}$ | | a | a |
| $\geq 0$ | $1$ | $0$ | $-i+1-2^{-i}$ | $1-2^{-i}$ | | a,c | b |
| $0$ | $3 \geq j \geq 1$ | $0$ | $0$ | $0$ | | | |
| $\geq 1$ | $2$ | $0$ | $-i+1$ | $0$ | | | |
| $1$ | $\geq 3$ | $0$ | $j - 3\frac{1}{3} + \frac{2^{3-j}}{3}$ | $1 - 2^{3-j}$ | | c | c |
| $\geq 2$ | $\geq 3$ | $0$ | $j - i - 2 + 2^{2-j}$ | $1 - 2^{2-j}$ | | a,b | a,b,c |
| $0$ | $\geq 4$ | $0$ | $j - 3\frac{1}{2} + 2^{2-j}$ | $1 - 2^{2-j}$ | | c | c |

## 3.3   Unblocked on Offensive Side and Blocked on Defensive Side

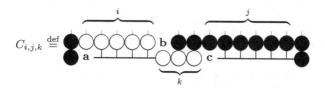

$$C_{i,j,k} \stackrel{\text{def}}{=}$$

**Table 12.** $C_{i,j,k}$: Unblocked on offensive side and blocked on defensive side.

| i | j | k | Mean | Temp | ish | Black | White |
|---|---|---|---|---|---|---|---|
| $1$ | $0$ | $\geq 1$ | $0$ | $0$ | | | |
| $\geq 2$ | $0$ | $\geq 1$ | $-i+3-2^{2-i}$ | $1-2^{2-i}$ | | b | b |
| $\geq 2$ | $1$ | $\geq 1$ | $-i+2\frac{1}{2}-2^{1-i}$ | $1-2^{1-i}$ | | a | a |
| $1,2$ | $\geq 1$ | $\geq 1$ | $j-2+2^{1-j}$ | $1-2^{1-j}$ | | c | c |
| $\geq 3$ | $2$ | $\geq 1$ | $-i+3-2^{1-i}$ | $1-2^{1-i}$ | | a | a |
| $\geq 2$ | $\geq 3$ | $\geq 1$ | $j-i+1+2^{3-i-j}-2^{2-i}$ | $1-2^{3-i-j}$ | | c | c |

A conjectured generalization of the last line of this table appears in Sect. 5. We omit the table of $C_{i,j,0}$, since $C_{i,j,0}$ is $-B_{j,i,0}$.

### 3.4    Unblocked on Both Sides

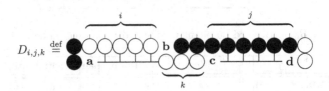

**Table 13.** $D_{i,j,k}$: Unblocked on both sides.

| i | j | k | Mean | Temp | ish | Black | White |
|---|---|---|------|------|-----|-------|-------|
| $\geq 2$ | $i$ | $\geq 1$ | $2^{2-i} - 1$ | $1 - 2^{2-i}$ | | a,b,c | a,b |
| $\geq 0$ | $i+1$ | 1 | 0 | 0 | | | |
| $\geq 1$ | $0,1$ | $\geq 1$ | $-i+2-2^{1-i}$ | $1-2^{1-i}$ | | a | a |
| $> j$ | $\geq 2$ | $\geq 1$ | $j-i-1+3\cdot 2^{1-j}-2^{1-i}$ | $1-2^{1-i}$ | | a | a |
| 0 | $\geq 2$ | $\geq 1$ | $j+k-3+2^{2-j}$ | $k+1-2^{2-j}$ | | b | b |
| $\geq 1$ | $i+2$ | $\geq 1$ | $j-i-1+2^{3-j}-2^{2-i}$ | $1-2^{3-j}$ | | b,c | b,d |
| 1 | $\geq 3$ | $\geq 1$ | $j-4+2^{3-j}$ | $1-2^{3-j}$ | | c,d | d |
| $\geq 1$ | $\geq i+3$ | $\geq 1$ | $j-i-1+2^{3-j}-2^{2-i}$ | $1-2^{3-j}$ | | b,c | d |

In Table 14, we assume $i \leq j$ without loss of generality.

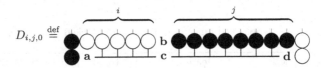

**Table 14.** $D_{i,j,0}$: Unblocked on both sides.

| i | j | k | Mean | Temp | ish | Black | White |
|---|---|---|------|------|-----|-------|-------|
| $\geq 0$ | $i$ | 0 | 0 | 0 | | | |
| $\geq 2$ | $> i$ | 0 | $j-i+2^{2-j}-2^{2-i}$ | $1-2^{2-j}$ | | d | c,d |
| 1 | $\geq 1$ | 0 | $j-3+2^{2-j}$ | $1-2^{2-j}$ | | a,b,c,d | c,d |

## 4    Three Adjacent Corridors with Gaps

We did not investigate all the cases of three adjacent corridors with two gaps but figured out the values of many cases in which both outer corridors are blocked and the length of the inner corridor is less than 4.

## 4.1  Blocked on Both Outer Corridors

We show the results using three tables according to the length of inner corridor. The table format is similar to the previous one and the rows in the tables are sorted in lexicographic order of $(j, i, k, l)$.

$$X_{i,j,k,l,m} \stackrel{\mathrm{def}}{=}$$

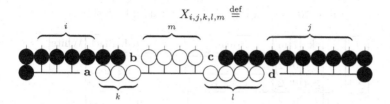

**Table 15.** $X_{i,j,k,l,m}$ $(m = 2)$: Blocked on both sides.

| i | j | k | l | m | Mean | Temp | ish | Black | White |
|---|---|---|---|---|---|---|---|---|---|
| 0 | 0 | 1 | 1 | 2 | $-\frac{1}{3}$ | $\frac{2}{3}$ | | b,c | b,c |
| 0 | 0 | $\geq 2$ | 1 | 2 | $-\frac{1}{4}$ | $\frac{3}{4}$ | | c | b,c |
| 1 | 0 | $\geq 1$ | 1 | 2 | $-\frac{5}{8}$ | $\frac{7}{8}$ | | c | c |
| 2 | 0 | $\geq 1$ | 1 | 2 | $-\frac{1}{8}$ | $\frac{7}{8}$ | | c | c |
| 3 | 0 | 1 | 1 | 2 | $\frac{37}{48}$ | $\frac{43}{48}$ | | a | a |
| $\geq 3$ | 0 | $\geq 2$ | 1 | 2 | $i - 2\frac{1}{4} + 2^{-1-i}$ | $1 - 2^{-1-i}$ | | a | a |
| $\geq 4$ | 0 | 1 | 1 | 2 | $i - 2\frac{1}{4}$ | $\frac{11}{12}$ | | a | a |
| 0 | 1 | 1 | $\geq 1$ | 2 | $-\frac{5}{8}$ | $\frac{7}{8}$ | | b | b |
| 0 | 1 | $\geq 2$ | $\geq 1$ | 2 | $-\frac{1}{2}$ | 1 | | b | b |
| 1 | 1 | $\geq 1$ | $\geq 1$ | 2 | $-1$ | 0 | | | |
| 2 | 1 | $\geq 1$ | $\geq 1$ | 2 | $-\frac{1}{2}$ | $\frac{1}{2}$ | | c | c |
| $\geq 3$ | 1 | 1 | $\geq 1$ | 2 | $i - 2\frac{9}{16}$ | $\frac{15}{16}$ | | a | a |
| $\geq 3$ | 1 | $\geq 2$ | $\geq 1$ | 2 | $i - 2\frac{1}{2}$ | 1 | | a | a |
| 0 | 2 | $\geq 2$ | 1 | 2 | 0 | 1 | | a | a |
| 1 | 2 | $\geq 2$ | 1 | 2 | $-\frac{1}{2}$ | $\frac{1}{2}$ | | b | b |
| 2 | 2 | $\geq 2$ | 1 | 2 | 0 | 0 | | | |
| 2 | 2 | $\geq 2$ | $\geq 2$ | 2 | 0 | 0 | | | |
| $\geq 3$ | 2 | $\geq 2$ | 1 | 2 | $i - 2$ | 1 | | a | a |
| 0 | $\geq 3$ | $\geq 2$ | 1 | 2 | $j - 2\frac{1}{8}$ | $\frac{7}{8}$ | | d | d |
| 1 | $\geq 3$ | $\geq 2$ | 1 | 2 | $j - 2\frac{9}{16}$ | $\frac{15}{16}$ | | d | d |
| 2 | $\geq 3$ | $\geq 2$ | 1 | 2 | $j - 2\frac{1}{16}$ | $\frac{15}{16}$ | | d | d |
| $\geq 3$ | $\geq 3$ | $\geq 2$ | 1 | 2 | $i + j - 4\frac{1}{8} + 2^{-2-i}$ | $1 - 2^{-2-i}$ | | a | a |
| $\geq 3$ | $\geq 3$ | $\geq 2$ | $\geq 2$ | 2 | $i + j - 4$ | 1 | | a,b,c,d | a,d |

$$X_{i,j,k,0,m} \overset{\text{def}}{=}$$

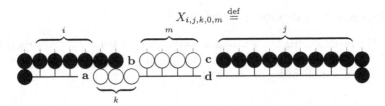

**Table 16.** $X_{i,j,k,0,m}$ ($m = 2$): Blocked on both sides.

| i | j | k | l | m | Mean | Temp | ish | Black | White |
|---|---|---|---|---|------|------|-----|-------|-------|
| 0 | 0 | 0 | 0 | 2 | 0 | 0 | | | |
| $\geq 1$ | 0 | $\geq 1$ | 0 | 2 | $i - 2 + 2^{1-i}$ | $1 - 2^{1-i}$ | | a | a |
| 0 | $\leq 1$ | $\geq 1$ | 0 | 2 | 0 | 0 | | | |
| $\geq 1$ | $\leq 1$ | 0 | 0 | 2 | $i - 1$ | 0 | | | |
| 1 | 1 | $\geq 1$ | 0 | 2 | $-\frac{1}{3}$ | $\frac{2}{3}$ | | d | d |
| 2 | 1 | $\geq 1$ | 0 | 2 | $\frac{1}{4}$ | $\frac{3}{4}$ | | d | d |
| $\geq 3$ | 1 | $\geq 1$ | 0 | 2 | $i - 2 + 2^{-i}$ | $1 - 2^{-i}$ | | a | a |
| $\geq j$ | $\geq 2$ | 0 | 0 | 2 | $i + j - 2\frac{1}{3}$ | $\frac{5}{6}$ | | a | a |
| 0 | $\geq 2$ | 1 | 0 | 2 | $j - 1\frac{1}{3}$ | $\frac{2}{3}$ | | d | d |
| 0 | $\geq 2$ | $\geq 2$ | 0 | 2 | $j - 1\frac{1}{4}$ | $\frac{3}{4}$ | | d | d |
| 1 | $\geq 2$ | $\geq 1$ | 0 | 2 | $j - 1\frac{5}{8}$ | $\frac{7}{8}$ | | d | d |
| 2 | 2 | 0 | 0 | 2 | $1\frac{3}{4}$ | $\frac{3}{4}$ | | a,d | a,d |
| 2 | $\geq 2$ | $1, 2$ | 0 | 2 | $j - 1\frac{1}{8}$ | $\frac{7}{8}$ | | d | d |
| 2 | 2 | $\geq 3$ | 0 | 2 | 1 | 1 | | d | d |
| 3 | $\geq 2$ | $\geq 2$ | 0 | 2 | $j - \frac{1}{8}$ | $\frac{7}{8}$ | | d | d |
| $3, 4$ | $\geq 2$ | 1 | 0 | 2 | $i + j - 3\frac{11}{48}$ | $\frac{43}{48}$ | | a | a |
| $\geq 4$ | $\geq 2$ | $\geq 2$ | 0 | 2 | $i + j - 3\frac{1}{4} + 2^{-i}$ | $1 - 2^{-i}$ | | a | a |
| $\geq 5$ | $\geq 2$ | 1 | 0 | 2 | $i + j - 3\frac{1}{4}$ | $\frac{11}{12}$ | | a | a |
| 2 | $\geq 3$ | $\geq 3$ | 0 | 2 | $j - 1\frac{1}{32}$ | $1\frac{1}{32}$ | | d | d |

In the bottom row of Table 16 the temperature is greater than one. The figure below describes this type of position. It seems to be an ordinary position like any other, but its temperature is $1\frac{1}{32}$.

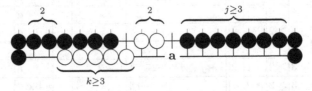

Figure 4 shows an example game tree which describes the crucial variation of this type of position. The root of the game tree is the game of $\{3 \,||||\, \{4|3\} \,|\, 2 \,||\, 1 \,|||\, 0\}$ and its mean value and temperature are $1\frac{31}{32}$ and $1\frac{1}{32}$, respectively.

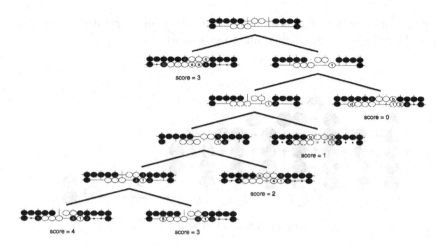

**Fig. 4.** Game tree of a surprisingly hot position.

$$X_{i,j,k,l,m} \stackrel{\text{def}}{=}$$

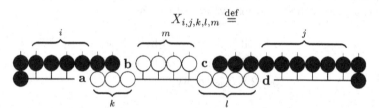

**Table 17.** $X_{i,j,k,l,m}$ ($m = 3$): Blocked on both sides.

| i | j | k | l | m | Mean | Temp | ish | Black | White |
|---|---|---|---|---|------|------|-----|-------|-------|
| 0 | 0 | 1 | $\geq 1$ | 3 | $-1$ | 1 | $\downarrow$ | b,c | |
| 0 | 0 | $\geq 2$ | $\geq 2$ | 3 | $-1$ | 1 | $\rightharpoonup$ | b or c | b or c |
| 1 | 0 | $\geq 1$ | $\geq 1$ | 3 | $-1\frac{1}{2}$ | 1 | $\rightharpoonup$ | c | c |
| 2 | 0 | $\geq 1$ | $\geq 1$ | 3 | $-1$ | 1 | $\rightharpoonup$ | c | c |
| 1 | 1 | $\geq 1$ | $\geq 1$ | 3 | $-2$ | 0 | | | |
| 2 | 1 | $\geq 1$ | $\geq 1$ | 3 | $-1\frac{1}{2}$ | $\frac{1}{2}$ | | c | c |
| 1 | 2 | $\geq 1$ | $\geq 1$ | 3 | $-1\frac{1}{2}$ | $\frac{1}{2}$ | | b | b |
| $\geq 3$ | 1 | $\geq 1$ | $\geq 1$ | 3 | $i - 3\frac{1}{2}$ | 1 | $\rightharpoonup \vert 0$ | a | a |
| 2 | 2 | $\geq 1$ | $\geq 1$ | 3 | $-1$ | 0 | | | |
| $\geq 3$ | 2 | $\geq 1$ | $\geq 1$ | 3 | $i - 3$ | 1 | $\rightharpoonup \vert 0$ | a | a |

# 5  Bicolored Multiple Corridors

We begin with the reference diagram on Page 27 of Berlekamp-Wolfe, shown in Fig. 5. It shows a position along the western edge of the board. We change the bottom three lines to the novel *half socket*. The figure is rotated 90 degrees counter-clockwise from the original one to save space.

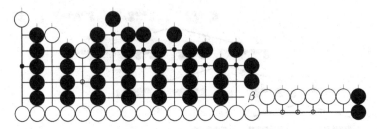

**Fig. 5.** Multiple Corridors.

The values (all marked and chilled) of $b_i$, $u_i$, and $s$ are as on page 27. As in the book, $s$ is the shortest unblocked corridor, of length $\log_2 \frac{8}{s}$. The new socket contains a point called $\beta$, whose value is taken as

$$1 \quad \text{if } \beta \text{ is occupied by White}$$
$$0 \quad \text{if } \beta \text{ is occupied by Black}$$
$$1/2 \text{ if } \beta \text{ is empty}$$

Below the half socket is a (new) white corridor, whose value is taken as $\sigma$, a negative number. The length of this corridor, measured by counting the sequence of empty points along the first column, is $\log_2 \frac{2}{|\sigma|}$.

Here is the conjectured value of the entire position, $v$:

**Case 1:** IF $\beta + \sum b_i \geq 1$, then
$$v = -1 + \beta + \sum b_i + \sigma + \sum u_i + s$$

**Case 2:** IF $\beta + \sum b_i \leq 1$, and $2|\sigma| \leq s/2$, then
$$v = \sum u_i + \sigma + s/2(1 + \beta + \sum b_i)$$

**Case 3:** IF $\beta + \sum b_i \leq 1$, and $2|\sigma| \geq s/2$, then
$$v = \sum u_i + s + |\sigma|(-3 + 2(\beta + \sum b_i))$$

**Case 4:** IF $\beta + \sum b_i \leq 1$, and there is no $s$ (and therefore no $u_i$) and $\beta = 0$ (full socket) then
$$v = 0^n| \maltese$$

All this may be easier to understand if stated in terms of a rather different generalization of page 27, in which the bottom four lines are instead replaced with a sequence of $k$ liberty-free sockets. Figure 6 shows a bottom when $k = 3$.

**Fig. 6.** Bottom when $k = 3$.

Begin by considering the simplest case, in which each $b_i = 0$ or 1. Then $\sum b_i$ is the number of liberties enjoyed by the portion of the white string adjacent to the $b$'s. Any liberties the white string has adjacent to the $u$'s are ephemeral; they will disappear at a higher temperature. If each $u_i$ is locally played canonically, then Black, if he is the first to move on that corridor, will play on the left side and take that liberty away. On the other hand, if White is the first to play on the corridor, then a black move at either end reverses the value, so w.l.o.g, we can assume that as soon as White plays on any $u_i$, Black responds locally immediately to reverse the local $u_i$ value and to take away that White liberty.

If there are $k$ sockets, and if $\sum b_i \geq k$, then every $u_i$ can be played canonically. However, if $\sum b_i < k$, (and each $b_i = 0$ or 1, and there are plenty of $u_j$'s), then only the longest $u_j$'s will decouple, each to its independent value, $u_j$. But Black will be able to play each of the $(k - \sum b_i)$ shortest $u_j$'s in sente. When Black plays on one of them, say $u$, changing its value to $u/2$, White must respond immediately by filling a socket. So each of the shortest $(k - \sum b_i)$ $u_j$'s has an effective value of $u_j/2$, while each of the longer $u_j$'s has an effective value of $u_j$.

The more interesting case, of course, is when the $b_i$'s are arbitrary. If $\sum b_i$ is an integer, then by canonical play within the $b_i$, we have miai, so that in the final result the $\sum b_i$ will be unchanged. The white string will then enjoy $\sum b_i$ liberties. And, as we have seen before, this implies that the requisite number of the shortest $u$'s will have their values divided by 2, because when a canonical Black eventually plays there, White will be compelled to respond by filling a socket. But when a canonical Black plays first on a longer $u_j$, a canonical White will respond on the opposite end of the same corridor.

Finally, if $\sum b_i$ lies between two integers, then there is an issue of whether the play will round it up or down. The number of $u_j$'s whose values need to be halved depends on the outcome. There will then be one critical $u_j$, called $s$. The values of longer $u_j$'s will be unaffected, and the values of all shorter $u_j$'s will be halved. The values of the $b_i$'s interact with the value of the critical $u_j$, resulting in a term

$$s/2((\sum b_i - \text{greatest integer in} \sum b_i) + 1)$$

The play of the $b_i$'s will determine whether this term becomes $s$ or $s/2$.

All of this is proved in the book in the special case in which $k = 1$. We think the case of multiple $k$'s was evaded because further complications arise if there are corridors emanating from the portions of White strings between different sockets. A typical Go player will consider such generalizations to be increasingly remote from anything he has directly encountered over the board.

If the white corridor at the bottom is the critical corridor, then the question to be resolved by playing the $b$'s is whether its value is $\sigma$ or $2\sigma$. When both this white corridor and unblocked black corridors exist, then the white corridor must compete with the unblocked black corridors to determine which is the critical one. The conjecture stated above asserts that which corridor is critical is determined by a comparison of lengths. We don't yet have a plausible intuitive explanation for this; before examining the evidence of several specific cases, we might have been inclined to compare the values of the white and black corridors rather than their lengths.

The half socket appears in earlier sections of this paper. Such positions look very realistic. They readily include the new $\sigma$ plus one other corridor. When there is an empty space ($\beta$) on the second column, someone may play on the first column opposite it. In some circumstances, this yields the last line of Table 12, which is a special case of our conjecture.

In most respects, the new $\beta$ behaves like a blocked corridor of length two. But when $-\log_2 |\sigma|$ is large, the half-point black mark associated with $\beta$ belongs superimposed on a white mark immediately to its southwest. It is easiest to assume that $\beta$ is not played until every other $b_i$ has become 0 or 1. That turns out to be one canonical line. When there is exactly one other $b_i = 1/2$, and no $u$'s or $s$, but $\sigma$ is present, the position has value $\sigma$. If Black makes two consecutive plays, on $b_i$ and on $\beta$, and White then fills, the new value is $2\sigma$, which is the same as Black could have obtained by extending the bottom invasion instead of attacking White prematurely. But if Black does attack, the strictly dominant order in which to play them is to begin by playing the $b_i$ to 0 rather than by playing the $\beta$ to 0.

We have found many real pro endgames in which two adjacent regions are not-quite independent. The most common reason is that there is some potential pressure against the group which separates these two regions. Although each move will fall into one region or the other, plays which affect eyes or liberties of the potentially-pressured group may have a higher temperature than they would if the regions were truly independent. *This situation is very common.* We now entertain visions of "Federation" software to help analyze it. We also think there is real hope of more theory to be discovered that might prove relevant to this problem.

We see our historical work on "multiple corridors", and the extension(s) discussed above, as a significant stepping stone in this direction. Although its direct applicability is too rare to excite very many Go players, we continue to be somewhat awed by its precision. It will certainly serve as a benchmark against which we can test any new theory or heuristics for evaluating a federation of two regions.

# References

1. Berlekamp, E., Wolfe, D.: Mathematical Go: Chilling Gets the Last Point. AK Peters (1994)
2. Berlekamp, E., Conway, J., Guy, R.: Winning Ways for your Mathematical Plays. Academic Press, London (1982)
3. Guy, R., Nowakowski, R.: Unsolved problems in combinatorial games. In Nowakowski, R., ed.: More Games of No Chance. Volume 42. Cambridge University Press (2002) 457–473

# New Winning and Losing Positions for 7×7 Hex

Jing Yang[1], Simon Liao[2], and Miroslaw Pawlak[1]

[1] Electrical & Computer Engineering Department,
University of Manitoba, Canada
{jingyang,pawlak}@ee.umanitoba.ca
[2] Business Computing Department,
University of Winnipeg, Canada
sliao@uwinnipeg.ca

**Abstract.** In this paper we apply a decomposition method to obtain a new winning strategy for 7x7 Hex. We also show that some positions on the 7x7 Hex board, called trivial positions, are never occupied by Black among all of the strategies in the new solution. In other words, Black can still win the game by using the strategies described in this paper even if White already has pieces placed on those positions at the start of the game. Considering the symmetry properties of a Hex board for both players, we also derive 14 losing positions for Black's first move on a 7x7 Hex board.

## 1 Introduction

Hex is an interesting board game that was invented in 1942 by Piet Hein, a Danish mathematician. The game was also reinvented independently by the American mathematician John Nash in 1948. Hex is played by two opponents, Black and White, where Black moves first. Each player owns the two opposite edges of the board that bear his color. The object of the game is to build a connected chain of pieces across opposite sides of the board. The Hex board is a hexagonal tiling of $n$ rows and $m$ columns, where $n$ is usually equal to $m$. Figure 1 shows an empty 7x7 Hex board. The rules of the game are relatively simple:

- The players take turns playing a piece of their color on an unoccupied hexagon.
- The game is won by the player that establishes an unbroken chain of his or her pieces connecting the player's sides of the board.

For example, Fig. 2 shows a Hex game in progress, in which it is Black's turn to play. If Black plays a piece at position "A", Black wins the game. However, if Black plays at any other position rather than "A", White can play a piece at position "A" and declare the win.

In 1949, John Nash proved that there is a winning strategy for the first player, but did not indicate what that play might be. A winning strategy for 7x7 Hex based on a decomposition method was declared in 2001 [1]. According to [2], this is the largest board size for which a solution has been found. In the solution,

J. Schaeffer et al. (Eds.): CG 2002, LNCS 2883, pp. 230–248, 2003.

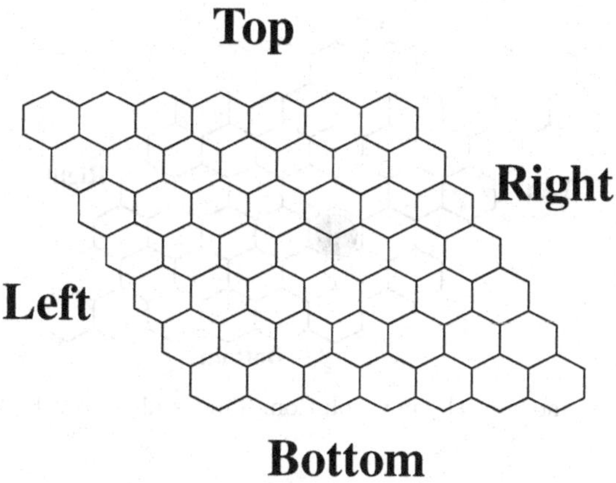

**Top**

**Right**

**Left**

**Bottom**

**Fig. 1.** An empty 7x7 Hex board.

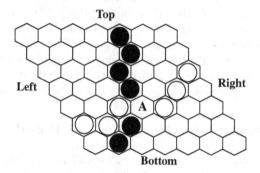

**Fig. 2.** Play on position A to win the game.

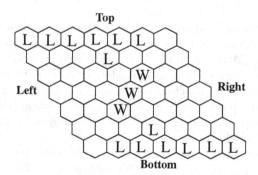

**Fig. 3.** Hexagons marked by W and L indicate the winning or losing opening positions. The empty hexagons remain unsolved opening positions.

the first Black piece is played at the center of the Hex board in order to take advantage of the symmetry properties. In this paper a new solution for 7x7 Hex

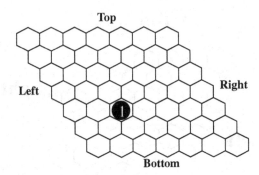

**Fig. 4.** The first move in the new solution. This is also LocalPattern 1 in [3].

is described that is based on the same decomposition method. This research also leads to the discovery of several losing first moves for Black, that is, if Black plays any of these moves White simply adopts Black's known winning strategy. As a result of this research, Fig. 3 shows all of Black's first moves that are known to lead Black either to win or to lose the game. The hexagons containing the letter "W" indicate that Black has a winning strategy if the first Black piece is played there. The hexagons containing the letter "L" indicate that if Black plays the first piece there he loses, that is, White has a winning strategy.

## 2    A New Solution for Hex on a 7×7 Board

In the already published winning strategy for 7x7 Hex [1], the first Black piece is played at the center of the board. In this paper we present a new solution for the 7x7 Hex game, in which the first piece is played at position ❶ as shown in Fig. 4.

The decomposition method is inspired by the concept of subgoals in AI planning. Finding winning strategies for a Hex game can be viewed as a Markov Decision Process (MDP). If a MDP can be broken up into several sub-MDPs, then there exists a parallel decomposition for the process. Each of the sub-MDPs has its own action space, which forms either a product or join decomposition of the original state space. Under a parallel decomposition, the sub-MDPs can be processed completely independently, and the original (global) MDP may be solved exponentially faster [4].

In a Hex game, the goal for Black is to form a connected chain from the top side to the bottom side of the board. This goal can be viewed as a sum of several subgoals. A subgoal may be "one Black piece is connected to another Black piece", "one Black piece is connected to Top", "one Black piece is connected to Bottom", "Top is connected to Bottom", or a combination of their OR/AND logical expressions. For example, "one Black piece is connected to another Black piece" OR "this Black piece is connected to Top" is one of the typical cases. The successes of achieving all of the subgoals in a game will lead to the success of accomplishing the goal of winning the Hex game. An important characteristic of

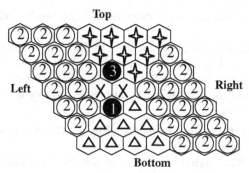

**Fig. 5.** ✦s establish LocalPattern 5, whose subgoal is to connect ❸ to Top; Local-Pattern 2 is composed of two ✗ s and its subgoal is to connect ❸ to ❶; LocalPattern 5 is marked by △s and its subgoal is to connect ❶ to Bottom. Black can win the game by playing ❸ if White plays in any of the ② positions in response to ❶.

a Hex game is that the success of a subgoal may be determined by only a small empty local region, which is called the influence region of the subgoal. If each of those subgoals has an influence region for the subgoal's success and all of the influence regions in a game are independent from each other, we can decompose the entire board into several local patterns. For example, in Fig. 5, the game is decomposed into three different local regions. The subgoal of the local pattern, whose influence region is covered by ✦s is to connect ❸ to the top row. The subgoal of another local pattern, whose influence region is over the two hexagons marked by ✗ , is to connect ❶ to ❸. The subgoal of the third local pattern is to connect ❶ to the bottom row, its influence region is marked by △s.

Obviously, the three influence regions shown in Fig. 5 do not overlap, and Black can win the game by forming a connected chain from Top to Bottom if all of the three subgoals are reached. If we can find a strategy on each of the local influence regions for Black to reach its subgoal, summing up all of the local winning strategies would form the winning strategy for Black to win the game. Since White can make a move in only one of the local pattern regions at a time, Black only needs to play the next piece following the strategy for the corresponding local pattern. In most cases, a local pattern can be further decomposed into smaller local patterns by applying the same decomposition technique, though there do exist some cases in which a parallel decomposition is not possible. A big advantage of the parallel decomposition process is that most local patterns appear repeatedly in different games. Therefore, it is possible to represent the whole winning-strategy tree by recursively using those local patterns. In fact, the new winning-strategy tree can be represented by 63 different local patterns [3]. Although the number of local patterns in the new winning strategy discussed in this research is higher than that of the previous solution [1], the new solution introduces some more interesting results, which are discussed in Sect. 3.

The new winning strategy is outlined in the Appendix. Figures 11 to 24, in combination with Fig. 5 above, show the winning move for Black against all possible White first move replies. The figures also show the local patterns involved. The remaining figures in the appendix further decompose the higher level patterns as follows:

- Figures 25 and 26 are used to portray Fig. 24, the strongest possible defense by White.
- Figures 27 to 35 describe one layer further the local pattern (LocalPattern 19) covered by ✗ s in Figs. 14, 15, and 16.
- Figures 36 to 42 describe one layer further the local pattern (LocalPattern 11) marked by ✗ s in Fig. 17.

The local patterns depicted in the Appendix are representative of the new solution. For the complete winning strategy, including all of the 63 local patterns, see [3]. In each figure, different local influence regions are labeled by different symbols, for example, ✗ , △, ✦, ♣, and ▲. A local influence region, combined with several played pieces, forms a local pattern. Each of the local patterns is indexed by a local pattern number in the form of "LocalPattern n", which is the same terminology as used in [3]. Each of the local patterns ensures the success of reaching a subgoal. Therefore, the sum of all local patterns' successes in a given game will lead to a win for Black. There is no overlap between the local influence regions in a game, and all White's defense moves in a local influence region must have been covered by the winning strategy of its corresponding local pattern. From a logical point of view, the winning strategy has been proved. Considering that the winning-strategy tree is very complicated with many local patterns, a computer program was developed for its verification. The verification program can generate all possible moves by White in a local pattern, and test the subgoal by playing with the winning strategy. For testing larger local patterns, a so called "virtual connection" technique [5] has been applied in order to eliminate the obvious winning cases by forcing a game to end early.

## 3   Losing Opening Moves for Black

Although there are winning strategies for the first player to win a Hex game, a "bad" opening move can lead to a loss. Beck proved that the acute corner moves in any $n$ by $n$ Hex is a first-player-loss [6]. Jack van Rijswijck outlines all winning strategies for 6x6 Hex [7]. In a solution for a Hex game, if there are some hexagons that Black never needs to occupy for winning the game, we define those hexagons as "trivial positions". In other words, even if White has some pieces on those "trivial positions", Black can still play with the strategies in the solution in order to win the game.

For example, in Fig. 6, all positions marked by ✗ s are "trivial positions", which are never needed by Black in the aforementioned solution. Black can win

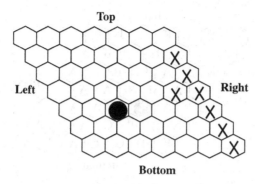

**Fig. 6.** All of **X** s are trivial positions. Even if they are all occupied by White's pieces, Black still can form a connected chain from Top to Bottom.

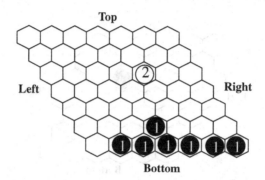

**Fig. 7.** ❶s are losing opening positions for Black. If Black plays its first move on one of ❶s, White can play ② and win the game. Note that, White will follow exactly what Black does in the new solution described in Sect. 2 except White's goal is to form a connected chain form Left to Right. If the board is reflected in the long diagonal and colours are interchanged, the situation will be the same as that of Fig. 6.

the game by following the winning strategies in the solution regardless of if those positions are occupied by White at the beginning of the game.

If the Hex board is turned over along the diagonal (from the left-top corner to the right-bottom corner), we can find that all of these marked positions in Fig. 6 then become Black's losing positions. These positions are marked by ❶s in Fig. 7. If Black occupies one of those positions with the first piece, White can play at ② and follow the winning strategy proposed in this paper to win the game.

## 4   Conclusions and Remarks

In this paper, we described a new winning strategy for 7x7 Hex where the first piece is played at ❶ in Fig. 8. In order to do so we decompose the game into 63

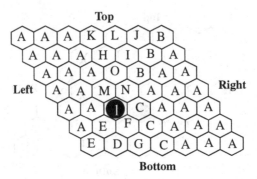

**Fig. 8.** White's next move on position As, Bs, Cs, ..., N, or O are discussed in Fig. 5, Fig. 11, Fig. 12, ..., Fig. 23, and Fig. 24, respectively. It is obvious that all White's defense moves are covered in the new solution.

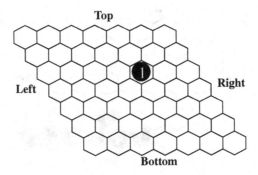

**Fig. 9.** Another winning position for Black due to the symmetry properties.

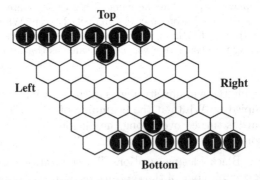

**Fig. 10.** All losing positions for Black derived from the new solution.

local patterns. Each local pattern will ensure a subgoal being achieved. Through Fig. 5, and Figs. 11 to 24 (position A to O in Fig. 8) we show that Black wins against all possible second level moves by White. We conclude that Black's ❶ in Fig. 8 is a winning move. Due to the symmetry properties of the 7x7 Hex board, we conclude that ❶ in Fig. 9 is also a winning move. The winning strategy described in this paper ensures the win with some simple coordinate transformations.

According to the "trivial positions" discovered in this research and the symmetry properties of a 7x7 Hex board, we also derived that there are 14 losing opening positions for Black, which are shown in Fig. 10. The newly discovered losing positions are especially valuable when the "swap rule", which gives the player to move second an option of swapping colors after Black's first move, is applied in a Hex game.

Figure 3 shows all solved winning and losing opening positions for 7x7 Hex. The empty hexagons in the figure still remain unsolved initial opening positions. However, with the decomposition method discussed in this paper, resolving those unsettled opening positions should be feasible. The decomposition method certainly can be applied to Hex games played on larger board, i.e., 8x8, 9x9, and beyond.

Parallel decomposition is a general approach, which has been applied successfully to AI planning [4, 8, 9]. Apart from Hex, it can possibly be applied to other games, for example, Go endgames. However, some modifications may be necessary because of the more complicated relationships between the subgoals and the final goal.

# References

1. Yang, J., Liao, S., Pawlak, M.: A decomposition method for finding solution in game Hex 7x7. In: International Conference On Application and Development of Computer Games in the 21st Century. (2001) 96—111
2. Stewart, I.: Hex marks the spot. Scientific American (2000) 100–103
3. Yang, J., Liao, S., Pawlak, M.: Another solution for Hex 7x7. Technical report, University of Manitoba, Department of Electrical and Computer Engineering (2002) http://www.ee.umanitoba.ca/~jingyang/TR.pdf.
4. Boutilier, C., Dean, T., Hanks, S.: Decision-theoretic planning: Structural assumptions and computational leverage. Journal of Artificial Intelligence Research 11 (1999) 1–94
5. Anshelevich, V.: The game of Hex: An automatic theorem proving approach to game programming. In: Seventeenth National Conference of the American Association for Artificial Intelligence (AAAI-00), AAAI Press (2000) 189–194
6. Beck, A., Bleicher, M., D.Crowe: Excursions into Mathematics. Worth Publishers (1969)
7. Rijswijck, J.: QUEENBEE's home page. http://www.cs.ualberta.ca/~queenbee/ (2000)
8. Singh, S., Cohn, D.: How to dynamically merge Markov decision processes. In Jordan, M., Kearns, M., Solla, S., eds.: Advances in Neural Information Processing Systems. Volume 10., The MIT Press (1998)
9. Meuleau, N., Hauskrecht, M., Kim, K., Peshkin, L., Kaelbling, L., Dean, T., Boutilier, C.: Solving very large weakly coupled Markov decision processes. In: Fifteenth National Conference of the American Association for Artificial Intelligence (AAAI-98), AAAI Press (1998) 165–172

# Appendix: Detailed Descriptions of the New Winning Strategy

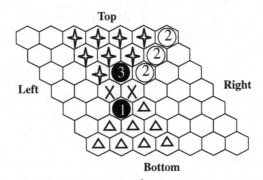

**Fig. 11.** The influence regions marked by ✦s, ✗s, and △s are discussed in Local-Pattern 5, LocalPattern 2, and LocalPattern 5, respectively.

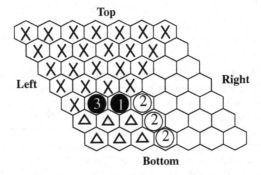

**Fig. 12.** ✗s and △s are covered by LocalPattern 20 and LocalPattern 6.

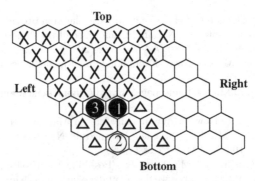

**Fig. 13.** The influence regions distinguished by ✗s and △s are explained by Local-Pattern 20 and LocalPattern 4.

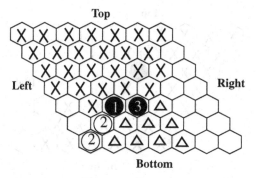

**Fig. 14.** The regions marked by X s and △s are clarified by LocalPattern 19 and LocalPattern 5. More discussions on LocalPattern 19 can be found in Fig. 27 to Fig. 35.

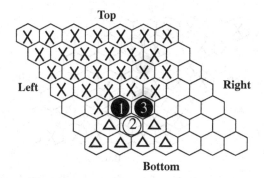

**Fig. 15.** The influence regions covered by X s and △s are discussed in LocalPattern 19 and LocalPattern 3.

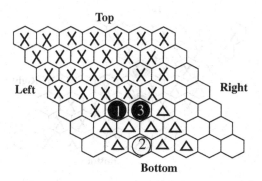

**Fig. 16.** The regions indicated by X s and △s form LocalPattern 19 and LocalPattern 4.

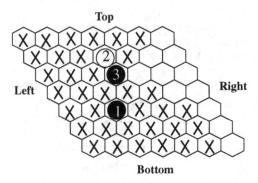

**Fig. 17.** ●, ②, ❸, and all positions marked by **X** s are explained by LocalPattern 11, which will be further discussed in Fig. 36 to Fig. 42.

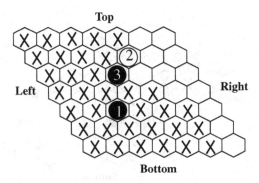

**Fig. 18.** LocalPattern 12 is slightly different from LocalPattern 11 shown in Fig. 17 in terms of pattern positions and winning strategy.

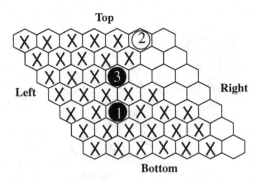

**Fig. 19.** The influence region marked by **X** s are explained by LocalPattern 14.

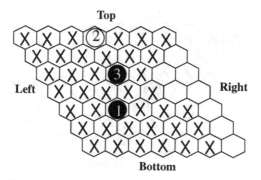

**Fig. 20.** This figure represents LocalPattern 15.

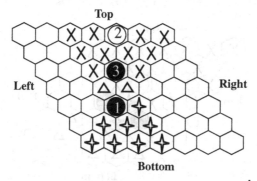

**Fig. 21.** The influence regions distinguished by X s, △s, and ✝s are explained in LocalPattern 9, LocalPattern 2, and LocalPattern 5, respectively.

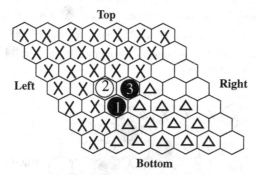

**Fig. 22.** Responding to ②, ❸ divides the board into two local patterns, LocalPattern 41 (marked by X s) and LocalPattern 21 (marked by △s).

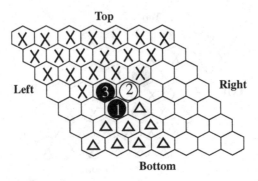

**Fig. 23.** After Black's ❸, the board can be split into two local patterns, LocalPattern 34 (marked by ✕ s) and LocalPattern 5 (marked by △s).

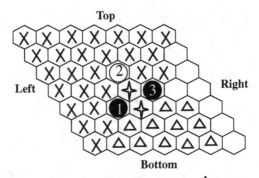

**Fig. 24.** The influence regions marked by ✕ s, △s, and ✚s are discussed in LocalPattern 22, LocalPattern 21, and LocalPattern 2, respectively. White's ② is the strongest possible defense move in the new solution. Black's ❸ would divide the board into three local patterns. LocalPattern 21 would guarantee Black's ❶ and ❸ to connect to Bottom. LocalPattern 22 is responsible for connecting either ❶ and ❸ to Top, or connecting Top to Bottom directly.

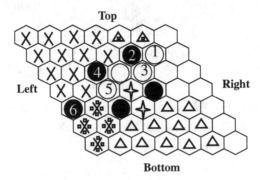

**Fig. 25.** Further development of Fig. 24. The influence regions marked by ✕ s, ◭s, ✚s, ✿ s, and △s are explained by LocalPattern 25, LocalPattern 2, LocalPattern 2, LocalPattern 13, and LocalPattern 21, respectively.

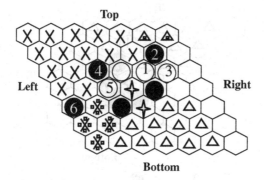

**Fig. 26.** Similar to Fig. 25, except the influence region marked by **X** s is covered by LocalPattern 30.

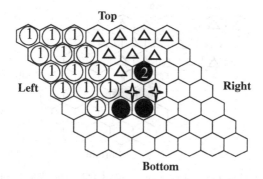

**Fig. 27.** Referenced to LocalPattern 19, whose influence region is distinguished by **X** s in Fig. 14, Fig. 15, and Fig. 16, respectively. Black's ❷ will ensure the connection to Top after White plays at one of the ① s.

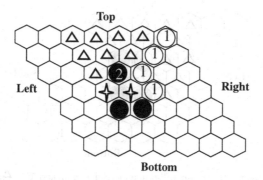

**Fig. 28.** Referenced to LocalPattern 19, if White plays at one of ①s, Black's ❷ will be able to ensure a connection to Top.

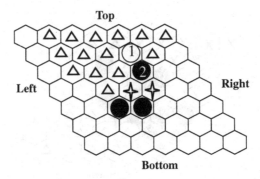

**Fig. 29.** If White plays at ①, Black's ❷ can ensure a connection to Top. The local influence region marked by △ is explained in LocalPattern 36.

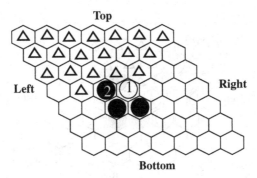

**Fig. 30.** This is another parallel case associated with the discussion of LocalPattern 19. If White plays at ①, Black's ❷ would ensure a connection to Top. The influence region distinguished by △ is explained in LocalPattern 34.

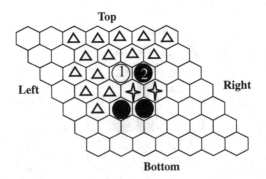

**Fig. 31.** LocalPattern 39, whose influence region is marked by △s, would ensure a connection between ❷ and Top.

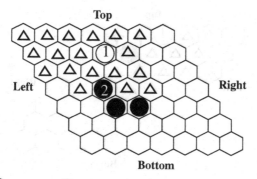

**Fig. 32.** Black's ❷ responds ① for making a connection to Top. The influence region is explained in LocalPattern 37.

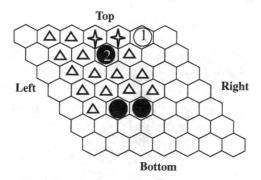

**Fig. 33.** If White plays at ①, Black can play ❷ to make a connection to Top. This case is discussed in LocalPattern 62.

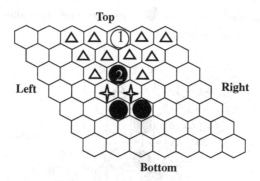

**Fig. 34.** If White plays at ①, LocalPattern 9, whose influence region is marked by ▲s, would ensure a connection between ❷ and Top.

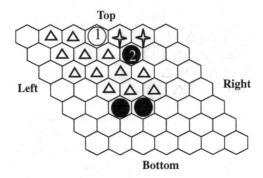

**Fig. 35.** The influence region distinguished by △s is explained in LocalPattern 38. This is the last case on the further descriptions of LocalPattern 19. From Fig. 27 to Fig. 35, all of White's defense moves in the influence region of LocalPattern 19 are covered.

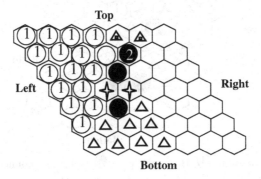

**Fig. 36.** This is the first figure to describe LocalPattern 11 one layer further. If White plays at any of ①s, Black can play at ❷ to win the game. The influence region marked by △s is illustrated in LocalPattern 5.

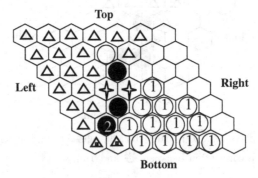

**Fig. 37.** If White plays at any of ①s, LocalPattern 52, whose influence region is marked by △s, will ensure ❷ a connection to Top.

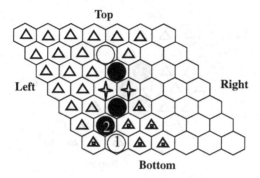

**Fig. 38.** Similar to the above case, Black is ensured a connection between ❷ and Top.

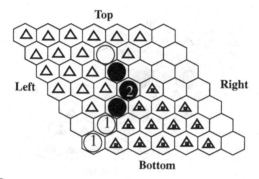

**Fig. 39.** Black's ❷ divides the board into two independent local regions. LocalPattern 57, marked by △s, ensures a connection to Top; LocalPattern 8, marked by ▲s, guarantees a connection to Bottom.

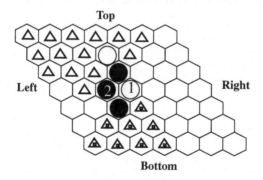

**Fig. 40.** Distinguished by △s, LocalPattern 36 would ensure a connection between ❷ and Top.

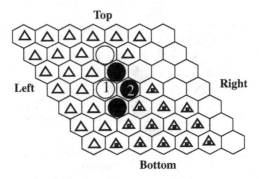

**Fig. 41.** If White plays at ①, Black can play at ❷ to win the game. Local regions marked by △s and ▲s are explained in LocalPattern 47 and LocalPattern 8.

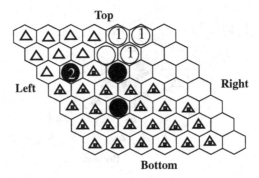

**Fig. 42.** The influence regions marked by △s and ▲s are discussed in LocalPattern 5 and LocalPattern 10, which would ensure a connection to Top and Bottom, respectively. This is the last case on the further discussion of LocalPattern 11. From Fig. 36 to Fig. 42, all of White's defense moves in the influence region of LocalPattern 11 are covered.

# Position-Value Representation in Opening Books

Thomas R. Lincke

ETH Zürich, Zürich, Switzerland
thomas@lincke.ch

**Abstract.** The standard position-value representation in search engines uses heuristic values, proven win, proven loss and, if applicable, proven draw. The advantage of this set of values is that propagation is efficient and that the propagated information is sufficient for selecting a move in the root position. In this paper we propose a new position-value representation which is suited for the use in opening books. In an opening book, propagation speed is not as important as in a search engine. Instead we want to have values which describe as accurately as possible the knowledge we have gathered about a position. As a solution we introduce three new value types: the *at-least-draw* and *at-most-draw*, which propagate additional information when the opening book contains positions with value `draw`, and *cycle-draw*, which propagates additional information when the opening book contains cycles which lead to a position repetition.

## 1 Introduction

The problem of position-value representation is similar for game search engines and opening books. In both cases, leaf nodes have either an exact or a heuristic value, and in both cases these values must be propagated to the start node.

However, the requirements for propagation speed and value accuracy differ. For a game search engine we want to use a position-value representation which allows for fast propagation, because we need to search as many nodes as possible. We are not interested in the value of the start node; we are only interested in the best move. For example the B* algorithm [1] terminates as soon as one move is proved to be better than any other move, possibly without calculating an exact value for that move.

An algorithm for automatic opening book construction repeatedly selects a leaf node of the book, evaluates and adds all successors of the leaf node, and propagates the new values towards the start position [2, 3]. Here we do not necessarily need the fastest propagation algorithm, because the book is constructed offline. But we need a good position-value representation because the values of successor nodes influence our selection of a leaf node (see Fig. 1), and to improve our selection of leaf nodes will improve the usefulness of the opening book in a tournament game.

We will show that when an opening book contains positions with a `draw` value or position-repeating cycles, then the standard propagation rules as used

J. Schaeffer et al. (Eds.): CG 2002, LNCS 2883, pp. 249–263, 2003.
© Springer-Verlag Berlin Heidelberg 2003

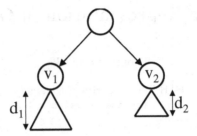

**Fig. 1.** For automatic opening book construction a leaf node is selected for expansion by following a path of 'best' moves from the start position. In [3] we proposed to calculate the expansion priorities of the successors as a function of the values of the successors and the depths of the sub-trees of the successors. In this simplified example we might expand the sub-tree for which $ep_i = d_i + \omega(max(v_1, v_2) - v_i)$ is minimal.

in many search engines represents the values of too many positions with the heuristic value 0. This leaves the opening book expansion algorithm clueless on which sub-tree should be selected for further expansion.

The question of how to mix heuristic and exact values in a search tree was also addressed in [4]. The author shows that, by propagating bounds on exact values, nodes are solvable even if the exact values of some successor nodes are unknown. He then extends alpha-beta search to propagate lower and upper bounds on the exact value in addition to the heuristic value.

Section 2 discusses the problem of incomparable values and how it is dealt with in search engines. In Sect. 3 we solve this problem for opening books by introducing two new value types, **at-least-draw** and **at-most-draw**, which are bounds on exact values. And in Sect. 4 we extend the idea of attributed values by introducing another value type **cycle-draw**.

## 2   Incomparable Values

The only values we would like to deal with are **win**, **loss** and **draw**. Unfortunately it is not always possible to determine the exact value of a position within a given amount of time. In that case we use so-called *heuristic values* to calculate an estimate of the exact value. A heuristic value is an integer $h$ in a range $h_{min} \le h \le h_{max}$. If a heuristic value is close to $h_{max}$ then we say that the corresponding position is likely to be a **win**. If a heuristic value is close to $h_{min}$ then we say that the corresponding position is likely to be a **loss**. If a heuristic value is close to 0 then we say that the corresponding position is likely to be a **draw**.

While the introduction of heuristic values allows us to estimate position values in a limited amount of time, it also gives rise to a new problem. We cannot, without information about the context of a game, tell whether a **draw** is better than a heuristic value or not: the **draw** and the heuristic values are *incomparable*.

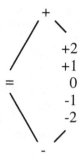

**Fig. 2.** Hasse diagram for a game-value set with $h_{min} = -2$ and $h_{max} = 2$. The draw value and the heuristic values are incomparable.

For example, imagine your are playing a game in the last round of a tournament. In your current position you have two moves: one leads to a draw, but for the other move you have only a heuristic value. Which one should you play? Now assume that a draw is sufficient to win the tournament. In that case it is obviously better to play safe and to choose the move to the position with a draw value. On the other hand, it might be that you will at least end up on second place, regardless of the outcome of the last game, but a win might allow you to win the tournament. In this case it is obviously better to play for "all or nothing" and to choose the move with the heuristic value.

## 2.1  Partially Ordered Sets

Because the draw value and the heuristic values are incomparable the set of game-position values forms a *partially ordered set* [5, Chapter 3]. A partially ordered set $P$ is a set together with a binary relation $\leq$ satisfying three axioms:

1. Reflexive: for all $x \in P$, $x \leq x$.

2. Anti-symmetry: if $x \leq y$ and $y \leq x$, then $x = y$.

3. Transitive: if $x \leq y$ and $y \leq z$, then $x \leq z$.

We use the obvious notation $x \geq y$ to mean $y \leq x$, $x < y$ to mean $x \leq y$ with $x \neq y$, and $x > y$ to mean $y < x$. We say two elements $x$ and $y$ of $P$ are comparable if $x \leq y$ or $y \leq x$; otherwise $x$ and $y$ are incomparable.

If $P$ is a partially ordered set, and if $x, y \in P$, then we say that $y$ covers $x$ if $x < y$ and if no element $z \in P$ satisfies $x < z < y$. The *Hasse diagram* of a partially ordered set $P$ is the graph whose vertices are the elements of $P$, whose edges are the cover relations, such that if $x < y$ then $y$ is drawn 'above' $x$. Figure 2 shows the Hasse diagram for a game-value set with $h_{min} = -2$ and $h_{max} = 2$.

**Table 1.** Simple max-propagation rules. $+$ means win, $-$ means loss, $=$ means draw, $h$ means heuristic value and $x$ means any type of value.

$$propagate(+, x) \quad \rightarrow +$$
$$propagate(-, x) \quad \rightarrow x$$
$$propagate(=, h) \quad \rightarrow max(0, h)$$
$$propagate(h_1, h_2) \rightarrow max(h_1, h_2)$$

**Table 2.** Max-propagation rules with draw-threshold $h_{draw}$.

$$propagate(+, x) \qquad\qquad \rightarrow +$$
$$propagate(-, x) \qquad\qquad \rightarrow x$$
$$propagate(=, h < h_{draw}) \rightarrow =$$
$$propagate(=, h \geq h_{draw}) \rightarrow h$$
$$propagate(h_1, h_2) \qquad\quad \rightarrow max(h_1, h_2)$$

## 2.2   Workaround for Search Engines

For search engines the problem of incomparability between draw and heuristic values is usually solved by treating draw and 0 as equal. This makes sense, because 0 is the heuristic estimate of draw, and because this can be implemented efficiently. Table 1 shows the propagation rules and propagate1() in Fig. 3 is the corresponding propagation function.

Another way to solve the incomparability problem is to provide an additional parameter $h_{draw}$, the draw-threshold, to the propagation function, where $h_{min} \leq h_{draw} \leq h_{max} + 1$. $h_{draw}$ represents a player's willingness to end the game with a draw and is chosen separately for every game, or even for every search. Whenever we compare a draw with a heuristic value $h$, we propagate draw if $h < h_{draw}$, and we propagate $h$ if $h \geq h_{draw}$. This means that the higher the value of $h_{draw}$, the higher is our preference for a draw. This effectively turns the partially ordered set into a totally ordered set (see Fig. 4).

Table 2 shows the propagation rules and propagate2() in Fig. 5 implements the corresponding propagation function. This solution is cheap and useful for search engines, but not usable for the construction of opening books, because at the time of construction we do not know the value of $h_{draw}$.

## 3   At-Least-Draw and At-Most-Draw

Figure 6 shows a min-max tree with four leaf nodes, two of them a draw. With the propagation rules of Table 1 the interior nodes A, B and C have value 0. There are two reasons why the interior nodes are misrepresented by the value 0.

First, the value 0 for node A implies that the outcome of a game from that node is completely open. But whatever move the min-player chooses, the max-player can always force a draw. In other words, the min-player can never win from node A.

```
int propagate1(int v₁,int v₂) {
    if ((v₁ == draw) && (v₂ != draw)) v₁ = 0;
    if ((v₂ == draw) && (v₁ != draw)) v₂ = 0;
    return max(v₁,v₂);
}/*propagate1*/
```

**Fig. 3.** Propagation function with proven draw. It is assumed that loss < draw < win, loss < $h_{min}$ and $h_{max}$ < win.

$$+$$
$$+2$$
$$+1$$
$$0$$
$$-1 \ = \ h_{draw}$$
$$=$$
$$-2$$
$$-$$

**Fig. 4.** Hasse diagram for a game-value set with a draw-threshold. This set is totally ordered.

```
int propagate2(int v₁,int v₂,int h_draw) {
    if ((v₁ == draw) && (v₂ >= h_min) && (v₂ <= h_max)) {
        if (v₂ < h_draw) return draw; else return v₂;
    }/*if*/
    if ((v₂ == draw) && (v₁ >= h_min) && (v₁ <= h_max)) {
        if (v₁ < h_draw) return draw; else return v₁;
    }/*if*/
    return max(v₁,v₂);
}/*propagate2*/
```

**Fig. 5.** Propagation function with draw-threshold. It is assumed that loss < draw < win, loss < $h_{min}$ and $h_{max}$ < win.

Second, although the values of the nodes B and C are the same, the sub-tree following B is more interesting for opening book expansion, because in C the max-player is more likely to move into the draw than he is in B. In other words node C is closer to being solved than node B, therefore B should get a higher priority for book expansion.

The first problem is solved by introducing a new value type, the at-least-draw value, denoted by '≥', and its inverse, the at-most-draw value, denoted by '≤'. We then define the 'max' value of a draw and a heuristic value to be at-least-draw. As shown in Fig. 7 the value of node A now correctly indicates that the min-player cannot win from that position.

The second problem is solved by introducing the attributed at-least-draw value, written ≥|h, where h is the value of the best heuristic successor ($h_{min}$ ≤

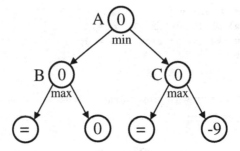

**Fig. 6.** Simple propagation in a min-max tree with **draw** values. All interior nodes evaluate to 0.

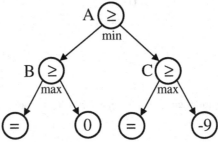

**Fig. 7.** The same min-max tree as in Fig. 6, now using **at-least-draw** and **at-most-draw** values. The value of node A is now bounded to indicate that the min-player cannot win.

$h \leq h_{max}$). For **at-most-draw** values, $\leq|h$ is defined analogously. Figure 8 shows that the values of nodes B and C are different now, and the values indicate that the sub-tree following node B is more interesting for opening book expansion than the sub-tree following node C.

Figures 9 and 10 show that the **at-least-draw** and **at-most-draw** values even help to solve additional nodes. With the simple propagation rule the value of the start position is a heuristic 0. With the new value types we are able to prove that the value of the start position is a **draw**.

Figure 11 shows the Hasse diagram of a value representation with attributed **at-least-draw** and **at-most-draw** values, and with $h_{min} = -2$ and $h_{max} = +2$. $\geq|h$ means that the player to move has a forced **draw**, but has the option to move into a position with value $h$ instead. $\leq|h$ means that the opponent has a forced **draw**, but has the option to move into a position with value $h$. The values $[\geq|h_{min}, \ldots, \geq|h_{max}]$ and $[\leq|h_{min}, \ldots, \leq|h_{max}]$ are continuous ranges of values between **draw** and **win** and **loss** and **draw** respectively, just in the same way as $[h_{min}, \ldots, h_{max}]$ is a continuous range of values between **loss** and **win**. Therefore these new values give us a more accurate estimate of the exact value than what we had before.

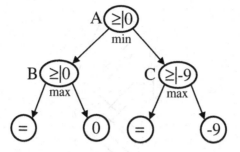

**Fig. 8.** The same min-max tree as in Fig. 7, now using attributed at-least-draw and at-most-draw values. The values of nodes B and C are different now and indicate that the sub-tree following node B is more interesting for opening book expansion.

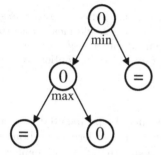

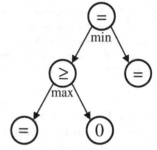

**Fig. 9.** With simple propagation the start position of this graph is not solved.

**Fig. 10.** With at-least-draw and at-most-draw values the start position of this graph is solved.

The term "at least draw" is sometimes used with a different semantics. For example chess positions with a king and a pawn against a king (KPK) are said to be "at least draw" for the player with the pawn: whatever he does, no mistake is bad enough to make him lose.

Our definition of at-least-draw is weaker. The value $\geq|h$ must be read as "draw exclusive-or $h$", and if a player chooses $h$ then he might still lose. However, the name at-least-draw is still justified, because whatever the exact value of the position with the current heuristic value $h$ will be, the exact value of $\geq|h$ will be either win or draw.

Although the problem of incomparability between draw and heuristic values is now solved, Fig. 11 shows that now two new types of incomparability call for a propagation heuristic:

– $\geq|a$ and $h$ are incomparable if $a < h$. This case is simple to solve. Remember that $\geq|a$ means that the player to move can force a draw, but has the option to move to a position with a heuristic value $a$ (and $a$ is the best heuristic value that is reachable). Therefore it is natural to propagate $\geq|max(a, h)$ in this case.

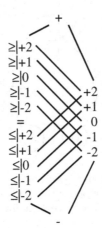

**Fig. 11.** The Hasse diagram for a value representation with attributed `at-least-draw` and `at-most-draw` values, and with $h_{min} = -2$ and $h_{max} = +2$. $\geq|h$ means that the player to move has a forced `draw`, but has the option to move into a position with value $h$ instead. $\leq|h$ means that the opponent has a forced `draw`, but has the option to move into a position with value $h$.

- $\leq|a$ and $h$ are incomparable if $a > h$. We know that the propagated value of $\leq|a$ and $h$ must be a heuristic value, because we cannot derive any bounds on the exact value from $\leq|a$ and $h$. Now $\leq|a$ means that the opponent can force a `draw`, but has the option to move to a position with a heuristic value $a$ (and $a$ is the best heuristic value that is reachable). Therefore when we compare $\leq|a$ and $h$, $a$ is an option for the opponent and $h$ is an option for the player to move, the two cannot be related to each other. To solve this problem we devised the following propagation heuristic: we assume that the opponent will always play the forced `draw` if $a \geq 0$, but will always avoid the `draw` if $a < 0$. The propagated value will be $max(h, min(a, 0))$.

Table 3 summarizes the propagation rules for a value representation with attributed `at-least-draw` and `at-most-draw` values.

## 4    Cycles

The graphs of many games are cyclic. The most common rule for handling position repetitions is that the game is declared to be a `draw`. See Table 4 for an overview of cycle handling in various games.

The first problem of cycle handling in opening books is cycle detection, see Fig. 12. Any time a move is added to the book, we have to find out if this new move closes a cycle, and update the values accordingly. Unfortunately it is impossible to detect a closed cycle locally, in general we have to search the whole graph to detect cycles.

**Table 3.** Propagation rules with `at-least-draw` and `at-most-draw` values. The last three rules define the propagation for incomparable values.

$$propagate(+,x) \quad\quad \rightarrow +$$
$$propagate(-,x) \quad\quad \rightarrow x$$
$$propagate(h_1,h_2) \quad\quad \rightarrow max(h_1,h_2)$$
$$propagate(=,\geq|a) \quad\quad \rightarrow \geq|a$$
$$propagate(=,\leq|a) \quad\quad \rightarrow =$$
$$propagate(\geq|a_1,\geq|a_2) \rightarrow \geq|max(a_1,a_2)$$
$$propagate(\geq|a_1,\leq|a_2) \rightarrow \geq|a_1$$
$$propagate(\leq|a_1,\leq|a_2) \rightarrow \leq|min(a_1,a_2)$$
$$propagate(=,h) \quad\quad \rightarrow \geq|h$$
$$propagate(h,\geq|a) \quad\quad \rightarrow \geq|max(h,a)$$
$$propagate(h,\leq|a) \quad\quad \rightarrow max(h,min(a,0))$$

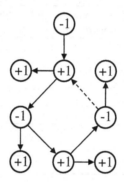

**Fig. 12.** [Negamax propagation] The cycle problem: when the dashed edge is added, a cycle is created. The cycle cannot be detected locally at the node where it was closed; instead a global search is necessary. Moreover, all the values in the cycle are locally consistent, but should be set to 0 because the positions are most likely **draw**.

## 4.1 Cycle-Draw

The second problem of cycle handling is cycle value propagation. Again our aim is to keep as much information about the value as possible. The simplest solution is to assign to every position in the cycle the heuristic value 0, see Fig. 13. However, this loses some information about the value of the position, as shown in Fig. 14: the same value is propagated for position A with a value almost identical to a 0 and for position B with a value almost identical to a **draw**. Therefore opening book expansion would select the sub-trees of nodes A and B with the same priorities even though the sub-tree of node A is more interesting for expansion than the sub-tree of node B.

As in the case with the `at-least-draw` and `at-most-draw` values we solve this by introducing a new attributed value type, the `cycle-draw`, denoted $0|+h_1|-h_2$, which has two attributes, $+h_1$ and $-h_2$. $+h_1$ is the smallest heuris-

**Table 4.** Cycle handling for some games.

| Game | Cyclic | Cycle Value | Cycle Breaking Moves |
|---|---|---|---|
| Abalone | yes | draw | capture |
| Amazons | no | | |
| Awari | yes | The remaining stones are split between the players, thus the configuration has value 0, the position value depends on the distribution of captured stones. | captures |
| Checkers | yes | draw | captures, conversions, checker moves |
| Chess | yes | draw | captures, conversions, pawn moves, castling |
| Chinese Checkers | yes | draw | none |
| Chinese Chess | yes | draw (Perpetual checks and perpetual threats to capture are forbidden.) | captures, forward moves by pawns |
| Connect-Four | no | | |
| Go | no | | |
| Go-moku | no | | |
| Hex | no | | |
| Nine Men's Morris | yes | draw | opening moves, captures |
| Othello | no | | |
| Qubic | no | | |

tic value with which the opponent can move out of the cycle. In other words the opponent either stays in the cycle and settles for a draw, or plays a move that has a value of at least $+h_1$ for the current player. $-h_2$ is the largest heuristic value with which the current player can move out of the cycle. In other words the current player either stays in the cycle and settles for a draw, or plays a move that has a value of at most $-h_2$ for the current player.

As the use of the + and − sign in the notation for the cycle-draw suggests, the first attribute is always strictly positive and the second attribute is always strictly negative. Otherwise one player would have a cycle-leaving move which is at least as good as any cycle-preserving move. Then the value of the cycle-leaving move is assumed to be better and the corresponding heuristic value is propagated. Figure 15 shows the same situation as Fig. 14, now using attributed cycle-draw values. The value of node A now indicates that this sub-tree should get a higher priority for book expansion.

When a game has a heuristic value range $h_{min}, \ldots, h_{max}$, the value ranges for the attributes of the cycle-draw are $1, \ldots, h_{max}$ for the first attribute and

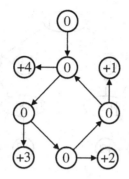

**Fig. 13.** [Negamax propagation] A cycle using the heuristic value 0 as cycle value.

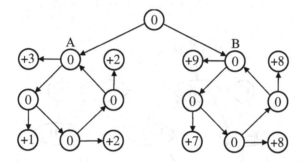

**Fig. 14.** [Negamax propagation] Using heuristic 0 as cycle value loses information. In this graph, the two successors of the start position look exactly the same and would be selected for expansion with the same priority. However, the penalty for leaving the right cycle is higher for both players, therefore the right cycle is closer to a **draw** than the left cycle, so the expansion priority should be higher for the left sub-tree.

$h_{min}, \ldots, -1$ for the second attribute. However it is possible that all the cycle-leaving moves for one or both players have **loss** values. We use $0| + | - h$ for the case where the opponent has only losing cycle-leaving moves, we use $0| + h| -$ for the case where the current player has only losing cycle-leaving moves, and we use $0| + | -$ for the case where both players lose if they play a cycle-leaving move. Obviously, the values $0| + | - h$, $0| + h| -$ and $0| + | -$ are equivalent to $\geq | - h$, $\leq | + h$ and **draw**, respectively. We conclude that **cycle-draw** values are a generalization of the previously introduced **at-least-draw**, **at-most-draw** and **draw** values.

Table 5 shows the propagation rules for a value representation with **cycle-draw** values.

### 4.2  Cycle-Draw and Tournament Play

Like for book expansion, the **cycle-draw** values are useful in tournament play to estimate how close we are in proving that a position is a **draw**. However, when

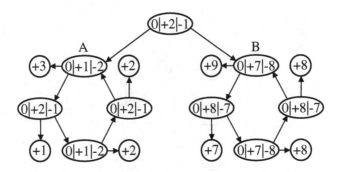

**Fig. 15.** [Negamax propagation] The same graph as in Fig. 14, but now with attributed `cycle-draw` values. The left sub-tree now gets higher priority for book expansion because the absolute value of the `cycle-draw` attributes is smaller.

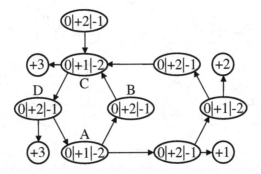

**Fig. 16.** [Negamax propagation] It is not possible to find the best cycle-leaving moves by only looking at the successor values. For example, if we are at position $A$, we cannot decide locally which move leads to the best cycle-leaving moves (the leaf nodes with values $+1$ and $+2$).

a position has the value $0| + h_1| - h_2$ it only means that a move *exists* such that the current player can leave the cycle with value $-h_2$. Whether such a move is reachable may depend on the moves of the opponent.

Assume for example that in Fig. 16 the first player in the start node decides that he is willing to leave the cycle with a penalty of $-1$, and moves into the cycle. But at position $A$, it is the opponent's move, and if the opponent moves to position $B$, the game is forced into a **draw**.

Note that we cannot simply conclude that position $A$ is at least a **draw** for the opponent; this depends on where the cycle was entered. Had the current player entered the cycle at position $A$, then a move to $B$ would not force a **draw**, but the first player could still leave the cycle with a penalty of $-3$ at position $C$. The dependency of the value of a node in a search tree on the sequence of moves on which that node is reached is also known as the GHI (graph history interaction) problem [6].

**Table 5.** Max-propagation rules with `cycle-draw` values. When compared to other types of values, the `cycle-draws` are treated like heuristic 0 values.

$$propagate(+,x) \qquad\qquad\qquad \rightarrow +$$
$$propagate(-,x) \qquad\qquad\qquad \rightarrow x$$
$$propagate(h_1,h_2) \qquad\qquad\quad \rightarrow max(h_1,h_2)$$
$$propagate(=,\geq|a) \qquad\qquad\quad \rightarrow \geq|a$$
$$propagate(=,\leq|a) \qquad\qquad\quad \rightarrow =$$
$$propagate(\geq|a_1,\geq|a_2) \qquad\quad \rightarrow \geq|max(a_1,a_2)$$
$$propagate(\geq|a_1,\leq|a_2) \qquad\quad \rightarrow \geq|a_1$$
$$propagate(\leq|a_1,\leq|a_2) \qquad\quad \rightarrow \leq|min(a_1,a_2)$$
$$propagate(=,h) \qquad\qquad\qquad \rightarrow \geq|h$$
$$propagate(h,\geq|a) \qquad\qquad\quad \rightarrow \geq|max(h,a)$$
$$propagate(h,\leq|a) \qquad\qquad\quad \rightarrow max(h,min(a,0))$$
$$propagate(0|+h_1|-h_2,0|+h_3|-h_4) \rightarrow 0|min(h_1,h_3)|max(-h_2,-h_4)$$
$$propagate(h,0|+h_1|-h_2) \qquad\quad \rightarrow max(h,0)$$
$$propagate(=,0|+h_1|-h_2) \qquad\quad \rightarrow \geq|0$$
$$propagate(\geq|a_1,0|+h_1|-h_2) \qquad \rightarrow \geq|max(a_1,0)$$
$$propagate(\leq|a_1,0|+h_1|-h_2) \qquad \rightarrow 0$$

We conclude that `cycle-draw` values improve our selection of leaf nodes for opening book expansion. But in a tournament game we still have to deal with the GHI problem, we need look-ahead and opponent modeling to find out which leaf nodes are reachable.

## 5 The New Value Representation in Real Games

The position-value representation we proposed in this paper has been implemented in an opening book construction tool and has been used for several games. The usefulness of the new value types is somewhat game-dependent: if a game contains no or only a few `draw` values, then using `at-least-draw` and `at-most-draw` will not make a difference. The game has to be cyclic to benefit from `cycle-draw` values.

### 5.1 At-Least-Draw, At-Most-Draw and Awari

We constructed an Awari opening book with 2.2 million positions. Thanks to the use of endgame databases with up to 40 stones the book contains 825,000 solved positions, 75,000 of them are `draw`. About 12,000 of these `draws` are only solved because of the use of `at-least-draw` and `at-most-draw` values. The amount of work required to solve these positions without the use of bounded values can be arbitrarily large. In addition to the solved positions the book contains about 90,000 positions with bounded values.

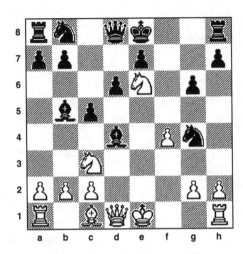

**Fig. 17.** Sax – Seirawan (Bruxelles, 1988): 1.e4 d6 2.d4 Nf6 3.Nc3 g6 4.f4 Bg7 5.Nf3 c5 6.Bb5 Bd7 7.e5 Ng4 8.e6 fxe6 9.Ng5 Bxb5 10.Nxe6 Bxd4 (**Diagram**)11.Nxd8 Bf2+ 12.Kd2 Be3+ 13.Ke1 Bf2+ 1/2–1/2.

## 5.2   Cycle-Draw and Chess

The game graph of chess contains many cycles. To show an example of cycle-draw values we added the game Sax–Seirawan (Bruxelles, 1988) to an opening book. Figure 17 shows the position after the 10th move of Black, and Fig. 18 shows the game graph starting from the same position. All nodes in the graph are labeled with their value. For better readability the graph contains only the best moves deviating from the game.

In this example White is unable to leave the cycle, annotated with the $+\infty$ and $-\infty$ values. Black can leave the cycle with values around $-40$, but that penalty would be too high because it is equivalent to losing about 4 pawns. We might consider this position as a draw for all practical purposes, just in the same way as humans consider positions like this one as drawn, and we don't need to consider the positions in the cycle for further book expansion.

## 6   Summary

In this paper we proposed three new position-value types: at-least-draw, at-most-draw and cycle-draw. The new value types propagate more information than what is possible with the standard value set used in search engines. For example, many positions which would be represented by a heuristic 0 in the standard value set now have bounded values or are explicitly identifiable as being part of a position-repeating cycle. This additional information enhances the selection of leaf nodes for the expansion of opening books.

The new position-value representation might also be useful for search engines. For example one might be able to prove that a good looking move with value 0

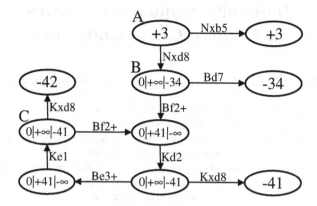

**Fig. 18.** Node A represents the position in Fig. 17. In every node only the best move deviating from the game is shown. Nodes B and C represent different positions because in B White still has castling rights. Inside the cycle White always has only one move. The best Black move leaving the cycle is always Kxd8, once with a value of -41 and once with a value of -42 (+10 being the value of a pawn).

is actually `at-most-draw` and should be avoided. However the new propagation rules are more complicated to implement and more expensive to use than the old rules.

# References

1. Berliner, H.: The B* tree search algorithm: A best-first proof procedure. Artificial Intelligence **12** (1979) 23–40
2. Buro, M.: Toward Opening Book Learning. International Computer Chess Association Journal **22** (1999) 98–102
3. Lincke, T.: Strategies for the automatic construction of opening books. In Marsland, T., Frank, I., eds.: Second International Conference on Computers and Games (CG 2000). Volume 2063 of Lecture Notes in Computer Science., Springer-Verlag (2001) 74–86
4. Beal, D.F.: Mixing heuristic and perfect evaluations: Nested minimax. International Computer Chess Association Journal **7** (1984) 10–15
5. Stanley, R.: Enumerative Combinatorics Vol. 1. Number 49 in Cambridge Studies in Advanced Mathematics. Cambridge University Press (1997)
6. Breuker, D., van den Herik, J., Uiterwijk, J., Allis, V.: A solution to the GHI problem for best-first search. In: First International Conference on Computers and Games (CG 1998). Volume 1558 of Lecture Notes in Computer Science. Springer-Verlag (1999) 25–49

# Indefinite Sequence of Moves
# in Chinese Chess Endgames

Haw-ren Fang[1,*], Tsan-sheng Hsu[2], and Shun-chin Hsu[3]

[1] Department of Computer Science, University of Maryland
hrfang@cs.umd.edu
[2] Institute of Information Science, Academia Sinica, Taiwan
tshsu@iis.sinica.edu.tw
[3] Department of Computer Science and Information Engineering,
National Taiwan University
schsu@csie.ntu.edu.tw

**Abstract.** In western chess, retrograde analysis has been successfully applied to construct 6-piece endgame databases. This classical algorithm first determines all terminal win or loss positions, i.e., those that are either checkmate or stalemate, and then propagates the values back to their predecessors until no further propagation is possible. The un-propagated positions are then declared draws.

However, in Chinese chess, there are special rules other than checkmate and stalemate to end a game. Therefore, some terminal positions cannot be determined by the typical retrograde analysis algorithm. If these special rules are ignored in the construction of the endgame databases, the resulting databases may contain incorrect information.

In this paper, we not only describe our approach in abstracting the special rules of Chinese chess and its consequent problems when retrograde analysis is applied, but also give a solution to construct complete endgame databases complying with the most important special rules.

## 1 Introduction

Retrograde analysis has been widely used to solve many problems. For example, it has been successfully applied to construct 6-piece endgame databases for western chess [1]. This classical algorithm first determines all terminal win or loss positions, i.e., those that are either checkmate or stalemate, and then propagates the values back to their predecessors until no further propagation is possible. The un-propagated positions are then declared as a draw.

In both western chess and Chinese chess, checkmate and stalemate are the two principal rules to end a game. They are also the foundations for propagation when retrograde analysis is applied. In western chess, if a game continues with a repeated series of moves it ends in a draw. On the contrary, in Chinese chess, such a game may result in a win/loss/draw depending on the characteristics of

---

* This work was mainly done while this author was with Institute of Information Science, Academia Sinica, Taiwan.

J. Schaeffer et al. (Eds.): CG 2002, LNCS 2883, pp. 264–279, 2003.

the indefinite move patterns involved. The rules deciding the game outcome in such cases are called *special rules* throughout this paper[1].

The most influential special rule is *checking indefinitely*. If only one player checks his opponent continuously[2], he loses the game. Therefore, the endgame databases of Chinese chess constructed by retrograde analysis may have errors if this special rule is not taken into account. Other special rules may also spoil the endgame databases in a similar way. It is known that endgames where only one side has attacking pieces are not affected by these special rules [3]. Using this fact, endgame databases with attacking pieces on one side only are constructed [3, 4]. This paper studies the problem of constructing Chinese chess endgame databases with both sides having attacking pieces. We believe this problem has not been tackled successfully before.

## 2   Notations and Rules of Chinese Chess

In Chinese chess, the two sides are called Red and Black. Each side has one King, two Guards, two Ministers, two Rooks, two Knights, two Cannons and five Pawns which are abbreviated as K, G, M, R, N, C and P, respectively. The pieces Rook, Knight, Cannon and Pawn are called *attacking pieces* since they can move across the *river*, the imaginary stream between the two central horizontal lines of the board. In contrast, Guards and Ministers are called defending pieces because they are confined in the domestic region[3]. Moreover, a side is called *armless* if it has no attacking pieces. A *position* in Chinese chess is an assignment of a subset of pieces to distinct addresses on the board with a certain player to move.

## 3   Special Rules in Chinese Chess

Retrograde analysis algorithms are widely used to construct the databases of finite, two-player, zero-sum and perfect information games [5]. In the endgame databases of western chess, terminal positions are either checkmate or stalemate positions. Furthermore, all unresolved positions at the end of the retrograde analysis can be safely declared as a draw (includes e.g. draws by 3-fold-repetition). In Chinese chess, on the other hand, repetitions may result in either a loss or a draw because of special rules. Resulting endgame databases using classical retrograde analysis may have errors if these special rules are neglected.

---

[1] The detailed special rules of Chinese chess are very complicated. Some minor rules differ in the Asian, Chinese, and Taiwanese versions, and might be revised as time goes on. Our discussions in this paper are all based on the rule book [2].

[2] In real games, each indefinite checking pattern is determined by the appearance of the same position for three times in a series of moves in which one side checks his opponent all the time.

[3] The information of Chinese chess such as notations and rules in English can be found in FAQ of the Internet news group rec.games.chinese-chess, which is available at http://www.chessvariants.com/chinfaq.html. More detailed rules can be found at http://txa.ipoline.com/.

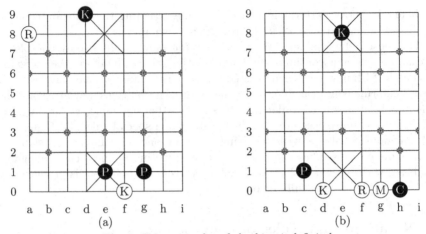

**Fig. 1.** Two examples of checking indefinitely.

### 3.1  The Rule of Checking Indefinitely

In Chinese chess, if only one player checks his opponent continuously *without ending*, then he loses. In this case, a game without ending is formally defined as having the same position occurring three times. This rule is called *checking indefinitely*. In real games, this means that a player loses if he cannot prevent his King from being captured without checking his opponent indefinitely.

As shown in Fig. 1(a), the Black side can move either Pg1-g0 or Pg1-f1 to checkmate the Red King. The Red side cannot avoid his King being captured without checking the Black King by making the move Ra8-a9. The checking continues endlessly with the moves Kd9-d8, Ra9-a8, Kd8-d9, Ra8-a9, etc. In Chinese chess, the Red side loses the game because he violates the rule of checking indefinitely. Note that if one of the two Black Pawns is removed from the board, the Red side can play Ra8-a1 to capture the Black Pawn and then easily win the game thereafter.

The example in Fig. 1(b) is most intriguing. The only logical move for the Red side that will not let him lose the game immediately is Rf0-e0. Then, if the Black side plays Ch0-e0 to capture the Red Rook, the game will end in a draw. However, if the Black side plays Ke8-f8 instead of Ch0-e0, the only move for the Red side is Re0-f0. The game continues cyclically with the moves Re0-f0, Kf8-e8, Rf0-e0, Ke8-f8, etc. The Black side wins because the Red side violates the rule of checking indefinitely. From this point of view, if one side can force his opponent to check him indefinitely, he wins.

In databases constructed by retrograde analysis that ignore this rule, a position in which either side cannot checkmate or stalemate the other, resulting in the game to continue endlessly, is regarded as a draw. Therefore, the positions in Fig. 1 are incorrectly recorded as draws, but should be treated as Black-to-win according to the rule of checking indefinitely. Ignoring the special rules may therefore result in potential errors in the constructed endgame databases.

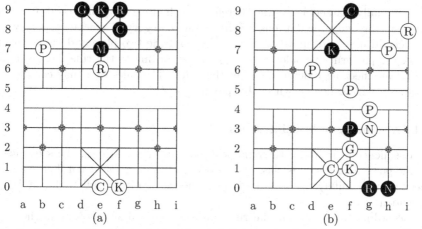

**Fig. 2.** An example of mutual checking indefinitely on the left and an example of non-mutual checking indefinitely on the right.

## 3.2   The Rule of Mutual Checking Indefinitely

The rule of checking indefinitely has an exception by another rule called *mutual checking indefinitely*, which means the two players check each other continuously without end. When this happens and the two players do not want to break the pattern, the game ends in a draw. As shown in Fig. 2(a), if the game continues cyclically with the moves Re6-e7, Cf8-e8, Re7-f7, Ce8-f8 Rf7-e7, Cf8-e8, etc., the game is treated as a draw since both sides check each other indefinitely.

The example in Fig. 2(b) is interesting as the game continues cyclically with the moves Pf5-e5, Pf3-e3, Pe5-f5, Pe3-f3, etc. The Black side only checks the Red side every two moves whereas the Red side checks all the time. This is not regarded as mutual checking indefinitely but treated as checking indefinitely. Thus, the Red side loses the game if he does not break the pattern.

## 3.3   Other Special Rules

In Chinese chess, there are rules of *chasing indefinitely*. The general concept is that one cannot chase his opponent's piece continuously without end. The term *chase* is defined similarly to the term check. The difference is that the prospective piece to be captured is not the King but another piece. The detailed rules of chasing indefinitely are very complicated. It is also possible that they may stain the endgame databases. In addition, there are special rules about other indefinite move patterns.

All these special rules can be classified into two categories. The first category is where, the game ends in a draw if a certain indefinite move pattern occurs and the two players do not want to break the pattern, e.g., the rule of mutual checking indefinitely. While constructing endgame databases by retrograde analysis, all the positions which are not propagated from checkmate or stalemate positions are eventually marked as draws. Therefore, the first category of special rules does not stain the endgame databases. The second category of special rules is when, if a certain indefinite move pattern occurs and the two players do not want to break the pattern, one side loses the game according to the rule. For example, the player checking indefinitely loses the game. The rules that may corrupt the databases will be dealt with in this paper.

### 3.4   Summary of Special Rules

In a sequence of moves with some position appearing 3 times, the sequence of all the plies of each side is classified as being *allowed* or *forbidden*. For example, checking indefinitely is forbidden[4]. A sequence of plies is allowed if it is not forbidden.

According to [2], the special rules are summarized as follows and these are applied to an indefinite move pattern:

1. If only one side checks the other indefinitely, the side who checks loses the game.
2. Otherwise, if the sequence of plies of one side is forbidden and that of the other side is not all forbidden, the side with forbidden sequence of plies loses.
3. Otherwise, the game ends in a draw.

With the above summary, both mutual checking indefinitely and mutual chasing indefinitely result in a draw. If one side chases his opponent all the time forming a forbidden sequence of plies and his opponent checks every other move with an allowed sequence of plies, the side which chases loses the game. In Chinese chess, the detailed special rules are very complicated. This summary is sufficient for the purpose of this paper.

## 4   Abstracting the Special Rules

A position is called a *state* in the graph representation. The state graph of Chinese chess is a finite, directed, bipartite and cyclic graph. The vertices are the states, and each directed edge indicates the corresponding ply (half-move) from one state to another, with the relationship of *parent* and *child* respectively. It is important to know the features of these indefinite move patterns in the state graph.

If we are given a series of moves, we can easily determine whether the rule of checking indefinitely is applied or not. However, to construct endgame databases, we assume both sides play perfectly and need to foresee which states must result in Red win or Black win because of the special rules. For this purpose, the term, indefinite move pattern for a special rule, needs to be defined rigorously.

---

[4] In the Asian and Taiwanese versions of the rules, each forbidden sequence consists of only plies of checking or chasing. In the Chinese version of the rules, it may contain another ply called *threatening to checkmate*. It means a ply that leads to a win if the opponent gave up the right of the next move, i.e., made a null move.

Note that the algorithms of retrograde analysis construct an endgame database by first assigning initial values, or *seeds*, to terminal positions that are checkmate or stalemate. These values are then propagated to the rest of the positions. Finally, when no propagation can be done, the unknown positions are marked as draws. Given initial values for a subset of positions $S$ in an endgame database expressed as a state graph $G$, let $G_S$ be the state graph with values assigned using a typical retrograde analysis algorithm before the final phase. That is, it contains win/loss/draw information for positions that are reachable using propagation from initial values from positions in $S$. We call $G_S$ the *propagated* state graph for $G$. A propagated database has the property that the children of every unknown vertex are either unknown, win or draw. Furthermore, an unknown vertex has at least one unknown child and has no loss child.

This paper addresses the special rule $R^*$ of non-mutual checking indefinitely. For the convenience of discussion, we assume there is an attacking side and that the other side is defending. The attacking side tries to win the game by forcing his opponent to violate the special rules. An example of the special rule for checking indefinitely, is when the defending side always checks the attacking side forming a sequence of forbidden plies.

We define an $R^*$ *primitive pattern* as a maximal (with respect to vertex addition) subgraph $G_S[R^*]$ of the propagated state graph $G_S$ with the following properties.

1. The out-degree of any vertex in $G_S[R^*]$ is at least 1.
2. Each state $s$ in $G_S[R^*]$ is marked unknown.
3. The edges from a vertex of the defending side to all of its unknown children are included in $G_S[R^*]$. A vertex of the defending side does not have children which are draws.
4. At least an edge from a vertex of the attacking side to one of its unknown children is included in $G_S[R^*]$.
5. The defending side checks the attacking side on every possible move within $G_S[R^*]$.

In an $R^*$ primitive indefinite move pattern, the attacking side can always keep the game in the pattern, and the defending side cannot quit this pattern without losing the game. Therefore, all the defending plies are forbidden. That is, all the attacking vertices are win, and all the defending vertices are loss. Our goal is to find such patterns efficiently using an algorithm.

Note that an indefinite move pattern does not need to have a strongly connected component. That is, an indefinite move pattern may not be periodic. For example in Fig. 1(a), we denote the sequence of moves Ra8-a9 Kd9-d8 Ra9-a8 Kd8-d9 by $A$, and Ra8-a9 Kd9-d8 Ra9-a8 Kd8-d7 Ra8-a7 Kd7-d8 Ra7-a8 Kd8-d9 by $B$. The game may continue as $ABAABAAAB\cdots$ without ending. The approach of periodicity does not meet our purpose. The defined primitive indefinite move patterns are the new seeds for finding the above patterns which may be non-periodic.

# 5    Constructing Complete Win-Draw-Loss Databases

We say some certain special rules are *harmless* to a certain database if the database maintains the correct win-draw-loss information when these special rules are ignored. An endgame database is called *complete* if it has correct information in concord with the special rules. In [3], a lemma is stated for the fact that special rules are harmless to the databases for endgames with exactly one armless side. Hence complete endgame databases are constructed using the classical retrograde analysis algorithm for 151 Chinese chess endgames, each of which has exactly one armless side.

In Chinese chess endgames, human experts believe the rule of checking indefinitely is much more influential to corrupt the databases than the other special rules. That is, special rules, other than checking indefinitely, need several more pieces on the board to corrupt the databases. In available literature, experts study the use of the special rules for chasing indefinitely with a total of at least three attacking pieces, four defending pieces and the two Kings. For example, in [2, page 82], the endgame with chasing indefinitely is KRGGKCPGG. Such an endgame contains at least $4.5 \cdot 10^{10}$ positions and is too large to be considered at this stage. It is possible that there are small games to demonstrate the effect of indefinite chasing. However, we cannot manually design any position to be stained by special rules other than checking indefinitely in our current study of computer endgames. Furthermore, in available literature, a meaningful game relevant to mutual checking indefinitely contains at least five attacking pieces, two defending pieces and the two Kings. For example in [2, page 67], the endgame with mutual checking indefinitely is KRCPKRCGM. Such an endgame contains more than $2 \cdot 10^{13}$ positions and is certainly too large to be considered at this stage. This is confirmed by our experimental results, in which no instance of mutual checking indefinitely is reported during our construction of the 43 important endgames.

In contrast, the rule of non-mutual checking indefinitely can be applied to many positions with only two attacking pieces and even no defending pieces. Therefore, we focus on solving the problems caused by non-mutual checking indefinitely in this paper.

Note that the classical retrograde analysis algorithm for constructing the win-draw-loss endgame databases, without considering the special rules for Chinese chess, consists of three phases: initialization, propagation, and the final phase. For details, see [3]. Our algorithm differs from the classical algorithm through the revision of the propagation phase to the following phase:

- {* The revised propagation phase. *}
  the original propagation phase
- **repeat**
  - find primitive non-mutual checking indefinite positions
  - the propagation phase of checking indefinitely
  **until** there is no unknown position being changed in the last iteration.

In the following sections, we describe our algorithm for finding primitive non-mutual checking indefinitely positions.

## 5.1   Finding Direct Checking Indefinitely Positions

In Chinese chess, a player can win the game in one of two ways. One is to checkmate or stalemate the opponent. The other is to force the opponent to violate the special rules, e.g., forcing his opponent to check or to chase indefinitely. In real games of Chinese chess, a player will break the special rules only when he knows that all the other moves will lead him to lose the game. A master can foresee the conditions a great number of moves ahead whereas an amateur may only be aware of what might happen in a few moves. Therefore, a master knows more about which move will lead him to win or lose a game than an amateur does. In real games, the occurrence of checking or chasing indefinitely depends on the knowledge of the two players or the information of the chess-playing system.

We call a win or loss position *relevant* to special rules if the winning side cannot win the game without these rules, i.e., he cannot checkmate or stalemate his opponent but can win the game by forcing his opponent to violate a special rule. This kind of position is special, since it is not noticed during the construction of databases by retrograde analysis ignoring the special rules. Note that in some positions, the winning side can win the game by either checkmate, stalemate his opponent, or by forcing his opponent to violate the special rules.

Similarly, a position is called *relevant* to checking indefinitely if the losing side can avoid his King being captured or stalemated by checking his opponent indefinitely. Only this kind of position is relevant to special rules in our current endgame databases with the assumption that the special rules, other than checking indefinitely, are harmless to our current computer endgame study.

With a set of propagated endgame databases, our algorithm to detect checking indefinitely consists of three phases: initializing candidates, pruning candidates, and the final phase. The first phase is to find prospective states relevant to checking indefinitely. The second phase is to prune unqualified candidates by an iterative process. The final phase is to update the win-draw-loss information of the resulting states.

In the phase of initializing candidates, the database is traversed twice. The first is to initialize the *win candidates* and the second is to initialize the *loss candidates*. In the first traversal, each unknown state is marked as a win candidate if the King of the next mover is in check. Note that all the moves/edges leading to the win candidates are checking moves. In the second traversal, each unknown state is marked as a loss candidate if its parent is a win candidate. The next mover of each win candidate is potential to win the game because he may be able to force the opponent to check himself in the next move. All loss candidates are potential to be loss states if its all children are win states or win candidates. Note that a win candidate may also be a loss candidate at the same time. These candidates with dual status are the ones possibly relevant to mutual checking indefinitely.

The phase of pruning candidates is composed of multiple traversals of the marked candidates until no further update on the candidates is possible. In the odd-numbered traversals, a loss candidate is pruned, i.e., no longer marked as a loss candidate, if it has at least one child that is neither a win candidate nor a win state. In the even traversals, a win candidate is pruned if it has no child as a loss candidate. The traversals end when no candidates can be pruned.

In the final phase, if there are candidates with dual status, they are the positions relevant to mutual checking indefinitely. However, we do not find any of them in our current experiments. Each win candidate has at least one loss candidate to keep the game in the pattern, and each loss candidate cannot quit the pattern without losing the game. Therefore, we mark the loss candidates loss and the win candidates win.

We call these newly marked win and loss states *direct* checking indefinitely since they immediately lead to checking indefinitely. They are new seeds for propagation. Thus, we have the propagation phase of checking indefinitely after that. The newly found win and loss states are called *indirect* checking indefinitely, because they may lead to checking indefinitely some plies later. For win-draw-loss information, the propagation of checking indefinitely are the same as the original propagation.

## 5.2 Splitting of State Graphs

In the design of endgame systems of western chess and Chinese chess, the state graph is usually split according to the total number of pieces on the board into multiple databases. In the graph representation, the vertices are the databases and the directed edges indicate the conversions of pieces. The graph representing the relations of the databases is non-cyclic; therefore, the bottom-up construction order can be clearly defined.

With the bottom-up construction order, each database has all its supporting databases before being constructed. This guarantees that the above algorithms work. In addition, the propagation might be from the states relevant to checking indefinitely in the supporting databases.

## 6    Infallible and Complete Endgame Databases

The Chinese chess endgame databases constructed by the algorithms in Sect. 5 contain complete win-draw-loss information. However, with only this win-draw-loss information available, one player may rove in the win states but never win the game by checkmate, stalemate, or forcing his opponent to check indefinitely. In this section, we show our approach to construct the infallible and complete endgame databases.

### 6.1    Position Values

Our algorithm for constructing endgame databases in distance-to-mate and distance-to-conversion metrics without taking care of special rules are discussed in [3]. The algorithm follows more or less a standard retrograde analysis algorithm devised for western chess. We use 0 to represent a draw. We use an odd number $i$ to denote win in $i$ plies, and an even number $j$ to denote loss in $j$ plies.

We now describe our scheme using positions that are marked with win or loss using the rule of non-mutual checking indefinitely.

To achieve infallibility, we set the position value of the win states of direct checking indefinitely to a special value $WinByDirectCheckingIndefinitely$. It is 255 in our implementation. And we set the position value of the loss states of direct checking indefinitely to a special value $LossByDirectCheckingIndefinitely$. It is 254 in our implementation.

The propagation phase of checking indefinitely starts after the phase of detecting checking indefinitely. However, it is a little different from the propagation of the states in distance-to-mate or distance-to-conversion metrics. Here we introduce the concept of *distance-to-check*, which is the distance to the win states by direct checking indefinitely. In our implementation, the position value in the distance-to-check metric is 255 minus the distance to win states of direct checking indefinitely. Each position in distance-to-check metric propagates its value minus 1 to all its parents, instead of plus 1 in distance-to-mate metric. For instance, the position value 250 in distance-to-check metric means 5 plies to the win state of direct checking indefinitely. Each loss state of direct checking indefinitely has position value 254. Note that in our implementation, all positive even position values indicate loss states and all odd position values represent win states in our indexing scheme in any metric. The only exception is 0 for draw positions.

In real games of Chinese chess, there is no preference in winning the game by using checkmate or special rules. In the design of Chinese chess endgame databases in concord with checking indefinitely, we assume that loss by checkmate or stalemate is worse than loss by breaking the special rules. That is, if one can win the game without using special rules, he will prefer this way of ending the game, i.e., checkmate or stalemate. Similarly, if one is losing the game, then losing by special rules is preferred against losing by checkmate. The assumption used here is reasonable since the use of special rules requires more knowledge.

The current position values are updated if the propagated values are better. By the zero-sum property, our assumption implies that win by checkmate or stalemate is better than win by forcing the opponent to check or chase indefinitely. Therefore, win position values of distance-to-mate are better than those of distance-to-check. Loss position values of distance-to-mate are better than those of distance-to-check. This assumption helps for propagation to update the current position values of their parents.

As mentioned in Sect. 5.1, some new states of direct checking may be detected when more information regarding the states is gained. Therefore, as shown in the algorithm in Sect. 5, the phases of detecting and propagation of checking indefinitely are repeated as multiple iterations until no position information can be gained. We call the positions with information found in the first iteration, *first* order checking indefinitely. Similarly, those with information found in the $k$th iteration, are called $k$th order checking indefinitely. In our implementation, the position value of the win states of the second order direct checking indefinitely is the largest odd integer less than minimum position value of the first order checking indefinitely, and the position value of the loss states of the second order direct checking indefinitely is 1 less than that of the win states. The propagation of second order checking indefinitely remains the same without changing. Similarly, third order checking indefinitely and its index scheme can be clearly defined, and so are the fourth order, fifth order, etc. The process continues until no more states relevant to checking or mutual checking indefinitely can be found. Because of technical problems in encoding the position values for multiple order checking indefinitely in one byte, we currently do not consider second or higher order checking indefinitely.

As state space is split into multiple databases, each resulting database has its own indices of checking indefinitely of different orders, except for the first order. Our data structure has a separate table to record these indices.

## 6.2    Infallibility of Our Databases

With the rule of checking indefinitely, the concept of shortest win can hardly be defined. Hence we concentrate on infallibility. With our databases, the infallibility can be achieved as follows:

1. Starting from a draw state, we may move to any child draw state. Note that this strategy never results in loss because of checking indefinitely.
2. Starting from a win state with position value in distance-to-mate metric, we can win the game by leading the game to a checkmate or stalemate state by playing the best possible next move.
3. Starting from a win state with position value in distance-to-check metric, we can lead the game to one of the following states and remain in win status:
   - a state of indirect checking indefinitely with the same order and shorter distance-to-check,
   - a state of direct checking indefinitely, with the same order,
   - a state relevant to checking indefinitely with higher order, i.e., smaller order index, or
   - a state with position value in distance-to-mate, i.e. irrelevant to checking indefinitely.

   Because the order and the distances cannot be infinite, we can eventually win the game by checkmate, stalemate, or the rule of checking indefinitely.

## 6.3    Modifications for Distance-to-Conversion

When checking indefinitely is taken into account, a database may have some positions in distance-to-check metric and the others in distance-to-conversion metric. We call it a database in *distance-to-mate/check* metric. When the position values in distance-to-mate and distance-to-check metrics are both encoded in 1 byte, the chance of overflow in representing the values is greater than those in pure distance-to-mate. Distance-to-conversion is introduced to reduce this potential problem. In this case, the propagated value of any win state in supporting databases is the smallest loss position value, and that of a loss state in supporting databases is the smallest win position value. The propagated value of a draw state remains a draw. Similarly, if a database has some positions in distance-to-conversion metric and the others in distance-to-check metric, we call it *distance-to-conversion/check* metric.

Note that the databases in distance-to-conversion/check metrics are still infallible. Starting from a win state in distance-to-conversion metric, we can always lead the game to some supporting database, or win by checkmate or stalemate. Starting from a win state in distance-to-check metric, we can always lead the game to the four types of states listed in Sect. 6.2 or a state in some supporting database. Because the order and the distances cannot be infinite, and the graph representation of databases are finite and acyclic, we can eventually win the game.

## 6.4   Verifying the Databases

We introduced new position values, other than those propagated from checkmate or stalemate, in our databases. We will assign numerical values to these position values such that they obey the following rules:

1. Win position values are better than draw position values, which is better than loss position values.
2. Win position values in distance-to-mate or distance-to-conversion metric are better than those in distance-to-check metric, whereas loss position values in distance-to-check metric are better than those in distance-to-mate or distance-to-conversion metric.
3. For win position values in distance-to-mate (respectively, distance-to-conversion) metric, the smaller the value, the shorter the distance to checkmate or stalemate (respectively, to reaching the supporting databases), and hence the better. For loss position values in distance-to-mate (respectively, distance-to-conversion) metric, the larger the value, the longer the distance to checkmate or stalemate (respectively, to reaching the supporting databases), and hence the better.
4. For the win states in distance-to-check metric, the higher the order, the better. For the loss states in distance-to-check metric, the lower the order, the better.
5. For the win states in distance-to-check metric with the same order, the larger the position value, the shorter the distance to the direct checking indefinitely, and hence the better. For the loss states in distance-to-check metric with the same order, the smaller the position value, the longer the distance to direct checking indefinitely, and hence the better.

To verify the correctness of a database with correct supporting databases, we just have to check if the position value is the best of the propagated values from all the children. The only exception is the win states of direct checking indefinitely. Each win state of direct checking indefinitely has at least one loss child of direct checking indefinitely and has no loss children in distance-to-check metric in higher order or in distance-to-mate metric. With the assumption that everything goes well in the phases of finding direct checking indefinitely, the verification is sufficient to prove the correctness and uniqueness of our databases. The verification program can detect program bugs or hardware errors which occur during the propagation.

# 7 Concluding Remarks

Retrograde analysis has been widely used and successfully applied in western chess. In Chinese chess, its application is confined because of these special rules. We successfully developed an algorithm to construct the endgame databases in concord with the rules of checking indefinitely. We list the statistics of 43 important endgame databases in the appendix. Our current code uses a large amount of main memory. To save memory usage, we need to encode the position values in one byte. Thus we currently do not consider second or higher order checking indefinitely, though our algorithm can handle them easily. We are working now in extending our code to deal with second and higher order checking indefinitely.

In the future, we plan to use knowledgeable encoding and querying of endgame databases discussed in [6] as an approach to extract win-draw-loss information and condense it in physical memory to improve the ability of present Chinese chess programs. When this has been realized, the level of Chinese chess computer systems might be close to that of human champions. The process to construct endgame databases which are too large to be constructed in present physical memory, is another subject. With successful development and implementation some open problems in Chinese chess endgames, such as the KRN-MKRMM endgame, may be solved. That would fertilize the field of Chinese chess endgame studies.

# References

1. Thompson, K.: 6-piece endgames. International Computer Chess Association Journal **19** (1996) 215–226
2. China Xiangqi Association: The Playing Rules of Chinese Chess. Shanghai Lexicon Publishing Company (1999) In Chinese.
3. Fang, H., Hsu, T., Hsu, S.: Construction of Chinese chess endgame databases by retrograde analysis. In Marsland, T., Frank, I., eds.: Second International Conference on Computers and Games (CG 2000). Volume 2063 of Lecture Notes in Computer Science. Springer-Verlag (2001) 96–114
4. Wu, R., Beal, D.: Fast, memory-efficient retrograde algorithms. International Computer Chess Association Journal **24** (2001) 147–159
5. van den Herik, J., Uiterwijk, J., van Rijswijck, J.: Games solved: Now and in the future. Artificial Intelligence **134** (2002) 277–311
6. Heinz, E.: Knowledgeable encoding and querying of endgame databases. International Computer Chess Association Journal **22** (1999) 81–97

# Appendix: Statistics of 43 Endgame Databases

In this paper, we focus on the databases which are potentially stained by the special rules of indefinite move patterns. Thus, we list only the statistics for the databases satisfying the following requirements.

1. Both sides have attacking pieces.
2. There is an attacking piece other than the pawn.
3. If there are only two attacking pieces, then neither of them can be a pawn.

As we mentioned in [3], if only one side has attacking pieces, the special rules of indefinite move patterns do not stain the databases. Therefore, in requirement 1, we list only the statistics for databases in which both sides have attacking pieces. The rule of checking indefinitely stains the positions in which the defending side cannot avoid his King being captured without checking the attacking side. This means the defending side must be powerful enough to check the attacking side indefinitely. In general, Pawn is considered as the weakest attacking piece in Chinese chess. Therefore, we add requirements 2 and 3. Experts believe that endgames not satisfying these requirements can hardly be stained by special rules.

Table 1 gives statistics for 43 endgame databases satisfying the above requirements. For each database, we report its number of legal positions, number of draw positions, maximal distances measured in plies, and distance-to-check rate for each database. The distance-to-check rate is the number of positions in distance-to-check metric divided by the number of legal positions. If no positions relevant to checking indefinitely exist, the maximal distance-to-check and distance-to-check rate are marked as *N/A*.

So far, we do not find any database with the positions of first order mutual checking indefinitely. Currently we only experiment on first order checking indefinitely. Therefore, the maximal distance-to-check value is the maximal distance measured in plies to the win states of first order direct checking indefinitely.

We construct a database in distance-to-conversion/check metric only if the position value cannot be encoded in 1 byte or it has some supporting databases in distance-to-conversion/check metric. Otherwise, the database is constructed in distance-to-mate/check metric. In our current experiments, the only two databases which have all supporting databases in distance-to-mate/check metric and cannot be indexed in 1 byte are KRKNGMM and KRKNCG. The databases above them are also constructed in distance-to-conversion/check metric as well, e.g., KRKNGGMM. In Table 1, if a database is in distance-to-conversion/check, the database name and the maximal distance-to-mate or -conversion has an ending character '*'. The only two databases in distance-to-conversion/check metric listed in the table are KRKNGMM and KRKNGGMM.

In distance-to-conversion/check metric, all win positions in the supporting databases propagate the same value to the positions in the constructed database, no matter they are distance-to-mate, distance-to-conversion, or distance-to-check. The loss positions perform the same propagation. Therefore, the maximal

**Table 1.** Statistics of 43 endgame databases.

| Database Name | # of Legal Positions | # of Draw Positions | Maximal Distance-to-Mate/Conv. | Check | Distance-to-Check Rate |
|---|---|---|---|---|---|
| KRNKRGG | 251181481 | 140966299 | 77 | 58 | 2.59% |
| KRNKRG | 138660209 | 62141914 | 79 | 62 | 7.22% |
| KRNKRM | 204263530 | 96645975 | 70 | 61 | 5.18% |
| KRNKR | 30118362 | 9279051 | 79 | 59 | 13.05% |
| KRPKRG | 84330363 | 48354000 | 32 | 24 | 1.17% |
| KRPKRM | 124050578 | 68111972 | 36 | 33 | 1.48% |
| KRPKR | 18443469 | 9383254 | 30 | 21 | 2.48% |
| KRKNN | 16300026 | 4672873 | 88 | 53 | 1.59% |
| KRKNC | 33568194 | 757635 | 116 | 73 | 1.84% |
| KRKCC | 17300976 | 1409308 | 82 | N/A | N/A |
| KRKNPGG | 168307887 | 2433079 | 152 | 81 | 1.02% |
| KRKNPG | 92456806 | 115809 | 60 | 14 | 0.64% |
| KRKNPM | 136200539 | 470847 | 72 | 19 | 0.66% |
| KRKNP | 20011890 | 20324 | 37 | 14 | 0.73% |
| KRKPPP | 103676439 | 1278141 | 54 | 15 | 0.43% |
| KRKPPGG | 52598998 | 947673 | 86 | 10 | 0.04% |
| KRKPPMM | 122221940 | 2214611 | 94 | 12 | 0.06% |
| KRKPPG | 28498574 | 53302 | 44 | 10 | 0.06% |
| KRKPPM | 41658907 | 107788 | 44 | 10 | 0.08% |
| KRKPP | 6084903 | 8238 | 34 | 10 | 0.10% |
| KNPKN | 21682338 | 5008874 | 154 | 43 | 2.01% |
| KNPKCG | 100076040 | 40958639 | 138 | 19 | 0.16% |
| KNPKCM | 149719630 | 67594106 | 154 | 36 | 0.39% |
| KNPKC | 22364304 | 6387677 | 46 | N/A | N/A |
| KCPKC | 22956705 | 20094132 | 41 | N/A | N/A |
| KRKRGG | 2997932 | 1840672 | 30 | 11 | 0.09% |
| KRKRG | 1628603 | 966498 | 19 | 11 | 0.19% |
| KRKRM | 2389472 | 1370793 | 19 | N/A | N/A |
| KRKR | 348210 | 193950 | 10 | N/A | N/A |
| KRKNGGMM* | 63684381 | 33638427 | 64* | 23 | 0.24% |
| KRKNGGM | 21879507 | 720797 | 104 | 6 | 0.33% |
| KRKNGMM* | 35195142 | 4726833 | 82* | 39 | 0.52% |
| KRKNGG | 3221138 | 6578 | 74 | 3 | 0.37% |
| KRKNGM | 12032732 | 27464 | 50 | 3 | 0.35% |
| KRKNMM | 7654095 | 135941 | 111 | 75 | 17.31% |
| KRKNG | 1762807 | 3275 | 40 | 1 | 0.39% |
| KRKNM | 2605497 | 5200 | 34 | 1 | 0.37% |
| KRKN | 380325 | 609 | 26 | 1 | 0.42% |
| KRKCGG | 3322727 | 1379102 | 48 | N/A | N/A |
| KRKCMM | 7913097 | 2969808 | 64 | N/A | N/A |
| KRKCG | 1820350 | 2462 | 36 | 3 | 0.01% |
| KRKCM | 2694400 | 91748 | 26 | N/A | N/A |
| KRKC | 393327 | 7479 | 16 | N/A | N/A |

distance-to-check value of a database may be different in distance-to-mate/check and distance-to-conversion/check metrics.

If the special rules are ignored during construction, the database may be stained by these rules. One thing we are interested in is the *stained rate*, which is defined as the number of positions relevant to checking indefinitely divided by the number of legal positions. For databases in distance-to-mate/check, it is the same as distance-to-check rate. The statistics in Table 1 excludes second or higher order checking indefinitely. For endgames with positions of second or higher order checking indefinitely, the listed distance-to-check rate is lower than the stained rate. For databases in distance-to-conversion/check metric, some positions in distance-to-conversion may be relevant to checking indefinitely, because the positions in distance-to-check metric in the supporting databases do not propagate the checking indefinitely information during the construction. Thus, the distance-to-check rate is usually lower than the stained rate.

# ORTS: A Hack-Free RTS Game Environment

Michael Buro

NEC Research Institute, Princeton
mic@research.nj.nec.com

**Abstract.** This paper presents a novel approach to Real-Time-Strategy
(RTS) gaming which allows human players as well as machines to compete in a hack-free environment. The main idea is to replace popular but
inherently insecure client-side game simulations by a secure server-side
game simulation. Only visible parts of the game state are sent to the
respective clients. Client-side hacking is therefore impossible and players
are free to choose any client software they please. We discuss performance issues arising from server-side simulation and present ORTS – an
open RTS game toolkit. This software package provides efficient C++
implementations for 2D object motion and collision detection, visibility
computation, and incremental server-client data synchronization, as well
as connectivity to the Generic Game Server (GGS). It is therefore well
suited as a platform for RTS related A.I. research.

## 1 Introduction

Real-time strategy (RTS) games have become very popular over the past couple
of years. Unlike classic board games which are turn-based, RTS games are fast
paced and require managing units and resources in real-time. An important element of RTS games is incomplete information: players do not know where enemy
units are located and what the opponents' plans are, unless they send out scouts
to find out. Incomplete information increases the entertainment value and complexity of games. The most popular RTS titles so far have been the million-sellers
WarCraft-II and StarCraft by Blizzard Entertainment and Age of Empires series
by Ensemble Studios. These games offer a wide range of unit and building types,
technology trees, multi-player modes, diverse maps and challenging single-player
missions. From the A.I. research perspective the situation looks ideal: playing
RTS games well requires mastering many challenging problems such as resource
allocation, spatial reasoning, and real-time adversarial planning. Having access
to a large population of human expert players helps to gauge the strength of
A.I. systems in this area – which currently leaves a lot to be desired – and in-
spires competition. The commercial success of RTS games, however, comes at a
price. In order to protect their intellectual property, games companies are disinclined to publish their communication protocols or to incorporate A.I. interfaces
into their products which would allow researchers to connect their programs to
compete with peers and human experts. Another reason for keeping game software closed is the fear of hackers caused by software design concessions. Due

J. Schaeffer et al. (Eds.): CG 2002, LNCS 2883, pp. 280–291, 2003.

to minimal hardware requirements that the game companies want to meet, the aforementioned commercial RTS games rely on client-side simulations and peer-to-peer networking for communicating player actions. This approach reduces data rate requirements but is prone to client hacking – such as revealing the entire playing field – and threatens the commercial success of on-line gaming as a whole. The ORTS project [1] deals with those two problems: it utilizes an open communication protocol – allowing players and researchers to connect any software client they wish – and implements a secure RTS game environment in which client hacking is impossible. ORTS is an open software project which is licensed under the GNU Public License to give users the opportunity to analyze the code and to share improvements and extensions.

The remainder of this paper is organized as follows: Sect. 2 presents the ideas behind server-side RTS simulations and takes a detailed look at information hiding and data rate requirements. Section 3 deals with implementation issues and discusses the ORTS kinematics model, visibility computation, and server-client data update. A summary and outlook section wraps-up the article.

## 2   A Server-Client RTS Game Model

Popular RTS games utilize the client-side game simulation described in Fig. 1. A detailed description of this approach – including various optimizations that mask client latency – can be found in [2]. Clients run game simulations and only transmit user actions to a central server or directly to their peers. At all times game states are synchronized and known to all clients, regardless of what is visible to the respective player. The client software hides information not meant for the respective player. An alternative approach is presented in Fig. 2. Here,

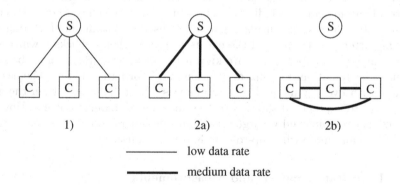

**Fig. 1.** Common client-side game simulation: clients first establish contact to a central server (1). When a game is created, all clients start simulating world changes simultaneously and send issued user commands either back to the server (2a), which broadcasts them to all other clients, or directly to their peers (2b) using a ring or clique topology. The data rate requirements are modest if the number of players and the number of commands they issue are small.

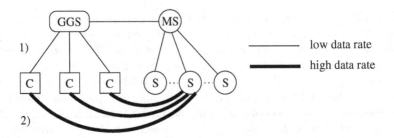

**Fig. 2.** ORTS server-side game simulation: clients and a master server (MS) are connected to a central server – GGS in this case – and several worker servers (S) are connected to the master server (1). When a game is created on MS, it schedules an idle worker server on which the game is to be simulated. This server then connects to the clients to send the individual game views and to receive unit commands (2).

a central server runs the simulation and transmits only the part of the game state the respective client is entitled to know. In the following two subsections we compare both models with regard to information hiding and data rate issues.

## 2.1    Information Hiding

Playing games with incomplete information at a competitive level on the internet demands that information can be physically hidden from clients. To see this one just needs to imagine a commercial poker server that sends out all hands to each client. Hacking and losing business is inevitable. Translated into the RTS world, information hiding rules out client-side simulations in theory. Regardless of how clever data is hidden in the client, the user is in total control of the software running at her side and has many tools available to reverse engineer the message protocol and information hiding measures, finally allowing her to reveal "secret" game state information during a game. [3] discusses means of thwarting such hacking attempts. Despite all the efforts to secure clients, hacks – which spoil the on-line game experience – are known for all popular RTS games based on client-side simulation. Perhaps encryption and information hiding schemes that change with every game and require analysis that takes longer than the game itself lead to a practical solution to the information hiding problem. However, uncertainty remains and we argue that game designers do not even have to fight this battle any more when up-to-date hardware is used.

## 2.2    Data Rate Analysis and Measurements

Synchronized client-side simulations require only player commands to be transmitted to peers, which keeps the data rate low if only a few commands are generated during each frame[1] and the number of players is small. Independent

---
[1] From here on a *frame* is a simulation round consisting of updating the game state, sending visible information to the clients, and receiving commands from clients.

of the chosen communication topology and player views during the game, the client input data rate $d^{(in)}$ – measured in bytes/frame – depends on the number of participating players and the rate they issue commands:

$$d_k^{(in)} = \sum_{i=1, i \neq k}^{n} d_i^{(out)}. \tag{1}$$

In the server-side simulation model the input data rate grows linearly in the number of objects the players see in the current frame:

$$d_k^{(in)} = D \cdot \#\text{objects visible to player } k, \tag{2}$$

where $D$ is the average data rate generated by one visible object. To compare the i/o requirements of both models we look at the extreme cases: 1) small vs. large number of players and unit commands per frame and 2) overlapping vs. disjoint player views. Client-side simulation has a lower input data rate requirement if a small number of players only generate few commands during each frame. Server-side simulation excels if the number of players and unit commands is high and the player views are mostly disjoint.

Data rates in the server-side simulation model can be decreased by incremental updates, compression, and partial client-side simulation of visible objects. The empirical results presented in Fig. 3 indicate that even without partial simulation it is possible to play RTS games at 5 frames/sec featuring hundreds of visible objects over conventional DSL or cable modem lines.

For the tests the motion of up to 1500 circular objects were simulated. In the initial configuration objects of diameter 16 were distributed evenly on an empty 800x800 playing field. Then object ownership was assigned randomly to

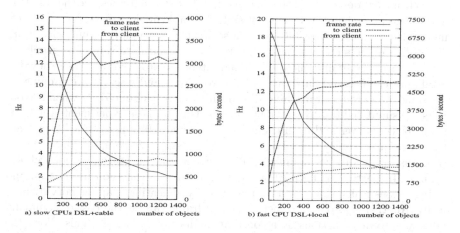

**Fig. 3.** Data and frame rates dependent on the number of objects. (a) Slow (500 MHz) client CPUs, "real-world" DSL and cable modem connection with both 80 msec ping time. (b) Fast (850+ MHz) client CPUs, DSL and local connection.

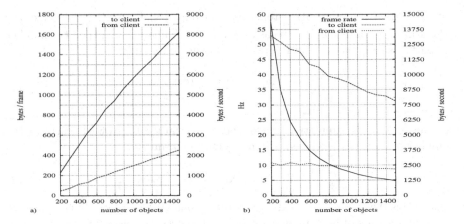

**Fig. 4.** Data and frame rates dependent on the number of objects: (a) Bytes per frame and bytes per second at a fixed simulation rate of 5 Hz. (b) Total frame and data rates measured on a dual Pentium-3/933 system for the entire simulation including object motion, collision test, (de)compression, and data transmission.

two players. During the game both players picked new random headings for their objects whenever they collided. All objects had constant speed of 4/frame and a sight range of 60 which ensured that a large fraction of enemy units was visible at all times (Fig. 7). All experiments were conducted on a dual Pentium-3/933 system under Linux. Fig. 4a) shows the generated data rates dependent on the number of objects when using message compression. On average approximately 1.2 bytes per visible object per frame is sent to each client and 0.6 bytes per own object per frame is returned to the server. The details of the utilized compression scheme are described in the next section. Fixed at 5 frames/sec the resulting data rates can be handled by current DSL and cable modem technology. Neglecting latency it is even possible to run a server for a two-player game with up to 400 visible objects on systems with 6 KB/sec upload data rate. The graphs in Fig. 4b) take server and client CPU load latencies into account. They show the total performance and data rates when two clients and a server are running on the same dual processor machine. The frame rate drops from 57 Hz when 200 objects are simulated to 5 Hz for 1500 objects. During these experiments the CPU loads for the server and both clients stayed around 60%/40%/40%. To increase the frame rate, latency caused by simulation, message (de)compression, and data transmission has to be minimized. The current implementation is not well optimized. In particular, data (de)compression and transmission use stream and string classes which slow down computation by allocating heap memory dynamically. Another approach for increasing the frame rate at the cost of command latency is to delay actions by a constant number of frames [2]. This allows the server to continue its simulation after sending out the current game views without waiting for action responses. Whether built-in command latency is tolerable is game dependent. Currently, this technique is not employed in ORTS.

To check how server-side simulation performs in conjunction with smaller channel capacities and higher latencies we ran two external clients on slower (500 MHz) machines which transmitted data over a 128/768-kBaud DSL and a 240/3200-kBaud cable modem line. The frame rates we measured are shown in Fig. 3. Apparently, latency caused by transmission and slower client side computation rather than the available data rate is the bottleneck in this setting. Nevertheless, we can conclude that playing RTS games in a hack-free environment featuring hundreds of moving objects is possible using up-to-date communication and PC equipment. It is important to note that the reported frame and data rates are lower bounds because in actual RTS games player views usually do not coincide (Fig. 7). On the other hand, data rates will increase if features are added and more object actions become available. However, the data rate increase is expected to be moderate because compression routines have access to all feature vectors and the entire action vector and can therefore exploit repetitions.

## 3   Implementation Issues

### 3.1   ORTS Kinematics

ORTS objects are circles placed in a rectangular playing field. Each object has a fixed maximal speed and can move in any direction or stop anytime. ORTS kinematics is simple: there is no mass, no acceleration, and no impulse conservation. When two objects collide they just stop. This basic model simplifies motion and collision computation but already covers the important motion aspects of RTS games. Object motion is clocked. In each time interval objects move from their current location to the destination location which depends on the object's heading and speed. The algorithm presented in Fig. 6 detects all collisions that occur in one time interval and moves the objects accordingly. The algorithm

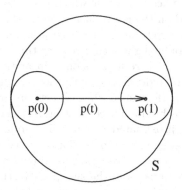

**Fig. 5.** Motion bounding sphere. A circular object moves from position $p(0)$ to $p(1)$. The motion bounding sphere S covers all points occupied by the object in time interval $[0, 1]$. Spheres are good approximations if the motion vector is short compared to the object radius – which is usually the case in RTS games.

first computes the motion bounding spheres for all objects (Fig. 5). Then it constructs the sphere intersection graph $G$ in which nodes represent bounding spheres and edges indicate intersections. In order to minimize the sphere distance computation – which consumes quadratic time if implemented naively – spheres are first assigned to grid sectors and then all distances between spheres in each sector are determined to form $G$. Motion and collisions can now be handled local to the connected components of $G$ because objects do not collide if their motion bounding spheres are disjoint. The algorithm then generates the sequence of collision times for each component of $G$ separately and moves the objects to their respective final positions. The central data structure is an augmented heap which gives access to the edge with the earliest collision time and allows adding, updating, and removing of edge-time pairs in logarithmic time. An additional mapping from edges to time allows to perform delete operations of pairs with unspecified collision time which is used in several placed throughout the subroutine. The algorithm starts by computing local collision times ignoring global effects at first. Then, starting with the earliest collision, it moves colliding objects to the collision location, stops the objects there, and updates the collision times between newly stopped objects and objects in their neighborhood. When all collisions are handled the remaining non-colliding objects are moved to their final location. Compared with the visibility computation we describe in the next subsection the running time for object motion and collision test is negligible.

### 3.2   Visibility Computation

In ORTS objects have circular vision. Enemy objects are reported to a player if it is in sight of at least one friendly object ("Fog of War", Fig. 7). Similar to object motion, a naive implementation leads to quadratic run time. Moreover, the sector approach we adopted for object motion is slow in case of large sight ranges because many objects fall into single sectors. The ICollide software package [4] implements a fast on-line collision test which can also be applied for visibility computation. It is based on the fact that axis-aligned rectangles intersect if and only if their projections onto the x and y axes overlap, and exploits that the order of projection intervals only slightly change over time. Although this algorithm is fast in sparse settings – where it takes only linear time in average – its worst case run time is quadratic independent of the actual number of intersections. To increase the worst-case performance when objects are crowded we make use of a well known line-sweep technique for computing rectangle intersections. Objects and vision areas are approximated by bounding squares. Square intersections are detected in a left-to-right sweep by maintaining two priority search trees [5] which contain the current set of vertical intervals of sight and object squares. Whenever a new square appears in the sweep its vertical projection is checked for intersections with the other type of squares and then added to the respective set of intervals. At exit of a square its vertical projection is removed from the respective priority search tree. Adding and removing intervals from a priority search tree takes $O(\log n)$ time, while reporting all $K_i$ intersections at time step

```
compute motion bounding spheres
compute sphere intersection graph G
forall connected components c in G {
  empty H  // pair-heap of edges and collision times
  empty S  // stopped nodes collected here
  forall active edges (u,v) in c {       // at least one moving node
    if (u and v intersect) { stop u and v at time 0 and add them to S }
    add ((u,v), NextCollisionTime(u,v,0)) to H
  }
  T:= 0                                   // current collision time
  forever {
    while (S not empty) {
      empty N                             // newly stopped nodes collected here
      forall u in S {
        forall neighbors v of u in G {
          if (v is moving) {
            d:= NextCollisionTime(u, v, T)
            if (d = 0) {
              stop v at time T            // instant collision
              add v to N
              remove ((u,v),?) from H
            } else if (T+d >= 0 && T+d <= 1)
              add ((u,v),T+d) to H        // update collision time
            else
              add ((u,v),2) into H        // no collision anymore
          } else
            remove ((u,v),?) from H       // remove inactive edge
        }
      }
      S:= N;
    }
    finished:= false;
    forever {                             // find next collision time
      if (H empty) { finished:= true; break; } // all done?
      retrieve ((u,v),t) with minimum t from H
      T:= t;                              // next collision time
      if (T > 1) { finished = true; break; }   // no more collisions?
      if (u or v is moving) break;        // edge active? ->
                                          //       handle collision
      remove ((u,v),?) from H             // remove inactive edge
    }
    if (finished) break;                  // component done
    if (u is moving) { stop u at time T; add u to S }
    if (v is moving) { stop v at time T; add v to S }
  }
}
```

**Fig. 6.** Pseudo-code for object motion and collision test. NextCollisionTime(u,v,t) returns the elapsed time until the next collision of objects u and v occurs after time t. If the objects do not collide during the current time interval the function returns a value greater than 1.

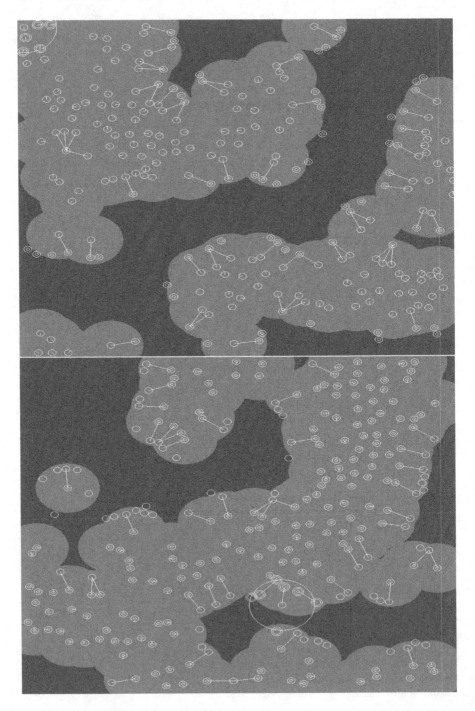

**Fig. 7.** Two views of a two player ORTS game. Straight lines represent attacks. Object sight ranges are indicated by filled ovals.

$i$ is possible in time $O(\log n + K_i)$. Including the initial sort of the x-coordinates the total worst-case run time amounts to $O(n \cdot \log n + K)$, where $n$ is the total number of objects and $K$ the number of square intersections.

### 3.3 Server-Client Data Update

A server-side simulation cycle consists of sending the respective game state views to all clients, waiting for their object action responses, executing all actions, moving objects, and resolving collisions. To ensure high frame rates, latency and required data rates have to be minimized. For slow data connections compression is essential because clients need to be informed about a potentially large number of visible objects. In ORTS each object has an associated numerical feature vector with the following components:

(Object ID, Owner, Radius, Sight Range, Min. Attack Range, Max. Attack Range, Speed, Attack Value, Replicating, Hit Points, Moving, Position)

Most of these values stay constant during simulation or vary only slightly. Before compression algorithms such as LZ77 [6] are applied it is therefore beneficial to reduce entropy by computing differential vector updates first. Compression and decompression increase CPU loads and latencies in the server and the clients. Decreasing the compression rate in favor of lower compression times may therefore result in a better overall performance. The better performance when using a fast CPU at the client side (Fig. 3b) indicates that compression induced latencies currently form the bottleneck in the ORTS implementation. We will deal with (de)compression speed optimizations and implement the command delay approach to increase the simulation rate in future ORTS releases.

## 4 Summary and Outlook

ORTS is a hack-free RTS game toolkit. Its open message protocol and available client software enable A.I. researchers to gauge the performance of their algorithms by playing RTS games in a secure environment based on server-side simulation. Even though popular RTS titles do not provide dedicated A.I. interfaces, programs can – in principle – be constructed to play those games by accessing the frame buffer and audio streams and generating mouse and keyboard events. However, only having indirect access to the game state and being restricted to a sector view imposed by popular GUIs slows down computation and communication by forcing the A.I. to switch focus often. It also limits the command rate considerably because only objects within the current sector can receive instructions. The server-side simulation discussed here removes these artificial limitations at the cost of higher data rates and latencies which, however, can be handled by modern hardware quite easily. Human players also benefit from the open game concept as they are no longer confined to static user interfaces and predefined low-level object behavior. Instead, players can utilize self-made or third-party client software for low- or mid-level object control –

**Fig. 8.** StarCraft User Interface (http://www.blizzard.com). The main screen area
shows a detailed view of a playing field sector. Its location is indicated on the "mini-
map" in the lower left part.

such as path-finding and multi-unit attack/defense. Improved A.I. frees players
from cumbersome hand-to-hand combat and lets them concentrate on strategic
decisions. Moreover, GUIs can be chosen freely because server-side simulations
are not prone to client-side hacking. For instance, players may want to replace
the usual detailed view of a single playing field sector (Fig. 8) by several low
resolution views which together cover much more space and allow to control
multiple sectors more efficiently.

ORTS provides basic RTS game functionality and can be extended easily.
Currently, objects are restricted to moving circles placed on a rectangular playing
field, there are no landmarks or resources, motion is acceleration free, objects
can replicate, have limited vision and just stop when they collide, and object
interaction is restricted to attacking objects in a given attack range. This simple
RTS game is already challenging and playing it well requires understanding
of multi-object attack/defense formations, scouting, and motion planning. We
are currently working on a machine learning approach for basic unit behavior
and think about similar ideas to train the other components of a hierarchical
command and control structure.

## Acknowledgment

The ORTS project has benefited from many fruitful discussions with Susumu
Katayama and Igor Durdanovic.

# References

1. Buro, M.: ORTS project. http://www.cs.ualberta.ca/~mburo (2002)
2. Bettner, P., Terrano, M.:  1500 archers on a 28.8: Network programming in Age of Empires and beyond. Gamasutra http://www.gamasutra.com/features/20010322/terrano_01.htm (2001)
3. Pritchard, M.: How to hurt the hackers: The scoop on Internet cheating and how you can combat it. Gamasutra http://www.gamasutra.com/features/20000724/pritchard_01.htm (2000)
4. Cohen, J., Lin, M., Manocha, D., Ponamgi, K.: I-Collide: An interactive and exact collision detection system for large-scaled environments. ACM International 3D Graphics Conference (1995) 189–196
5. McCreight, E.: Priority search trees. SIAM Journal on Computing **14** (1985) 257–276
6. Ziv, J., Lempel, A.: A universal algorithm for sequential data compression. IEEE Transactions on Information Theory **23** (1977) 337–342 Implemented for instance in zlib http://www.gzip.org/zlib.

# Causal Normalization:
# A Methodology for Coherent Story Logic Design in Computer Role-Playing Games

Craig A. Lindley and Mirjam Eladhari

Zero Game Studio, The Interactive Institute, Sweden
{craig.lindley,mirjam.eladhari}@interactiveinstitute.se

**Abstract.** A common experience in playing computer role-playing games, adventure games, and action games is to move through a complex environment only to discover that a quest cannot be completed, a barrier cannot be passed, or a goal cannot be achieved without reloading an earlier game state and trying different paths through the story. This is typically an unanticipated side effect caused by the player having moved through a sequence of actions or a pathway different from that anticipated by the game designers. Analogous side effects can be observed in traditional software engineering (referred to as data coupling and control coupling), in database design (in terms of unnormalized relations), and in knowledge base design (in terms of unnormalized truth-functional dependencies between declarative rules). In all cases, good design is a matter of minimizing functional dependencies, and therefore coupling relationships between different parts of the system structures, and deriving system design from the minimized dependency relationships. We propose a story logic design methodology, referred to as causal normalization, that minimizes some forms of causal functional dependency within story logics and therefore eliminates some unintended forms of causal coupling. This can reduce the kind of unexpected dead ends in game-play that lead to player perceptions of poor game design. Normalization may not be enough, however. Extending the principle of minimal coupling, we propose an object-oriented approach to story logic, and relate this to principles of normalization and game architecture.

## 1 Introduction

The study of games and game-play has historically been concerned with the study of competitive systems, associated with economic theory more strongly than with play. Traditional board games and puzzle games typically model competitive situations in a very abstract way, involving little or no story context, game world, or characterization. It is only with the advent of computer games that the distinctions between games/game-play and narrative have become unclear, and the study of games has shifted focus more strongly towards games as a type of fiction. Computer games span a range of forms, varying from the highly narrative, to the highly non-narrative. This range of perspectives, from

J. Schaeffer et al. (Eds.): CG 2002, LNCS 2883, pp. 292–307, 2003.

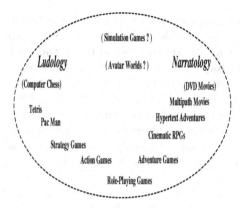

**Fig. 1.** Games fall within a continuum from the ludological to the narratological.

the ludological[1] to the narratological[2] is depicted in Fig. 1. At the ludological extreme are computer implementations of traditional board games, and abstract game forms that rely upon the active dynamics of a computer implementation, but have little or no function in terms of representing a fictional world. At the narratological extreme are highly story-oriented productions, from multi-path movies to hypertext stories and adventures.

The more dominant computer game forms lie in a continuum between these extremes, using different approaches for the integration of narrative and patterned game-play. This may lead to a perceived tension between game-play and story in computer role-playing games (see, for example, [2]), although this ultimately amounts to a matter of style and taste; players will gravitate to the games that satisfy their preferences in terms of narrative framing and its relation to the core game-play experience (or the game-play gestalt, as an essentially nonnarrative interaction pattern [3]). There are game forms more concerned with simulation, or realizing multi-user, on-line avatar worlds, that strain the definition of games, although similar tensions between game play and story also occur in these systems.

More consistently frustrating for players than the game-play/narrative tension is the experience of moving through a rich game world, completing the tasks, meeting the challenges, completing the quests, etc., only to come to a point that is unpassable since some unknown critical action has not been performed at a previous point in the game. This may result in the player needing to experiment with reloading past game states in order to try to discover the "correct" sequence that needed to be completed, restarting the game, or floundering around with no idea of what went wrong or how to move forward. If the frustration level is too high, the player will stop playing the game, leaving her with the impression

---

[1] Ludological means pertaining to the scientific study of games, i.e., ludology, from the Latin ludus, meaning play [1].

[2] Narratological means pertaining to the scientific study of narrative, i.e., narratology. See http://www.narratology.net.

that it was too hard, impossible to understand, too tedious to resolve, and/or badly designed. This is a problem of story logic. Story logic becomes more explicit in the movement from the ludological pole towards the narratological pole of Fig. 1. However, story logic problems are most prevalent in the intermediate zones, where story design is obscured by game-play and simulation.

In this paper we consider the extent to which problems in story logic can be attributed to design characteristics that can be analyzed in software engineering terms, i.e., in terms of dependency relationships among story elements. We present a simple example of a stereotypical game quest, together with two undesirable outcomes that have actually been experienced by the authors in real game play. We present a semi-formalized representation of the causal relationships involved in the quest and the problematic situations. We then review the concepts of coupling, functional dependency, and normalization theory as they have been used in the history of software engineering methodologies, and explore the applicability of these concepts to story logics. A number of principles for normal forms for story logics are presented. We also consider the applicability of object-oriented concepts to story telling, and what this means in terms of system architectures for games.

## 2    The Dead End – Errors in Game Logic

The forms of games in which stories dominate the player experience are branching narratives based upon a hypertext model (e.g., multi-path movies, http://www.brilliantdigital.com/solutions/movies). At the other extreme, strong game-play can be experienced in very abstract games having no story at all, such as traditional board and puzzle games. Problems in game logic of the kind discussed here lie in intermediate forms, where game designers wish to impose a specific series of plot points in order to create particular story structures. Game players are given high levels of freedom in interaction, especially in relation to the exploration of the landscapes and architectures of a game world.

A common strategy for imposing a specific story sequence within a highly interactive game is to make progress in the game conditional upon completing a specific sequence of actions or plot points. This is where design problems may arise. Consider, for example, the following clichéd scenario. The player plays the part of a fantasy protagonist (the player character, or PC) moving through a medieval world inhabited by various helpful or enemy non-player characters (NPCs). The designers have created a quest: an ailing wizard will give the player a key to an underground cave system in return for killing an old enemy dragon that the wizard has failed to destroy in time before his own death, and which therefore now threatens the local town. This is programmed into the game. However, as a function of the virtual geography of the game, the player's interactive possibilities for traversing this geography, and the way the quest is imposed upon the player, several story outcomes are possible. First, the outcome intended by the designers:

1. The player meets the wizard and is given the quest. The player follows the wizard's instructions, finds, battles and defeats the dragon, returns its head to the wizard, and is rewarded with the key. The player can now continue in the game by seeking and entering the underground cave system to further her higher level quest.

As a simple example of a design problem, however, we consider the case when the player has enough freedom in the environment to go to the dragon's lair before going to the wizard's lair. This occurs in part due to the game designers' attempts to simulate a world, since one solution for the dependency problem (generally undesirable for players) is to restrict freedom of movement in the world to enforce the required sequence of events. One design solution for imposing the intended story without restricting freedom of movement is to not instantiate the dragon until the wizard has been encountered. This leads to the following possible outcome:

2. The player goes into an empty lair (no dragon yet). The player goes on to receive the wizard's instructions. The player is now looking for a dragon in a lair, but does not go back to the lair because it was previously found to be empty. The player searches through all reachable but previously unexplored parts of the terrain. No dragon is found. Either the player must revisit all previously visited areas of the map just in case one of them was the lair which is now by chance inhabited, or will give up, having no options to go anywhere new, without understanding why the dragon is not to be found.

To avoid this, designers allow the dragon to be in its lair before the player character visits the wizard, leading to another possible outcome:

3. The player goes into the dragon's lair, battles and defeats the dragon. The player then goes on to meet the wizard and is given the quest. However, the quest cannot be completed, because the dragon no longer exists. The player must reload a game state prior to the point of defeating the dragon, and go through the battle again, this time after visiting the wizard. If no suitable state has been saved, the player must restart the game, or stop playing.

Of course, there are solutions that avoid these outcomes. For example, to avoid outcome 2, the wizard can explain where the lair is, and the designers can hope that this can be related to the player's memory of the lair if it has already been visited. This can however detract from the fun element of finding the lair as part of the quest, and also raises the question of why the player didn't run into the dragon along the long and winding route from the lair to the wizard. The solution violates the expected existential logic of the world for the sake of a specific story sequence. Outcome 3 can be avoided by having the wizard reward the player's action of killing the dragon even though the action was performed before the player was instructed to do it, so it is no longer necessarily a quest. This is a matter of weakening the imposition of the designer's desired story sequence, for the sake of a more plausible simulation of a world.

While these solutions are possible, they and the situation leading to them raise the question of whether there is a more general and coherent method for understanding and resolving this kind of problem in story logic. Here we propose two methods. First, we consider the analysis of causal dependencies in the game logic, including the notion of causal coupling, and a design methodology based upon the minimization of causal coupling by causal normalization. This approach is appropriate when specific story structures (such as quests) are desired as an intrinsic part of the game form. The second approach, that of object-oriented story telling, is desirable when the world is to function more as a simulation, in which stories are an emergent and retrospective phenomenon.

## 3    Causal Modeling for Game Logics

The story example above can be represented in the following way. We use the notation:

$$E1(P \text{ meets } W \text{ and receives } Q) \rightarrow E2(P \text{ goes to } L)$$

to represent a causal relationship, where:

E1 and E2 denote events 1 and 2, respectively.

P refers to the player.

W refers to the wizard.

Q refers to the quest instruction.

L refers to the dragon's lair.

$\rightarrow$ is a causal relationship, where the event(s) on the left hand side of the arrow causes the event on the right hand side of the arrow.

We have not completely formalized this notation, nor adopted an existing causal logic. This level of formalization is sufficient for the analysis presented here, i.e., as a tool for the analysis of patterns of causal dependency.

Using this notation, we present outcome 1 above in terms of the following sequence of causal dependencies desired by the game designers:

Sequence 1:
  E1(P meets W and receives Q)
  $\rightarrow$ E2(P goes to L)          where L denotes the dragon's lair
  $\rightarrow$ E3(P meets D)          where D denotes the dragon
  $\rightarrow$ E4(P defeats D)
  $\rightarrow$ E5(P returns victoriously to W)
  $\rightarrow$ E6(P receives R)          where R denotes the reward.

A crucial issue in game design is whether or not to impose these kinds of causal relationships as rules that the player must obey. This becomes very complex, since a decision to impose causal rules raises the need for desirable formal properties, such as soundness, completeness, decidability and consistency (see [4]). The undesirable outcomes 2 and 3 above result from the lack of these properties for the causal system expressed in Sequence 1. For instance, the system is incomplete in the sense that E2 can be true without being derived from (or caused by) E1. The presence of the player as an active causal agent in the game world, and the function of that world as a simulation, make it impossible to formalize all possible simulated causal relationships in that world. Thus, a formal approach to proving desirable behavior is generally not feasible.

Examining outcome 2 above, in which the user encounters the lair without the dragon prior to encountering the wizard, we find the causal sequence:

Sequence 2:

E2

$\rightarrow$ E1

$\rightarrow$ confusion!

Outcome 3 involves the sequence:

Sequence 3:

E2

$\rightarrow$ E3

$\rightarrow$ E4

$\rightarrow$ E1

$\rightarrow$ E2

$\rightarrow$ E5

$\rightarrow$ dead end!

Since these problems arise from undesirable patterns of causal dependency, it may be feasible to apply systematic methods from software engineering practice, based upon dependency analysis, as an aid to story logic design.

## 4    Coupling, Dependency, and Normalization in Software Engineering

The analysis of dependencies underlies methodologies for system development within a variety of programming and development paradigms. This includes structured development (analysis and design) for procedural software systems [5], normalization of relational database systems [6, 7, 8, 9], and normalization of rule-based knowledge systems [10, 11]. Structured development for software systems is based upon an analysis of the data flow relationships within an application, as captured by hierarchical data flow diagrams (DFDs; see [12, 13]). A data flow diagram is a representation of the data within a system, and how data flows between different transforming processes. Structured software development methodology has traditionally used DFDs to represent data flow as part of the analysis of a system, and the resulting DFDs have then been used as a basis for hierarchically defining program modules. In developing this approach, Yourdon and Constantine [5] articulate the concept of coupling, as the degree to which one functional module of a system must know about another, which then amounts to the likelihood that modifications to one module will effect the operation of another in some way. Coupling can be further classified into data coupling and control coupling, where data coupling involves a data dependency between modules (modifying a data value in one module changes the data outputs of another), and control coupling involves a control dependency (the behavior of one module influences the control sequencing of another). A good structured design amounts to creating a system with a minimum of coupling between modules, so that future modifications to a module will have a minimum impact upon the operation of the rest of the system. Structured analysis and design techniques focus on data flow relationships, and seek to minimize data functional dependencies between modules by defining systems having a structure that reflects data dependency.

Database normalization involves constructing relations for relational databases that reflect the functional dependencies within the data domains. A functional relation from a domain A to a domain B means that a value within domain A uniquely determines a value within domain B. Values within domain B can have more than one determinant in A, but each value in A has only one dependent value in B. Database normalization is a process of eliminating redundancy and inconsistent dependencies within relational database designs by following the patterns of functional dependency within the data domains [9]. This can be seen to be a similar process to the minimization of coupling in structured analysis and design (or identical at an abstract level), the difference being that in pure database systems values are explicitly represented rather than being calculated dynamically.

Normalization theory is extended into rule-based systems by Debenham [10, 11], in this case dealing with the same or similar kinds of functional dependencies, but expressed in terms of abstract declarative relations, instead of database tuples. These dependencies are truth-functional dependencies, and normalization amounts to the minimization of truth-functional coupling. A simple example is the separation of repetitive premise subsets into distinct rules, analogous to Codd's first normal form for database systems. For example, consider the simple propositional rules:

$$\text{Rule 1: A, B, C, D, E :- F}$$
$$\text{Rule 2: G, H, C, D, E :- I}$$

where capitalized letters represent simple propositions, and :- represents logical implication. The occurrence of the subset of premises {C, D, E} in both rules suggests an interdependency between the propositions within the subset. This creates an update hazard, since any change to this interdependent set must be reflected everywhere that it occurs. Rules 1 and 2 are therefore truth-functionally coupled in the sense that the {C, D, E} subset represents a common meaning, which becomes ambiguous if the expression of that meaning becomes inconsistent in different rules. To avoid this, the rules can be normalized by extracting the subset as a new rule, and replacing the subset by the head of the new rule in rules 1 and 2, giving the new rule set:

$$\text{Rule 1: A, B, J :- F}$$
$$\text{Rule 2: G, H, J :- I}$$
$$\text{Rule 3: C, D, E :- J}$$

The meaning of the {C, D, E} subset is now encapsulated within Rule 3, and changes to the subset only have to be made in one place. As with structured software design and database normalization, the representation structures reflect the functional dependencies within the system.

Object-oriented software development methodologies have superseded many of the earlier methodologies, as a more coherent and universal method of addressing the standing issues of minimizing modular coupling and providing a principled approach to system development [14]. Object-based approaches provide a consistent methodology through all phases of software development, since objects identified during analysis may provide the foundation for objects in the design and implementation of systems. An object encapsulates both data and control, and provides what should be well-defined interfaces through which other modules can use their functionality. Object-based systems typically also use the concept of inheritance, allowing system constructs to be defined as classes at various levels of abstraction, with lower-level constructs inheriting features, data, and/or functions (methods) from higher abstraction levels. An object is then an instance of a class, having its own internal data (state information), and interfaces defined as methods by which other objects can interact with it. Ideally, a system composed of a set of interacting objects has minimal control and data coupling between its elements.

In the next section we examine the meaning of principles of dependency analysis for story logics. The issue of object-orientation in story structure is examined in the section after that.

## 5    Causal Normalization for Games

Examination of Sequence 1 together with Sequences 2 and 3 shows that these outcomes result from dependent and independent relationships that are not clearly represented in Sequence 1. In particular, outcome 2 results from a dependency between E1 and E3. That is, the player can only meet the dragon if she has first encountered the wizard. Outcome 3 results from a dependency between E4, E5 and E1; the player can only return to the wizard after killing the dragon and receive a reward if the wizard has been visited before the dragon was killed. In both cases, the ability to enter the sequence at E2 undermines the intended story logic.

This kind of causal influence resembles control and data coupling phenomena in software engineering, and unnormalized relationships in databases and rule-based systems. In all cases, there are dependencies that cut across the intended, explicit, or modeled dependencies of the system. For story logics we can refer to this as causal coupling, informally understood as a causal relationship that is excluded from a high-level causal model of the story logic. If causal coupling is ignored, Sequence 1 could be represented by a sequence of separate causal steps, as follows:

Sequence 4:

E1(P meets W and receives Q)   → E2(P goes to L)

E2(P goes to L)                → E3(P meets D)

E3(P meets D)                  → E4(P defeats D)

E4(P defeats D)                → E5(P returns victoriously to W)

E5(P returns victoriously to W) → E6(P receives R)

If each step is treated as a causal rule within the system, then the occurrence of a cause event must be followed by the occurrence of an effect event. This allows Sequence 1 to be side-stepped to different degrees, due to the nature of the game world as a simulation in which the traversal of the world by the player character, or the player character's effect within the world (e.g.. via magic), is not constrained in terms of this causal rule set. For instance, the player might remotely defeat the dragon by magical or other indirect means, without ever having met either the wizard or the dragon. Then E4 is satisfied, and by the steps E4 → E5 and E5 → E6, the player receives the reward from the wizard.

If the designers wish to impose the strategies that lead to outcomes 2 and 3, we can explicitly represent what were the hidden dependencies between E1, E3, and E5 in Sequence 1 by modifying the causal steps of Sequence 4 as follows:

Sequence 5:

E1(P meets W and receives Q)   → E2(P goes to L

E1(P meets W and receives Q)

         and E2(P goes to L)   → E3(P meets D)

E3(P meets D)                  → E4(P defeats D)

E1(P meets W and receives Q)

         and E4(P defeats D)   → E5(P returns victoriously to W)

E5(P returns victoriously to W) → E6(P receives R)

Now it is possible to see that the causal relations expressed within the second rule include a hidden relation within the causes analogous to that addressed by Boyce-Codd Normal Form (BCNF) in database theory [9]. This is because E3 is caused by E1 and E2, while E2 is an effect of E1. The hidden dependency creates precisely the kind of anomaly observed in outcome 2, that if E2 occurs without E1, there is no specified outcome. Similarly in rule 4, if E4 occurs without E1, there is no specified outcome, although in this case there are no dependencies between E1 and E4.

Modeling the previously hidden dependencies clarifies the existence of undesirable game states. It can also be asked if there is a methodology analogous to normalization that can be applied to causal models of this kind that might prevent or help to prevent these anomalies from occurring. Applying the principle of BCNF to the second causal step of Sequence 2, we could break it down into the first two separate steps of Sequence 4:

$$E1(\text{P meets W and receives Q}) \rightarrow E2(\text{P goes to L})$$
$$E2(\text{P goes to L}) \rightarrow E3(\text{P meets D})$$

These relationships are normalized in a form analogous to BCNF, eliminating interdependencies between the causes within a single relationship. Now we are back in the situation of no longer imposing the logic that leads to our earlier outcome 2. It appears that the imposition of a desired story sequence creates the unnormalized story structure responsible for the undesirable outcome.

This analysis suggests that it may be possible to define a general set of normal forms for the causal relationships in story logics. Assuming a representation of causal relationships that links a set of causes to a specific effect, such normal forms for story logics should at least:

- extract recurrent subsets of causes representing independent events as separate cause-event relations (an analog of Codd's first normal form for relational databases),
- eliminate irrelevant causes from cause sets (an analog of Codd's second normal form),
- separate multiple effects of a common set of causes into multiple relations, one for each effect (an analog of Codd's fourth and fifth normal forms), and
- eliminate interdependencies between causes within any single relation (an analog of BCNF).

Developing these ideas into a more precise, extensive, and formalized list of normal forms for causal relations is beyond the scope of this paper. Such a task will depend upon settling upon a specific representation for story logics. This should be able to be done for any explicit representation of causal dependencies in stories, and applying the above normal forms to the analysis of those dependencies. Using causal normalization, it should be possible to eliminate story logic anomalies for games in which the story logic covers all possible traversals of the game world. These are the games close to the narratological pole of Fig. 1.

# 6   Normalization Methodology and System Architecture

Database normalization theory derives from the relational formalization of database functionality. Relational databases are designed in accordance with this model, so the abstract methodology has a deep relationship to the operational semantics of a relational database. Applying a normalization method to

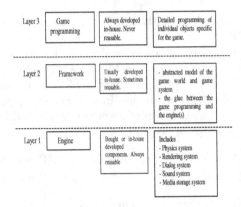

**Fig. 2.** Layers of text in the code level in computer games.

the story logic of a computer game requires a similar mapping from a representation that is convenient to normalize, to the semantics of that representation in terms of the simulated events and player experiences of the game world. What is required, therefore, is a mapping between different levels of interpretation of "the game." For a story-driven computer game there are three levels that internally form text layers and structures:

1. The *code level*, consisting of engines, a game framework and game programming. These together define the mechanics, the virtual geographical structure of the game world, and the conditions for the overall story and its deep structure.
2. The *narrative level*, consisting of the overall story, deep structure and the specific story carrying objects, which in turn can manifest the story, possible side quests, and internally independent stories.
3. The *discourse level*, consisting of the sequential order that is created between the parts of the narrative simultaneously with the players movement through the game. It is at this level that the surface structure of a game text can be monitored.

In these terms, given a game engine constituting the code level, the narrative level may be a primary concern for the game designer (who might implement the level as data to be inserted within the code level). The discourse level is the game play experience and the experience of the game as a story on the part of a player. Causal normalization is a narrative level methodology to help to ensure that the data entering the code level creates a coherent story experience at the discourse level. Effective normalization must be treated as an issue of defining a coherent and usable narrative level methodology together with a clear migration path to the semantics of narrative representation within the code level, just as a normalized logical relational database model has a clear mapping to table structures within a relational database (notwithstanding pragmatic variations in implementation). Before this is possible, it is necessary to devise clear representations for the narrative level and its semantics.

In general, what we have referred to here as the code level of a game can itself be subdivided into three levels, as shown in Fig. 2. The lowest layer of the code level is the engine, consisting of very general functions, such as the rendering system interface, animation interfaces, collision detection, terrain or portal management, a dialog system, and media storage and access. Above this is the framework for the game, which is the level of abstract representations of game structures, such as game agent classes, behavior controller classes, an event management system, and a communication (i.e., message passing) system. The engine may be general across many game genres, while the framework may be more genre-specific. On top of these levels is the specific game programming, consisting mostly of data and instance definitions for realizing a specific game. These layers together present the media that to the player is the game.

How this architecture is built and where the borders are between the layers is different from game to game, from developer to developer, and from genre to genre. It also depends upon the technical platforms and environments of the game.

# 7   Object-Oriented Story Telling and the Minimization of Causal Coupling

For story logics within highly interactive game worlds, where issues of story do not totally dominate the world simulation functions of a game, the concept of normalization is not as clear as, for example, the case of database systems. Within these worlds, story logics are generally not complex enough to justify a full causal logic, and story structures are often sparse in relation to the size of a game world and the overall cognitive density of the game-play experience. It is in this kind of world that cases like that of rule 4 of Sequence 5 above cause a problem beyond the scope of normalization. In this case, there is a straightforward stipulation that the player cannot receive the reward without visiting the wizard before killing the dragon. This may be the designer's interpretation of the personality of the wizard; the player must act as desired, or miss out on the prize that will unlock unexplored areas of the game world. In general this would be a perverse and undesirable discovery for the player, and we need a better method for reducing such chains of dependency for more flexible game play. This can be accomplished by pursuing object-oriented story telling, as a strategy for designing game entities in terms of story potential, rather than imposing causal dependencies.

Object-oriented story telling is an approach in which all objects in the world have integrity and contain their own stories, functions, possible developments, and possible responses to actions conducted by other objects that influence them. That an object has integrity means that the information available in the object is only available through the object, and all information retrieval or data access is implemented by objects. For object-oriented story telling, this may function as follows.

If a player object, controlled by the player, comes close to a non-player character (NPC) object in the game, communication between the player and the NPC is partly defined by the characteristics of the player object, and partly by the characteristics of the NPC object. Depending upon what it has been through earlier in the game, the player object can ask questions governed by the events that have become the history (recorded past) of the player object. The information that the NPC object in its current state can give is dependent on its own history, the location in which the player object encounters it, the time of day it is in the game, etc. Thus the content of the dialog and amount of information transferred from the NPC is dependent on a combination of conditions emerging from the meeting between the two objects. By maintaining the integrity of the object, false or confusing causalities need not occur. Actions of the NPC that may be undesirable from an overall story perspective can be avoided by encapsulating knowledge within appropriate objects. For instance, the existence or not of a dragon does not need to be conditioned upon remote interactions that have nothing directly to do with the dragon. An isolated action or state variable that the player object carries can directly correspond with an opportunity to activate a specific response.

This situation corresponds exactly to that discussed in the case of Sequence 4 above, and requires the designers to abandon the imposition of pre-specified sequences. If the player character goes to the wizard after killing the dragon and without having received the quest, she is nevertheless entitled to the reward. This follows from the simulated intent of the wizard to reward the act of killing the dragon with a key, without making knowledge of this intent a cause within the game world.

This kind of object-oriented approach means that it is unnecessary to create an overall story structure having a large number of conditions for which the internal relations must be correct in order to activate the specific response. The system governing the story logic will be more immune to the kind of causal logic problems discussed above. The advantage of this from a story perspective is that it is possible to construct an NPC and define exactly how it should behave according to its characteristics, the operations that can be performed on it, and the internal conditions set for releasing information to a player character. The advantage from a game-play perspective is that this NPC and other objects will seem more natural and intelligent, since there are no false casual relations conditioning their behavior.

# 8    Mixed Forms: Object-Oriented Story Telling and the Imposed Quest

In a highly simulation-based game, built according to principles of object-oriented story telling, a quest or a story becomes a history of interaction, as suggested by Oliver [15]. Rimmon-Kenan [16] derives a definition of story from Genette's concept of histoire [17], but stresses the chronological aspect of the term: " 'Story' designates the narrated events, abstracted from their disposition

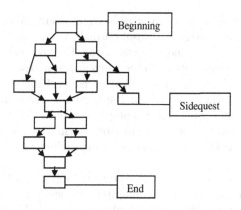

**Fig. 3.** A network structure depicting casual relations between events.

in the text and reconstructed in their chronological order, together with the participants in these events." In a story-driven computer game in the genre of the adventure game, and within the high-level structure of other genres such as role playing and action games, there is a chronological order in which the events occur within a particular player's experience. This order depends, however, on the nature and implementation of the story generating structures at the code level, which are usually not strictly linear and contain more or less possible chronological variations in what Anna Gunder would call the omni-story [18]. The omni-story in turn contains all possible real stories, that is, all possible chronologically ordered sequences of events. This must be regarded as a combination of possible variations both in the chronological ordering of events, and in the necessity or contingency of occurrence of events.

In a simulation-based game, the omni-story is vast and effectively unknowable. Massively multi-player on-line role-playing games (MMORPGs), for example, contain unlimited story potential. However, a role for a player character is still typically understood in narrative terms, providing purposes for the character in the game world, generally in the form of quests. A picture of the events that can occur and their possible causal order in a story between a start and an ending might look like the directed network structure shown in Fig. 3.

Each square in Fig. 3 represents an event. The arrows may represent casual relations between events. In order for an event to happen, the events that are represented by the boxes that have an immediate above connection must have happened. In a model like this, all existing casual relations are important. Thus there are only three existing hierarchical levels:

1. cogent relations leading to the end of the story,
2. relations that are only cogent for experiencing a side-quest, and
3. events that are not cogent at all and thus not represented in the model.

In these terms, the player's freedom to move their character beyond the structures of the predefined narrative reflect the simulation functions of the game and game world, representing a realm of non-cogent events from the perspective of the designed narrative patterns. Causal normalization is applicable to the narrative model, regardless of the non-cogent events. But to such a simple causal map must be added the complexity of the contingency or necessity of causal relations, and relations of joint sufficiency and joint necessity. Only then can a causal map represent possible variations both in the need for and order of occurrence of causally related events. This greatly complicates both the design process of narrative structures, and the processes of story normalization, suggesting that for simulation-based worlds, object-oriented story telling methods are much easier to handle.

In a game that uses object-oriented story telling, a high-level narrative model could be interesting as tool for planning possible story experiences (or as a tool for analyzing the game). But any such plan should only be regarded as a picture of a subset of story experiences possible within the game world. It should not be imposed upon the player or specified as an *a priori* set of dependencies between game objects.

## 9   Conclusion

Problems of story logic encountered in computer game play are a consequence of a lack of coherent game development methodologies. As discussed in this paper, the problem of defining a coherent game development methodology can draw from principles of software engineering. However, developing complete solutions must involve the development of production environments in which clear methodological principles have a coherent translation into designs and implementations that preserve the qualities of good designs. For games with highly constrained narrative possibilities, causal (or story logic) normalization provides a methodology for avoiding dead ends or confusing situations in stories. For highly interactive, simulation-based game worlds, however, it appears that the idea of imposing predefined story sequences, even branching sequences, must be largely, if not entirely, abandoned. Instead, we require object-oriented methods for encapsulating interesting behavior and states of game entities constituting a deep and non-sequential structure of story semantics.

## References

1. Huizinga, J.: Homo Ludens: A Study of the Play-element in Culture. Routledge & Kegan Paul Limited, London (1949)
2. Aarseth, E.: Cybertext: Perspectives on Ergodic Literature. Johns Hopkins University Press (1997)
3. Lindley, C.: The gameplay gestalt, narrative, and interactive storytelling. In: Computer Games and Digital Cultures Conference. (2002)
   http://www.interactiveinstitute.se/zerogame/pdfs/CGDClindley.pdf.
4. Frost, R.: Introduction to Knowledge Base Systems. MacMillan (1986)
5. Yourdon, E., Constantine, L.: Structured Design: Fundamentals of a Discipline of Computer Program and System Design. Prentice-Hall (1979)

6. Codd, E.: A relational model of data for large shared data banks. Communications of the ACM **13** (1970) 377–387
7. Codd, E.: Normalized data base structure: A brief tutorial. In Codd, E., Dean, A., eds.: ACM SIGFIDET Workshop on Data Description, Access, and Control. (1971) Also published in/as: IBM, Report RJ935 Nov. 1971.
8. Codd, E.: Further normalization of the data base relational model. In Rustin, R., ed.: Data Base Systems. Prentice-Hall (1972) Also IBM Research Report RJ909.
9. Date, C.: An Introduction to Database Systems. 3rd edn. Addison-Wesley (1981)
10. Debenham, J.: Knowledge Systems Design. Prentice Hall (1989)
11. Debenham, J.: Knowledge Engineering: Unifying Knowledge Base and Database Design. Springer Verlag (1998)
12. De Marco, T.: Structured Analysis and Systems Specifications. Yourdon Inc. (1978)
13. Gane, C., Sarson, T.: Structured Systems Analysis. Prentice-Hall (1979)
14. Booch, G.: Object-Oriented Analysis And Design With Applications. 2nd edn. Benjamin Cummings (1994)
15. Oliver, J.: Polygon destinies: The production of place in the digital role-playing game. In: Computational Semiotics for Games and New Media. (2001) http://www.kinonet.com/conferences/cosign2001/.
16. Rimmon-Kenan, S.: Narrative Fiction: Contemporary Poetics. Taylor & Francis Books Ltd (1998)
17. Genette, G.: Narrative Discourse — An Essay in Method. Cornell University Press (1983) Translation by J. Lewin.
18. Gunder, A.: Berättelsens spel. Berättarteknik och ergodicitet i Michael Joyces afternoon, a story. Human IT **3** (1999) 27–127 http://www.hb.se/bhs/ith/3-99/ag.htm.

# A Structure for Modern Computer Narratives

Clark Verbrugge

School of Computer Science, McGill University, Canada
clump@cs.mcgill.ca

**Abstract.** In order to analyze or better develop modern computer games it is critical to have an appropriate representation framework. In this paper a symbolic representation of modern computer narratives is described, and related to a general model of operational behaviour. The resulting structure can then be used to verify desirable properties, or as the basis for a narrative development system.

## 1 Introduction and Overview

In order to analyze or better develop modern computer games it is critical to have an appropriate representation framework. Existing frameworks are demonstrably inadequate in this respect. In this paper a symbolic representation of modern computer narratives is described, and related to a general model of operational behaviour. The resulting structure can then be used to verify desirable properties of computer narratives, or as the basis for a narrative development system.

### 1.1 Motivation

Many modern computer games build gameplay based to varying degrees on computer narrative—game progression is defined through a narrative sequence of events. For some genres, particularly "adventure" and to a slightly lesser extent "role-playing" games (RPGs), the gameplay consists almost entirely of building/following a narrative, and a properly-constructed narrative structure is paramount.

As any avid game player is aware, narrative development is an imperfect process; mistakes are evident in unsatisfying plot holes and non-sequiturs. More severe consequences can include unwinnable situations, game unbalancing effects, or even program failure. A rigorous system for developing and describing narratives can aid in reducing or eliminating these problems.

### 1.2 Narratives and Graphs

In its simplest form, a narrative is a sequential presentation of events. States, important points or actions are necessarily described in order over the time of presentation—a linear plot graph, for example. In this sense a narrative is a total ordering of events.

J. Schaeffer et al. (Eds.): CG 2002, LNCS 2883, pp. 308–325, 2003.

However, the presentation of a narrative is often regarded as secondary to its internal semantics. While all narratives will unfold sequentially over time, most people understand the events to be "actually" ordered by one or more internally-consistent ordering frameworks. Internal, narrative time, for instance, can allow for two events to occur concurrently ("meanwhile...."); the relating of these two events is then an interleaving of the narrative's necessary ordering.

**Plot DAGs.** Internal order-relations, such as narrative time or physical causality, produce a "plot DAG" (Directed Acyclic Graph). The plot DAG is a graph representation of the significant events/states in a narrative ordered by the internal system: each node represents a state, and directed edges between nodes represent necessary precedence. Since events in a traditional narrative do not repeat, the overall structure is acyclic. A simple example for the ubiquitous door locking "puzzle" found in adventure and RPG games is shown in Fig. 1.

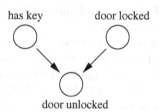

Fig. 1. A DAG structure describing a simple open door task. Multiple incoming edges form an "AND" relation—the door must be locked *and* the key possessed before it can be unlocked.

The above example illustrates a fundamental representation gap: in the plot DAG model, the door cannot be (infinitely) unlocked and re-locked, since that would imply a cycle. Narratives in computer games also tend to incorporate at least some amount of choice, and this is also not representable in the DAG system. Incoming edges in our plot DAG are combined as an "AND" relation, and so we cannot represent the possibility of the door perhaps also being unlocked by picking the lock. In a typical usage, the designer may avoid these issues by including this information as side-notes to their design, or they may attempt to abstract further (collapse the entire situation to just "get door unlocked"). With such limitations, however, plot DAGs are clearly an unsatisfying solution to the problem of representing computer narratives.

## 1.3   Roadmap

Section 2 discusses related approaches. Section 3 begins the presentation of our formalism. The initial, Petri Net derivation is explained in Sect. 3.1, and developed into a hypergraph model in Sect. 3.2. The final structure, along with path and other dynamic actions is defined in Sect. 3.3.

Section 4 discusses the use of this formalism for discovering various properties important to narrative development. The model is then applied to a more complex example in Sect. 5, where we show how our model and approach can be used to describe the initial portion of an actual computer narrative. Directions for further research are then discussed in Sect. 6, and conclusions drawn in Sect. 7.

## 2   Related Work

There are remarkably few academic computer science studies of modern computer games, and we are aware of no other formal attempts at representing computer game narratives. The concept of plot DAGs as a design tool for narrative computer games was discussed during the mid-1990's in the usenet discussion group `rec.arts.int-fiction` [1, 2, 3] and in an on-line trade journal [4]. The idea is sometimes credited to the "Oz" group's work on interactive drama [5]. Of course with DAG models cycles and choice cannot be represented, and must be attached as external information.

Approaches to analyzing *text* narratives, however, do exist. For example, Burg *et al.* use constraint logic programming to analyze time-relations in a William Faulkner short story [6]. High-level narrative development frameworks have been described by Strohecker and Brooks [7, 8, 9]; these systems are aimed at all aspects of (perhaps interactive) story development, and not the specifics of representation.

Low-level game "construction kits" also exist, automating the repetitive and/or stock tasks necessary to actually produce a working computer game. These are also typically not representation-oriented; the popular TADS [10] system, for instance, represents narrative implicitly through the control structure of a high-level object-oriented language.

Our hypergraph model is derived by simplifying a form of Petri Net. Similar Petri Net models have been used in modelling "workflow," the abstract relations and precedence requirements of tasks in a business environment [11]. Van der Aalst's workflow model is based on an enrichment of Petri Nets (transitions are augmented by the source of the firing impetus), and naturally does not discuss or define properties important to narratives per se.

## 3   Formalisms

Our interest is in a simple formalism that can model normal narrative progression (internal-dependencies), including cycles and choice. To keep our formalism simple, we do not attempt to explicitly model time[1]. Mechanics of player inter-

---

[1] E.g., a game action which must be completed in a specific real-time interval: "you have 30 seconds to defuse the bomb...". Real-time activities do occasionally appear in computer narratives, but at least in narrative-centric games like adventure games, real-time situations are not well-liked [12].

action (how actions and choices are actually presented and resolved) and other extra-narrative facets are also not considered.

The model we develop is derived from a form of Petri Net. The next subsection gives a quick introduction to Petri Nets, and is followed by an explication of our system.

## 3.1 Petri Nets

**Definition 1.** *A* Petri Net *is a 5-tuple,* $(P, T, E, W, M)$, *where* $P$ *is a set of* place *nodes,* $T$ *is a disjoint set of* transition *nodes* $(P \cap T = \phi)$, *and* $E \subseteq (P \times T) \cup (T \times P)$ *is a set of directed edges going between places and transitions. Edges are weighted* $(W : E \to \mathbb{N})$, *as are places via a "marking" or token-assignation function* $(M : P \to \mathbb{N} \cup \{0\})$.

Central to the model is the concept of a transition-node being "ready to fire," and the marking transformation that occurs through actual firing. In order for transition $t$ to be ready to fire, there must be enough tokens in each incoming place $p$ to satisfy the edges weights: $W((p, t)) \leq M(p)$. Actual firing conceptually removes tokens from the incoming places according to edge weights, and adds tokens to outgoing places according to edge weights. Note that firing is atomic; substeps in the firing of one transition cannot be interleaved with other transition firings. Also note that a place attached as both input and output to a firing transition will have its marking changed according to the difference between output weights and input weights.

**Definition 2.** *A 1-Safe* Petri Net *has edge weights of 1 in all cases,* $\forall (x, y) \in E, W((x, y)) = 1$, *and guarantees that markings are always either 1 or 0. If* $M_0 \xrightarrow{t_0} M_1 \xrightarrow{t_1} \ldots$ *is a sequence of markings generated by firing transitions* $t_0, t_1, \ldots$, *then:*

$$\forall i \geq 0, \ \forall p \in P, \ M_i(p) \in \{0, 1\}.$$

A variation on 1-Safe firing rules further requires all output places of a transition to be empty in order for it to fire. This does not affect expressive power, and for simplicity we will sometimes illustrate based on firing rules of either form. We will, however, explicity permit self-loops, and also assume that our nets are "$T$-restricted"—transitions do not map from or to empty place sets. The reader is referred to other texts (e.g., [13]) for more comprehensive introductions to Petri Nets.

1-Safe Petri Nets model finite state systems, and are sufficient to represent all narrative structures in which we are interested (formal expressiveness results can be found, e.g., in [14, 15]). For instance, an expanded version of the simple lock "puzzle," discussed in Sect. 1.2 is illustrated as a Petri Net in Fig. 2. Transitions are drawn as small lines, places as circles, and the marking by the presence/absence of a black dot in each place. To unlock the door, the door must be currently locked and the key has to be available; firing the "Unlock door" transition changes the door state to "door unlocked," but does not affect the presence of the key.

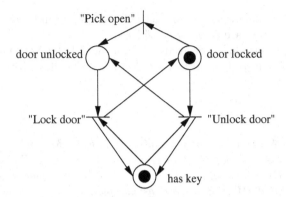

**Fig. 2.** A 1-Safe Petri Net describing a simple lock "puzzle" in a narrative.

Note that as opposed to the DAG example in Fig. 1, here the unlocking of the door is reversible—the graph is cyclic. Choice has also been incorporated; the lock is (un)lockable with the key, *or* it can be unlocked by picking the lock (firing the transition labelled "Pick open").

A Petri Net representation is quite flexible, and allows one to model more complex behaviours than a plot DAG. However, the bipartite structure and complex transition (firing) behaviour of Petri Nets adds unnecessary syntactic baggage to our model. In Fig. 2, for instance, edges from the lock/unlock transitions are needed to restore the token within the "has key" place—this represents the fact that (un)locking the door does not alter the possession of the key. In essence, the "has key" place functions here not as a transformable state, but as required *context* for other actions. The next section introduces a simplified representation that formalizes our requirements without requiring explicit token movement.

### 3.2   Hypergraphs

Our formalism is based on 1-Safe Petri Nets, and is a kind of *directed hypergraph:*

**Definition 3.** *A hypergraph is a graph $(V, E)$, with the property that edges (hyperedges) can connect more than just 2 vertices: $E \subseteq \mathcal{P}(V)$.*

**Definition 4.** *A directed hypergraph is a directed graph $(V, E)$, with the property that (hyper)edges can connect more than one tail vertex to more than one head vertex: $E \subseteq (\mathcal{P}(V) \times \mathcal{P}(V))$. The tail : $E \to \mathcal{P}(V)$ and head : $E \to \mathcal{P}(V)$ functions extract the input or output sets from a given hyperedge.*

A directed graph has directionality assigned to each connection between a hyperedge and a vertex. Note that some authors define a directed hypergraph by designating a single, distinct head for each hyperedge: $E \subseteq (\mathcal{P}(V) \times V)$ (e.g., see [16]); we are not following that pattern.

Directed hypergraphs can model the structure of 1-Safe Petri Nets: a correspondence can be built between transitions and hyperedges, and nodes can be identified. A restriction exists in that multiple, identical Petri Net transitions (i.e., transitions with the same input and output place sets) cannot be represented; this does not affect expressive power. Note that every directed hypergraph can be trivially transformed into a 1-Safe Petri Net if we assume the variant firing rules discussed in Sect. 3.1.

In order to model narratives succinctly, nodes that are both source and target of a hyperedge, *contexts* for other actions, should be abstracted out. This motivates the *context hypergraph* as a labelled, restricted variation of a directed hypergraph.

**Definition 5.** *A* context hypergraph *is a 6-tuple,* $(V, E, C, L, L_V, L_E)$ *where:*

$(V, E)$ *forms a directed hypergraph with the property* $\forall h \in E$, $head(h) \cap tail(h) = \phi$.
$C \subseteq (E \times \mathcal{P}(\mathcal{V}))$ *is such that* $(h, N) \in C \Rightarrow \forall n \in N$, $n \in tail(h)$.
$L$ *is a finite set of labels.*
$L_V : V \rightarrow L$ *is a node labelling function.*
$L_E : E \rightarrow L$ *is a hyperedge labelling function.*

A context hypergraph distinguishes nodes that function as *context* connections between a hyperedge $h$ and some subset of tail($h$). In the base hypergraph these are simple tail connections for a hyperedge; by identifying them with the $C$ relation it becomes possible to use them for context purposes. An example is shown in Fig. 3.

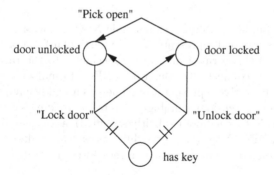

**Fig. 3.** A context hypergraph model of the same structure as Fig. 2. Context relations are marked by a line with two short lines intersecting it (e.g., the two lines connected to the "has key node").

## 3.3   Narrative Flow Graphs

Context hypergraphs form the syntactic basis for our model. We still need the concept of starting and ending nodes, and of course we need algorithms for discovering interesting and useful properties of our system. A simple context hypergraph is not yet sufficient for this. The *Narrative Flow Graph* defines the final representation.

**Definition 6.** *A* Narrative Flow Graph *(NFG) is a 4-tuple: $(H, a, w, \ell)$, where $H = (V, E, C, L, L_v, L_e)$ is a context hypergraph, $a \in V$ is an identified starting, source node with no context connectivity:*
$$\forall h \in E, \ a \notin head(h) \ \wedge \ \forall (h, N) \in C, \ a \notin N$$
*and $w, \ell \in V$ ($w, \ell \neq a$) are identified ending, sink nodes:*
$$\forall h \in E, \ w, \ell \notin tail(h)$$
*The $w$ and $\ell$ nodes must not be simultaneously reachable:*
$$\forall h \in E, \ |\{w, \ell\} \cap head(h)| \ \leq 1$$

The addition of specific starting and ending nodes allows for paths to be defined in the structure. The initial node, $a$, represents axiomatic precedence— all initial conditions are directly connected to $a$. Symmetrically, $w$ and $\ell$ represent termination, either by *winning* or *losing* respectively. The final condition above ensures that a specific hyperedge does not lead to both a win and a loss at the same time.

For narrative games, the concept of winning and losing is important. For narratives *per se*, however, termination remains critical, but the distinction between winning and losing is unnecessary. In these cases $w$ and $\ell$ can be equated, producing a simplified NFG:

**Definition 7.** *A* Simple Narrative Flow Graph *is an NFG $(H, a, w, \ell)$ such that $w = \ell$.*

The lock example is extended to a simple NFG in Fig. 4.

NFGs are a simplified representation of a particular form of 1-Safe Petri Net. The advantage of this formalism is in its specificity to the task. Through designated nodes and connectivity constraints NFGs formalize a general structure appropriate for narrative representation. There is an additional minor benefit in the representation of self-loops: ambiguity as to the readiness of such a transition in a 1-Safe Petri Net (when not explicitly specified) is eliminated when viewed as a context in an NFG. Note that these differences do not alter the close relation to Petri Nets, and so results in that area remain trivial to transfer.

### 3.4    Traversals

In a regular graph, a path or traversal can be represented as an alternating sequence of nodes and edges, terminating at a destination node. A simple path is a path without any cycles (repeated nodes).

For hypergraphs, due to the branching of hyperedges, traversal is more complicated, and (following [16]) simple paths are most easily represented as minimal sub-hypergraphs connecting two sets of vertices. For example, in Fig. 4, one minimal *hyperpath* from {AXIOM} to {END} would include nodes {AXIOM,

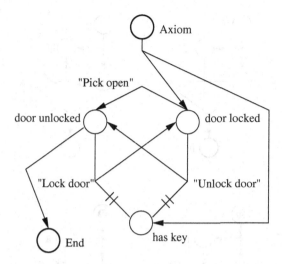

**Fig. 4.** A simple NFG of the same structure as Fig. 3. The "game" begins with the door locked and the key available. Once the door is unlocked (either by using the key or by picking), the door can be re-locked, or the game can end. There is no win or lose in this game, just termination.

DOOR LOCKED, DOOR UNLOCKED, HAS KEY}, along with the UNLOCK DOOR, axiomatic, and terminal hyperedges.

While we can define traversals using hyperpaths, it may not be that a given hyperpath is a realisable traversal—due to their Petri Net origins, there is an implicit, necessary ordering in a traversal of our hypergraphs, and this is lost in a sub-hypergraph representation. In Fig. 5, for instance, a sub-hypergraph representing a traversal from $A$ to $E$ is shown (as is its Petri Net equivalent). The version on the left, however, is not actually realisable as a game—$C$ is required for $X$, but it is also required for $E$. Context relations exist to permit such paths; the version on the right connects $C$ to $X$ using a context edge, akin to a bidirectional connection to a transition in a Petri Net, and so there is no difficulty in revisiting $C$ before moving to $E$.

In order to verify path-based properties in our model, then, a simple sub-hypergraph model of traversal is insufficient. We need to consider the (non)existence of contexts. This is captured through the following definition of *flow*:

**Definition 8.** *Let* $H = (V, E, C, L, L_v, L_e)$ *be a context hypergraph, and let* $X, Y, Q \subseteq V$ *be non-empty subsets of nodes. Let* $E_{Q \to Y} \subseteq E$ *be a set of hyperedges between* $Q$ *and* $Y$:
$$h \in E_{Q \to Y} \Rightarrow tail(h) \subseteq Q \land head(h) \subseteq Y$$
*There is a flow* $(X, Y)$ *representing a realisable traversal from* $X$ *to* $Y$ *if either* $X = Y$, *or if there exists a flow* $(X, Q)$ *and a non-empty* $E_{Q \to Y} \subseteq E$ *such that:*

1. $\forall q \in (Q/Y), \exists h \in E_{Q \to Y}. q \in tail(h)$, *and*
   $\forall y \in Y, y \in Q \lor \exists h \in E_{Q \to Y}. y \in head(h)$

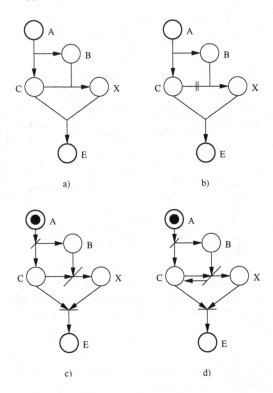

**Fig. 5.** Two small hypergraphs and their Petri Net equivalents. Graph a) does not represent a realisable path—in order to reach $X$ a traversal must pass through $C$, but having done so, since $C$ is also required for $E$, $E$ cannot be reached. This is reflected in the Petri Net equivalent, c), where no sequence of firings will result in a token in the $E$ place. In graph b), $C$ is connected to $X$ using a context relation; this does not "consume" $C$ and so the path is realisable; its Petri Net equivalent, d), can clearly have transitions fired to put a token in $E$.

2. For each $q \in Q$, let $E_q = \{h \in E_{Q \to Y} . q \in tail(h)\}$.
   Let $E_{q/C} = \{h \in E_q . \exists (h, \{q\} \cup N) \in C\}$.
   Then $|E_q| - |E_{q/C}| \le 1$, and if $q \in Y$ then $|E_q| - |E_{q/C}| = 0$.
3. For each $h_1, h_2 \in E_{Q \to Y}$, $head(h_1) \cap head(h_2) = \phi$, and
   for each $h \in E_{Q \to Y}$, $head(h) \cap Q = \phi$.

Note that given a pair of node sets $(X, Y)$, there may be many possible flows from $X$ to $Y$. In such cases it is convenient to describe a flow by *unfolding* it as sequences of node sets, $(X = X_0, X_1, \ldots, X_n = Y)$ for some finite or infinite $n$.

*Flows* are defined recursively, between two sets of nodes. Condition 1 of Definition 8 ensures that a flow progresses by following hyperedges. Condition 2 shows the underlying Petri Net semantics: a particular node can be "used" at most once to progress to the next node set, and all other uses must be defined as *context* uses. If a particular node actually remains in a node set, then all uses must be *context* uses. Condition 3 guarantees 1-Safety.

Despite the apparent complexity of its specification, this is a relatively simple condition to recognize (or compute). In Fig. 5 b), for instance, the node sets involved in a flow between A and E would {A}, {B,C}, {C,X}, and {E}. Sequences of node sets forming flows are analogous to sequences of Petri Net markings forming reachable states.

It will also be useful to have a notion of the "distance" between two node sets forming two ends of a flow. This notion is dependent on avoiding cycles, and hence is developed through the following definitions:

**Definition 9.** *A flow* $(X, Y)$ *unfolded as* $(X = X_0, X_1, \ldots, X_n = Y)$, $n \geq 0$ *is simple if it does not contain any cycles:* $\forall i \neq j$, $X_i \neq X_j$. *If each set of hyperedges between node sets consists of exactly one hyperedge* $(|E_{X_i \to X_{i+1}}| = 1)$ *then the flow is* sequential.

**Definition 10.** *Let* $F = (X = X_0, X_1, \ldots, X_n = Y)$, $n \geq 0$, *be a finite, simple flow. Then* $F$ *has* length $n$, *expressed* $|F| = n$.

Note that not all length 1 flows may be performed in a game as atomic or single actions—in our definition a length 1 flow may be comprised of concurrently following many independent hyperedges. *Sequential* length 1 flows, however, do indeed always follow one hyperedge.

Length is defined by the particular flow between two node sets. Since there may be many ways of reaching $Y$ from $X$, it is not by itself sufficient to describe the "distance" between these node sets. The following definitions supply terminology for describing both minimal and maximal distances.

**Definition 11.** *Given a pair of node sets* $(X, Y)$ *and an NFG, $N$, the* distance from $X$ to $Y$, *expressed* $D_N(X, Y)$, *is the smallest length over all possible simple flows in $N$ between $X$ and any $Y'$, where $Y \subseteq Y'$. The* sequential distance, *expressed* $d_N(X, Y)$ *is the smallest sequential length over all possible simple flows between $X$ and $Y'$.*

**Definition 12.** *Given a pair of node sets,* $(X, Y)$ *and an NFG, $N$, the* separation *between $X$ and $Y$, expressed* $s_N(X, Y)$, *is the length of the largest simple flow* $(X = X_0, X_1, \ldots, X_n)$ *such that* $\forall i \leq n$, $X_i \cap Y = \phi$. *Separation is always sequential.*

## 4 Narrative Properties

NFGs can be used as a design tool to describe narratives; Sect. 5 illustrates such a usage. Narratives, however, must also satisfy semantic properties that are not necessarily trivially apparent in every syntactically-correct NFG. Below some interesting properties and associated analysis/verification strategies are discussed.

## 4.1  Pointlessness

In some narrative games there is the possibility of the game persisting after the player has performed actions that make the game as a whole unwinnable. For example, in the adventure game *The Hitchhiker's Guide to the Galaxy*, "...if you didn't pick up the junk mail at the very beginning of the game, it was unwinnable at the very end" [12]. Since these games usually take considerable time to play through (often on a scale of days to weeks) such problems can be particularly vexing for players.

Ensuring a narrative reaches LOSE quickly if it cannot reach WIN is a desireable narrative property. This can be viewed as a form of reachability or path-length problem in the NFG formalism; below it is expressed in terms of the separation.

**Definition 13.** *An NFG $N = (H, a, w, \ell)$ is $p$-pointless if for all flows $(\{a\}, F)$ either there exists a flow $(F, \{w\} \cup Z)$ for some $Z$, or $s_N(F, \{\ell\}) \leq p$.*

The parameter $p$ defines how "quickly" one must reach the LOSE state if the game is not winnable. A small value of $p$ ensures a quick termination of a failed game.

Verifying this property in general involves determining reachability, a well-considered problem in Petri Nets [15]. There are a variety of efficient solutions available (e.g., [17]).

## 4.2  Narrative Progress

It is possible to build simple semantics for a DAG-based model by describing narrative progression as following a partial order on DAG subgraphs. A subgraph ordering, however, is too coarse for our purposes. First, it does not allow us to distinguish between progression towards the *win* as opposed to the *lose* nodes (and vice versa). Second, two identical subgraphs, equated in a subgraph ordering, may not be at all equivalent with respect to reaching a goal due to the flow mechanics.

Instead, we base our semantic interpretation of movement through a narrative on the state space of the NFGs and flows.

**Definition 14.** *Given an NFG, $N = ((V, E, C, L, L_v, L_e), a, w, \ell)$, and two node sets $X, Y \subseteq V$, we define $(N, X) \leq_w (N, Y)$ iff $d_N(X, \{w\}) \leq d_N(Y, \{w\})$.*

This is trivially a partial order: reflexive, transitive, and anti-symmetric. Distance is defined as the length of the shortest flow from a given node set to any superset of the target set. For NFGs, reaching the win state ($w$) is most important, and so our partial order is based on the distance to the $w$ node. Of course other orderings based on reaching other states are possible (e.g., a lose ordering: "$\leq_\ell$").

A partial ordering allows logical positions within a narrative to be compared: is player 1 "ahead of" player 2? Particular "actions" (flows of length 1) can also be categorized: actions forming increasing functions in this domain are known to increase the proximity to narrative conclusion.

# 5   An Extended Example

The examples presented so far are of a simple nature. In this section, the initial portion of an old, but nevertheless paradigmatic narrative-based computer game is described: *The Count*, one of the well-known Scott Adams adventure games [18]. Note that although resource constraints prevent discussion of the entire game, these initial scenes illustrate non-trivial narrative requirements.

## 5.1   Game Introduction

This game, and indeed all the Scott Adams games, function in discrete time steps furthered by user interaction. At each time step the player is presented with a command line, which allows the player to (potentially) perform a narrative movement. Like many computer narratives, *The Count* is a first-person narrative, where the player is conceptually one of the game characters, and hence is modelled as one of the game objects.

Since there are many subtasks and many potential player actions, the actual NFG is both large and non-planar. We do not address graph drawing concerns in this paper, and so use an *ad hoc* graph structuring to organize and present the model. Also note that the particular structure we present is only one system for doing so. Our presentation was developed experimentally, and we do not claim it is canonical in any way. Equivalence of different models is a non-trivial topic reserved for future work.

At least one node is defined for each object. Important objects in this example include you (the player), a sheet, and the end of the sheet (logically separate from the sheet). These are abbreviated in diagrams as **Y, S,** and **E** respectively. There are of course many other objects in the game; the majority, though, are not important to the initial scenes and so are not modelled here.

The state space of an object often includes its location within the game. Thus each object is represented by several mutually-exclusive states indicating its logical location in the game (which *room* it resides in). Relevant rooms in this example include the bed (starting location), the bedroom, the window ledge and the hall, as well as the player inventory. These are abbreviated **B, Br, L, H,** and **I**. Further state divisions of objects will be introduced as required. Also note that not all object-room combinations are allowable states (e.g., **Y-I** is nonsensical).

Application of location states for objects can be seen in the effects of the user *take* and *drop* actions. These require the player to be in the same location or for the object to be in the inventory; example NFG fragments are shown in Fig. 6. There will be such a substructure required for each place from or to which each object can be moved.

## 5.2   Game Tasks

Tasks within the game are basically organized according to location. Thus, as in the *take* and *drop* examples, the **Y-B, Y-Br, Y-H,** and **Y-L** nodes will

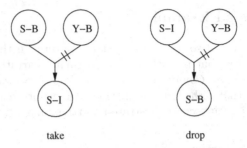

take                          drop

**Fig. 6.** NFG structure for representing effect of *take* and *drop* actions. On the left, while the player and the sheet are in the bed it is possible to move the sheet into the inventory. On the right is the symmetric situation—the sheet can be dropped into the bed provided it is in the inventory, and the player is also in the bed. Notice the context connection to **Y-B** in each case—taking or dropping an item does not affect the players' location.

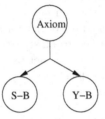

**Fig. 7.** Initial configuration of the extended example.

frequently be required context connections when representing state transformations. In the sections below, the tasks and state space available in two rooms are described.

The player begins lying in bed. Not surprisingly, the bed also contains the sheet (the complete set of connections to the axiom would also include the intial states of all objects, in all rooms). A portion of the initial game situation is illustrated in Fig. 7.

**The Bed.** The bedroom does not contain complex tasks. The only possibilities are to take or put objects here, and for the player to get up. One further possibility is to sleep. Sleeping three times terminates the game as a loss (player turns into a vampire, as in Fig. 8). Note that sleeping in any room includes an immediate movement back to the bed; this can be easily modelled, but is not shown in Fig. 8.

**The Bedroom.** The bedroom is by far the more complex of the rooms in this example. As well as the state of objects already mentioned, the room itself contains a window, which can be in open (**Wo**) or closed (**Wc**) state (initially

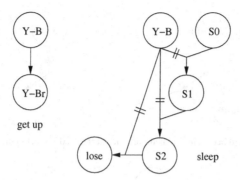

**Fig. 8.** Bed fragment. Here the player can "get up" and move out of the bed into the bedroom (left fragment), or they can sleep (right fragment). Having slept 3 times, they lose. In order to track this, 3 different "having slept" states are required (**S0, S1, S2**). This could also have been modelled by separate sleep actions, as opposed to sleep state nodes.

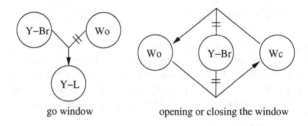

**Fig. 9.** Bedroom fragment. On the left is a structure allowing moving from the bedroom to the ledge, provided the window is open. On the right the window can be opened or closed.

closed). Figure 9 shows an NFG fragment for opening/closing the window and moving out to the ledge through the open window.

A more complex state structure is in the way the sheet and the bed interact. It is possible to tie the sheet to the bed; this transforms the bed (**Be**) into one with a sheet tied to it (**Bs**), and creates the end-of-the-sheet object (**E**) in place of the sheet. The end can then (later) be used as a makeshift rope when on the ledge, through the open window. Tying the sheet can be done whether the sheet is in the bedroom itself, or in the player's inventory (a multi-room context)—see Fig. 10 .

Further complexity is evident in how the tied sheet acts if certain other actions are taken. If the player takes the end of the sheet and ventures into the hall the sheet becomes "untied," restoring the sheet into the players inventory. This situation is shown in Fig. 11. As well, if the player sleeps, the tied sheet resets in a similar fashion (though the sheet appears in the bed, not the inventory); this is not shown in Fig. 8.

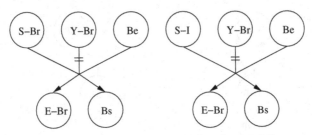

tie sheet to bed – sheet in bedroom  tie sheet to bed – sheet in inventory

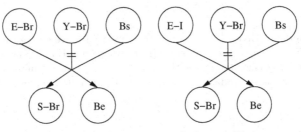

untie sheet – end in bedroom        untie sheet – end in inventory

**Fig. 10.** Tying (top) and untying (bottom) the sheet. Tying can only be done in the bedroom, provided the sheet is in the inventory or bedroom itself. Untying is more complex and not all situations are shown; the sheet can be untied only when *you* are in the bedroom (**Y-Br** as context), but the end (**E**) can be in the bedroom, bed, ledge or inventory. This can be handled by further structures like the bottom two.

## 5.3  Counting Time

Although we have explicitly excluded the modelling of real time, there are several narrative events based on counting. For instance, after 29 moves, a door-bell rings elsewhere in the castle. After 64 turns the sun sets, and the player finds themselves back in bed the next morning (as if they had slept). Static counts such as these can be explicitly modelled in our formalism by the appropriate number of state nodes, as we show for the sleep count (Fig. 8), or less-explicitly as a non-deterministic connection between normal behaviour and the post-count behaviour.

## 5.4  Pointlessness

The game fragment presented does not have a conclusive narrative goal. By adding winning and losing conditions, it becomes possible to discuss how *p*-pointless the narrative is. For example, Fig. 12 defines a win if the edge of the sheet is left on the ledge, and a loss if you go to the ledge without the sheet tied at all.

With this configuration, the aggregate NFG is 1-pointless—it is always possible to reach a win, unless in a condition where losing is the only possible action.

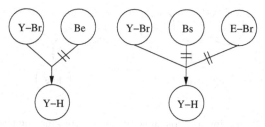

leaving without the sheet tied    leaving without the end

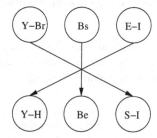

sheet resets if you leave with the end

**Fig. 11.** Leaving the bedroom for the hall. If the sheet is not tied, the movement is straightforward (upper left). If the sheet is tied but the end is not in the inventory, it remains behind (upper right; this requires several similar structures replacing **E-Br** with **E-B** and **E-L**). If the end of the sheet is in the inventory, the sheet becomes untied, and the end is replaced by the original sheet.

The actual game would have a $p$-pointless value larger than 1, since sleep (also leading to a loss) is forced at several points, potentially leaving insufficient moves available to reach a win.

## 6  Future Work

A number of issues remain to be addressed. Our formalism is an attempt to provide an initial infrastructure for narrative modelling and analysis; further analysis of complex computer narratives will expand and tune this model.

Further investigation of other formalisms, including variant forms of Petri Nets (e.g., coloured and/or timed Petri Nets) is warranted. 1-Safe Petri Nets can easily express the simple precedence relations (including cycles and choice) that form basic computer narratives. Other formalisms, however, may allow for the expression of concepts more efficiently (e.g., counting), or which cannot be expressed at all in our formalism (e.g., real-time counters). The increase in complexity of such representations would have to be balanced against other factors, including readability and representative modularity/cohesion—any expansions of the model should be careful to avoid encompassing more than just the game narrative.

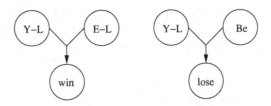

**Fig. 12.** Artificial winning and losing conditions. Dropping the sheet end on the ledge is a win; being there without having tied it at all is a loss. Other graph fragments allow for movement from the ledge back to the bedroom if neither condition is true.

The NFG fragments illustrated in Sect. 5 comprise a very small portion of the game. While these initial scenes are as complex as any subsequent scene, it is clearly more compelling to demonstrate the ability to represent and analyze a complete narrative.

The examination or use of larger examples raises other issues as well. An entire game would result in a very large, complex graph, and actively using such a graph for narrative development would be awkward for a human being. A higher-level environment automating many of the repetitive aspects of the formalism (e.g., implicitly modelling take/drop actions, etc.) is desireable for practical use.

An aspect we are actively working on is the overall semantics induced by narratives. In Sect. 4 some interesting narrative properties were discussed, and a partial order semantics introduced. We are developing these ideas further, investigating how various game actions and potential playability problems or goals may be understood within a semantic framework.

## 7   Conclusions

Traditional, text narratives can be simply represented through a straightforward DAG structure. Narratives in computer games, however, require a more complex presentation system, including the ability to represent both cycles and narrative choice, two things that are not possible with a DAG format.

In this paper we presented an alternative representation framework, Narrative Flow Graphs, derived from a limited form of Petri Net. This framework addresses the existing representation gap, while providing a syntactically simple narrative description format. Using this format, we defined some simple properties pertinent to narrative development, and demonstrated the application of our formalism to a realistic example.

## References

1. Baggett, D., Lee, F., Munn, P., Ewing, G., Noble, J.: Plot in interactive works (was re: Attitudes to playing (longish)). Discussion thread in `rec.arts.int-fiction` archives (1994)

2. Olsson, M., Ewing, G.: Plot DAGs "undo", and finite automata (was: Notes on "Annoy"). Discussion thread in rec.arts.int-fiction archives (1994)
3. Baggett, D., Plotkin, A.C., Arnold, J., Shiovitz, D., Clements, M.: Plot DAGs revisited (was re: Game design in general). Discussion thread in rec.arts.int-fiction archives (1995) http://bang.dhs.org/if/raif/pre1997/thrd14.html#07050.
4. Forman, C.E.: Game design at the drawing board. XYZZY News (1997) http://www.xyzzynews.com/xyzzy.4e.html.
5. Mateas, M.: An Oz-centric review of interactive drama and believable agents. Technical Report CMU-CS-97-156, School of Computer Science, Carnegie Mellon University (1997) http://www-2.cs.cmu.edu/afs/cs.cmu.edu/project/oz/web/papers/CMU-CS-97-%156.html.
6. Burg, J., Boyle, A., Lang, S.D.: Using constraint logic programming to analyze the chronology in "A Rose for Emily". Computers and the Humanities **34** (2000) 377–392
7. Brooks, K.M.: Do story agents use rocking chairs? The theory and implementation of one model for computational narrative. In: The 4th ACM International Conference on Multimedia. (1996) 317–328
8. Brooks, K.: Programming narrative. In: The 1997 IEEE Symposium on Visual Languages. (1997) 380–386
9. Strohecker, C.: A case study in interactive narrative design. In: The Conference on Designing Interactive Systems: Processes, Practices, Methods, and Techniques. (1997)
10. Roberts, M.J.: TADS the text adventure development system. Software, see http://tads.org (1987)
11. van der Aalst, W.M.P.: The application of Petri nets to workflow management. Journal of Circuits, Systems and Computers **8** (1998) 21–66
12. Adams, E.: The designer's notebook: "Bad game designer, no twinkie!". Online article from Gamasutra (1998) http://www.gamasutra.com/features/designers_notebook/19980313.htm.
13. Reisig, W.: Elements of Distributed Algorithms: Modeling and Analysis with Petri Nets. Spring-Verlag (1988)
14. Cheng, A., Esparza, J., Palsberg, J.: Complexity results for 1-safe nets. Theoretical Computer Science **147** (1995) 117–136
15. Esparza, J.: Decidability and complexity of Petri net problems—An introduction. In: Lectures on Petri Nets I: Basic Models. Springer-Verlag (1998)
16. Ausiello, G., Italiano, G.F., Nanni, U.: Optimal traversal of directed hypergraphs. Technical Report TR-92-073, International Computer Science Institute, University of California at Berkeley (1992) http://citeseer.nj.nec.com/ausiello92optimal.html.
17. Pastor, E., Cortadella, J., Pena, M.A.: Structural methods to improve the symbolic analysis of Petri nets. In: The 20th International Conference on Application and Theory of Petri Nets. (1999) 26–45 http://citeseer.nj.nec.com/pastor99structural.html.
18. Adams, S.: Scott Adams grand adventures (S.A.G.A). http://www.msadams.com (2003)

# Tackling Post's Correspondence Problem

Ling Zhao

Department of Computing Science University of Alberta, Canada
zhao@cs.ualberta.ca

**Abstract.** Post's correspondence problem (PCP) is a classic undecidable problem. Its theoretical unbounded search space makes it hard to judge whether a PCP instance has a solution, and to find the solutions if they exist. In this paper, we describe new application-dependent methods used to efficiently find optimal solutions to individual instances, and to identify instances with no solution. We also provide strategies to create hard PCP instances with long optimal solutions. These methods are practical approaches to this theoretical problem, and the experimental results present new insights into PCP and show the remarkable improvement achieved by incorporating knowledge into general search algorithms in the domain of PCP.

## 1 Introduction

Post's correspondence problem (PCP for short) was invented by Emil L. Post in 1946 [1], and soon became a highly cited example of an undecidable problem in the field of computational complexity [2]. Bounded PCP is NP-complete [3]. PCP of 2 pairs was proven decidable [4], and recently a simpler proof using a similar idea was developed [5]. PCP of 7 pairs is undecidable [6]. Currently the decidability of PCP of 3 pairs to PCP of 6 pairs is still unknown.

As the property of undecidability shows, there exists no algorithm capable of solving all instances of PCP. Therefore, PCP is mainly discussed in the theoretical computer science literature, for example, to prove the undecidability of other problems. Only recently did researchers begin to build PCP solving programs [7, 8, 9]. Richard J. Lorentz first systematically studied the search methods used to solve PCP instances and the techniques to create difficult PCP instances. These ideas were implemented and generated many good results [7]. Our work was motivated by his paper and can be regarded as an extension and further development of his work. As his paper is frequently cited in the following sections, we use the term *Lorentz's paper* to denote it for convenience.

In the past 20 years, search techniques in Artificial Intelligence (AI) have progressed significantly, as exemplified in the domains of single-agent search and two-player board games. A variety of search enhancements developed in these two domains have set good examples for building a strong PCP solver, and some of this research can be directly migrated to PCP solvers after a few application-dependent modifications. On the other hand, the characteristics of PCP lead to many unique search difficulties, which has prompted us to develop new techniques and to discover additional properties of PCP that can be integrated

J. Schaeffer et al. (Eds.): CG 2002, LNCS 2883, pp. 326–344, 2003.

into the solver. This work clearly demonstrates the significance of combining application-dependent knowledge with general search frameworks in the domain of PCP, and some of the ideas may be applied to other problems.

This paper mainly discusses three directions for tackling Post's correspondence problem. The first two directions focus on solving solvable and unsolvable instances respectively, i.e., finding optimal solutions to solvable instances efficiently and effectively, and identifying unsolvable instances. Six new methods were invented for the aim of solving instances, namely, the *mask method, forward pruning, pattern method, exclusion method, group method* and *bidirectional probing*.

The third direction of our work is concerned with creating hard instances that have very long optimal solutions. With the help of the methods developed in the above two directions, we built a strong PCP solver that discovers 199 hard instances whose shortest solution lengths are at least 100. Currently, we hold the hardest instance records in 4 PCP subclasses.

The paper is organized as follows. We begin by introducing the definition and notation in Sect. 2, as well as giving some simple examples. Section 3 describes six methods that are helpful to solve instances. Section 4 explains how to create difficult instances, and Sect. 5 contains the experimental results and related discussions. Finally, Sect. 6 provides conclusions and suggestions for future work.

## 2    Post's Correspondence Problem

An instance of Post's correspondence problem is defined as a finite set of pairs of strings $(g_i, h_i)$ $(1 \leq i \leq s)$ over an alphabet $\Sigma$. A solution to this instance is a sequence of selections $i_1 i_2 \cdots i_n$ $(n \geq 1)$ such that the strings $g_{i_1} g_{i_2} \cdots g_{i_n}$ and $h_{i_1} h_{i_2} \cdots h_{i_n}$ formed by concatenation are identical.

The number of pairs in a PCP instance[1], $s$ in the above, is called its *size*, and its *width* is the length of the longest string in $g_i$ and $h_i$ $(1 \leq i \leq s)$. *Pair i* represents the pair $(g_i, h_i)$, where $g_i$ and $h_i$ are the *top* string and *bottom string* respectively. *Solution length* is the number of selections in the solution. For simplicity, we restrict the alphabet $\Sigma$ to $\{0, 1\}$, as we can always convert other alphabets to their equivalent binary format.

Since solutions can be stringed together to create longer solutions, in this paper we are only interested in *optimal solutions*, which have the shortest length over all solutions to an instance. The length of an optimal solution is called the *optimal length*. If an instance has a fairly large optimal length compared to its size and width, we use the adjective *hard* or *difficult* to describe it.

An instance is *trivial* if it has a pair whose top and bottom strings are the same. It is obvious that such an instance has a solution of length 1. We call an instance *redundant* if it has two identical pairs. In this case, it will not influence the result if one of the duplicated pairs is removed. For brevity, we assume the instances discussed in this paper are all nontrivial and non-redundant.

---

[1] In the following, we use the name *instance* to specifically represent *PCP instance*.

To conveniently represent subclasses of Post's correspondence problem, we use $PCP[s]$ to denote the set of all instances of size $s$, and $PCP[s, w]$ for the set of all instances of size $s$ and width $w$. Then the following relations hold:

$$PCP[s, w] \subset PCP[s] \subset PCP$$

We use a matrix of 2 rows and $s$ columns to represent an instance in $PCP[s]$, where string $g_i$ is located at position $(i, 1)$ and $h_i$ at $(i, 2)$. PCP (1) below is an example in $PCP[3, 3]$.

$$\begin{pmatrix} 100 & 0 & 1 \\ 1 & 100 & 00 \end{pmatrix} \tag{1}$$

## 2.1   Example of Solving a PCP Instance

Now let us see how to solve PCP (1). First, we can only start at *pair 1*, since it is the only pair where one string is the other's prefix. Then we obtain this result:

Choose *pair 1*:   $\dfrac{\underline{100}}{1}$

The portion of the top string that extends beyond the bottom one, which is underlined for emphasis, is called a *configuration*. If the top string is longer, the configuration is *in the top*; otherwise, the configuration is *in the bottom*. Therefore, a configuration consists of not only a string, but also its position information: top or bottom.

In the next step, it turns out that only *pair 3* can match the above configuration, and the situation changes to:

Choose *pair 3*:   $\dfrac{100\underline{1}}{100}$

Now, there are two matching choices: *pair 1* and *pair 2*. By using *the mask method* (described in Sect. 3.1), we can avoid trying *pair 2*. Then, *pair 1* is the only choice:

Choose *pair 1*:   $\dfrac{1001\underline{100}}{1001}$

The selections continue until we find a solution:

Choose *pair 1*:   $\dfrac{1001\underline{100100}}{10011}$    Choose *pair 3*:   $\dfrac{10011001\underline{1001}}{1001100}$

Choose *pair 2*:   $\dfrac{100110010\underline{10}}{1001100100}$    Choose *pair 2*:   $\dfrac{1001100100100}{1001100100100}$

After 7 steps, the top and bottom strings are exactly the same, and the configuration becomes empty, which shows that the sequence of selections, *1311322*, forms a solution to PCP (1). When all combinations of up to 7 selections of pairs are exhaustively searched, this solution can be proven to be optimal.

## 2.2   More Examples

Some instances may have no solution. For example, the following PCP (2) is unsolvable, which can be proven through the *exclusion method* discussed in Sect. 3.2.

$$\begin{pmatrix} 110 & 0 & 1 \\ 1 & 111 & 01 \end{pmatrix} \tag{2}$$

An analysis of the experimental data we gathered shows that in PCP subclasses of smaller sizes and widths, only a small portion of instances have solutions, and a much smaller portion have long optimal solutions. PCP (3) is such a difficult instance with optimal length of 206. It is elegant that this simple form embodies such a long optimal solution. If a computer performs a brute-force search by considering all possible combinations up to depth 206, the computation will be enormous. That is the reason why we utilize AI techniques and new methods related to special properties of PCP to prune hopeless nodes, and thus, accelerate search speed and improve search efficiency.

$$\begin{pmatrix} 1000 & 01 & 1 & 00 \\ 0 & 0 & 101 & 001 \end{pmatrix} \tag{3}$$

The optimal solution to an instance may not be unique. For instance, PCP (4) below has 2 different optimal solutions of length 75.

$$\begin{pmatrix} 100 & 0 & 1 \\ 1 & 100 & 0 \end{pmatrix} \tag{4}$$

Now let us take a look at PCP (5). It is clear that *pair 3* is the only choice in every step, and as a consequence, configurations will extend forever and the search process will never end. This example shows an unfortunate characteristic of some instances: *the search space is unbounded.* This special property suggests that we cannot rely on search to prove some instances unsolvable. Several new methods such as the *exclusion method*, which helps to prove the unsolvability of PCP (5), were invented and are presented in the next section.

$$\begin{pmatrix} 100 & 0 & 1 \\ 0 & 100 & 111 \end{pmatrix} \tag{5}$$

# 3   Solving PCP Instances

Intuitively, solvable and unsolvable instances should be treated separately, but for PCP instances, many methods are valuable to both types of tasks. Therefore, we do not divide these methods into two sections, but discuss them together in this section instead.

Lorentz's paper introduced some general search techniques that can be used in a PCP solver:

1. Depth-first iterative-deepening search [10] works well on problems with exponential search space and provides a satisfactory trade-off between time and space.
2. A cache table can be used to prune revisited nodes (similar to what transposition tables do in game playing programs).
3. System-level programming techniques such as tail recursion removal result in satisfactory improvement of the running time.

Lorentz's paper also introduced the concept of *filters*, which consist of simple rules to identify unsolvable instances. In addition, configurations of an instance are not of fixed size, thus it will be very inefficient to use the standard memory allocation functions provided by an operating system. Therefore, we implemented specialized routines for configuration allocation, and this work resulted in about 15% improvement in speed. Please refer to [7, 9] for details of these ideas.

In the following subsections we describe six new methods to tackle PCP instances. They can be roughly categorized into three classes:

1. Pruning configurations: the mask method, forward pruning, and pattern method.
2. Simplifying instances: the exclusion method and group method.
3. Choosing the easier direction: bidirectional probing.

These methods are all closely related to the properties of PCP, and four of them including the mask method, pattern method, exclusion method and group method are very specific and cannot work for any given instance. However, they do improve the performance of the solver dramatically and enable it to answer some unresolved questions in Lorentz's paper.

### 3.1   Pruning Configurations

Configurations are a vital feature for solving PCP instances. The difficulty encountered in solving an instance is often presented by its vast number of configurations that need to be examined. But in many cases, a large portion of configurations can be pruned without examining their descendants, and thus, a significant percentage of search effort can be saved. Especially, the unbounded search space may be reduced to a finite one, making it possible to prove the unsolvability of some instances.

In order to make it easy to explain, we first introduce some definitions concerning configurations. The *reversal* of a configuration is generated by reversing its string. The *turnover* of a configuration is generated by flipping its position from top to bottom or vice versa. A configuration is *generable* to an instance if it can be generated from the empty configuration by a sequence of selections of pairs in the instance. Similarly, a configuration is *solvable* to an instance if it can lead to the empty configuration by a sequence of selections.

**Mask Method.** The mask method deals with pruning all configurations in the top or in the bottom. Before going into detail of the powerful and seemingly magical method, we first introduce the concepts of critical configuration and valid configuration.

**Definition 1:** A *critical configuration* in an instance is a non-empty configuration that can be *fully matched* by a pair: the resulting configuration is empty, or can be *turned upside-down* by a pair: the resulting configuration changes its position from that of the previous one.

**Definition 2:** A configuration is *valid* to an instance if it is generable and solvable.

Critical configurations are critical because they constitute an indispensable step for a configuration in general to reach a solution. For any configuration in the top, a necessary condition for it to lead to a solution is that there must exist a critical configuration in the top in the solution path. This property can be justified by the fact that a solvable configuration must change its position somewhere (we define the position of the empty configuration to be neither top nor bottom). A configuration being valid also means it occurs in a solution path. As a result, if no valid critical configuration in the top exists in an instance, then all configurations in the top can be pruned: the instance has a *top mask*. Similarly, a *bottom mask* means there is no hope of reaching a solution once the configuration is in the bottom.

To check if an instance has masks, we need to find all possible critical configurations, and then test their validity. The first step can be simply done by enumeration. The second step, validity testing, can also be automated, because there is a nice relation between an instance and its *reversal*, which is generated by reversing all the strings in the original instance: the question as to whether one configuration is generable to an instance is equivalent to the question as to whether the turnover of the reversal of the configuration is solvable to the reversal of this instance. This property can be easily proven by using the definition of PCP. It needs to be clarified that though the process of checking configurations solvable can be automated, it may not always terminate without other stopping conditions imposed. In fact, this process is undecidable in general.

The following will explain how these steps work to discover the top mask in PCP (6), whose reversal is PCP (7).

$$\begin{pmatrix} 01 & 00 & 1 & 001 \\ 0 & 011 & 101 & 1 \end{pmatrix} \tag{6}$$

$$\begin{pmatrix} 10 & 00 & 1 & 100 \\ 0 & 110 & 101 & 1 \end{pmatrix} \tag{7}$$

At first, we need to find all critical configurations in the top. Since they could either be fully matched or turned over by at least one pair, any matched pair must have a longer bottom string. In this instance, the matched pair could only be *pair 2* or *pair 3*. It is not hard to find that only one critical configuration in the top exists, i.e. 10 in the top, which can be fully matched by *pair 3*. Secondly, we check whether the 10 in the top can be generated by PCP (6), or equivalently, whether 01 in the bottom in PCP (7) can lead to a solution. However, in PCP (7), the configuration of 01 in the bottom cannot use any pair to match. Thus, PCP (6) has a top mask.

For some instances, the mask method is an effective tool to find optimal solutions. Take PCP (6) for example. It has a top mask, so we forbid the use of *pair 1* at the beginning, and can only choose *pair 3*, which helps quickly find the unique optimal solution of length 160. If this fact was not known, *pair 1* would be chosen as the starting pair and a huge useless search space would be explored before concluding that its optimal length is 160. That is the reason why Lorentz's solver found solutions to two instances but could not prove their optimality [7]. Both can be decided with the assistance of the mask method.

The mask method can also prove the unsolvability of many instances. For example, if an instance has some masks that forbid all possible starting pairs, it has no solution.

*GCD Rule.* The step to prove critical configurations invalid can be strengthened by the *GCD* (Greatest Common Divisor) *rule*: let $d$ be the greatest common divisor of the length differences of all pairs, then the length of every valid configuration must be a multiple of $d$ [2]. Consider the following PCP (8) as an example. The GCD of all length differences is 2.

$$\begin{pmatrix} 111 & 001 & 1 \\ 001 & 0 & 111 \end{pmatrix} \tag{8}$$

Although in PCP (8) we can find a critical configuration of 0 in the bottom, which can be turned upside-down by *pair 2*, it is invalid because its length is not a multiple of 2. As a result, this instance has a bottom mask, revealing that the starting selection must be *pair 2*. Finally, with a few steps of enumeration, we can prove that PCP (8) has no solution.

The above example uses the difference of length. We can similarly use the difference of the number of 0 or 1 elements. For example, if the difference of the number of occurrences of 0 in the two strings of any pair has a GCD $d$, then the number of 0's occurring in any valid configuration must be a multiple of $d$.

**Forward Pruning.** Similar to heuristic search algorithms such as $A^*$ and $IDA^*$ [11, 10], heuristic functions can be used to calculate lower bounds of the solution length for a configuration (for an unsolvable instance, its solution length is defined to be infinity). A heuristic value of a configuration is an estimate of how many more selections are needed before reaching a solution. When the heuristic value of one configuration added to its depth exceeds the current depth threshold in the iterative deepening search, this configuration definitely has no hope of reaching a solution within that threshold. Hence we can reject it

---

[2] This idea was separately mentioned by R. Lorentz and J. Waldmann in private communications.

$$11A \implies 1A0\vdots11 \implies 11B$$

**Fig. 1.** Deduction of a prefix pattern in PCP (9).

even if its depth is still far away from the threshold. Since the heuristic function is admissible, the pruning is safe and does not affect the optimality of solutions.

One simple heuristic value of a configuration in the top (bottom) can be calculated by its length divided by the maximum length difference of all pairs that have their bottom (top) strings longer. This heuristic is based on the balance of length, and similarly, we can calculate heuristics on the balances of elements 0 or 1.

More complex heuristics can be developed analogous to the pattern databases used to efficiently solve instances of the 15-puzzle [12]. We can pre-compute matching results for some strings as the prefixes of configurations and use them to calculate a more accurate estimate of the solution length than the simple heuristics.

**Pattern Method.** If a configuration has a prefix $\alpha$ and any possible path starting from it will always generate a configuration with the prefix $\alpha$ after some steps, and no empty configuration occurs in the middle, then this prefix cannot be completely removed whatever selections are made. The significance of this property is that any configuration having such a prefix never leads to a solution. This observation essentially comes from the goal to shrink configurations to the empty string. If there is a substring that will unavoidably occur, it is impossible for configurations to transfer to the empty string.

The pattern method is a powerful tool to prove instances unsolvable. The following example illustrates how it is effective in proving that PCP (9) has a prefix pattern of 11 in the top, and as a result, PCP (9) is proven unsolvable.

$$\begin{pmatrix} 011 & 01 & 0 \\ 1 & 0 & 100 \end{pmatrix} \tag{9}$$

For a configuration of $11A$ in the top, where $A$ can be any string, the next selection can only be *pair 1*. Thus a new configuration $1A011$ in the top is obtained. Now let us focus on how the substring 0 right after the $A$ is matched. The matched 0 can be supplied either by the only 0 in the bottom string of *pair 2*, or by the last 0 in the bottom string of *pair 3*. Whichever it is, after 0 is matched the substring 11 right behind it will inevitably become the prefix of a new configuration. So a configuration $11A$ in the top will definitely transfer to another configuration $11B$ in the top after a number of steps. The prefix cannot be removed, showing that any configuration in the top with a prefix of 11 will never lead to a solution. The procedure to find a prefix in PCP (9) is presented in Fig. 1.

The dotted vertical line in the figure partitions configurations into two parts: left part and right part. If a configuration still has the chance to lead to a

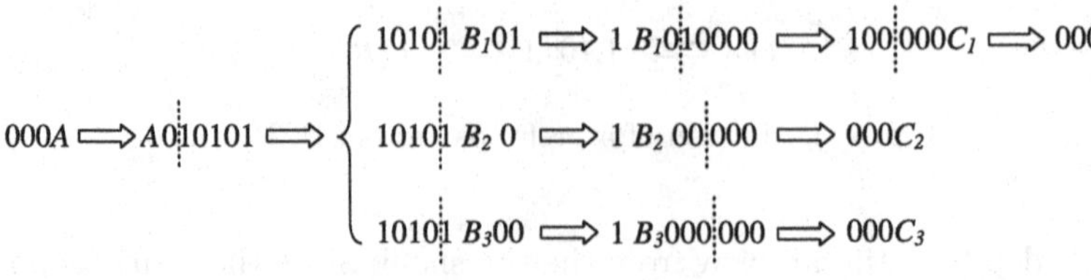

$$000A \Longrightarrow A0\vdots 0101 \Longrightarrow \begin{cases} 1010\vdots 1 \; B_101 \Longrightarrow 1 \; B_1 0\vdots 10000 \Longrightarrow 100\vdots 000C_1 \Longrightarrow 000D_1 \\[2ex] 1010\vdots 1 \; B_2 \; 0 \Longrightarrow 1 \; B_2 \; 00\vdots 000 \Longrightarrow 000C_2 \\[2ex] 1010\vdots 1 \; B_3 00 \Longrightarrow 1 \; B_3 000\vdots 000 \Longrightarrow 000C_3 \end{cases}$$

**Fig. 2.** Deduction of a prefix pattern in PCP (10).

solution, its left part from the line must be matched exactly during a number of selections. Therefore, the dotted vertical line works as a border: matching must stop at one side of it and continue on the other side; no substring is matched across the border.

It can be proven that PCP (9) has a bottom mask. So with the help of the prefix pattern derived above, we can exhaustively try all possible selections and prune any configuration in the top with a prefix of 11. When this process terminates, the unsolvability of PCP (9) is proven.

It is quite intuitive to discover the pattern in PCP (9), yet to find similar patterns in other instances may not be simple. For example, Fig. 2 illustrates the procedure to detect the prefix pattern of 000 in the top in PCP (10), which is indispensable for the unsolvability proof of this instance.

$$\begin{pmatrix} 01 & 0 & 00 \\ 0 & 100 & 10 \end{pmatrix} \tag{10}$$

## 3.2    Simplifying Instances

Through the analysis of specific instances, it is possible to simplify them by removing some pairs or replacing substrings with simpler ones. This simplification significantly reduces the search work, and makes it possible to solve some instances.

**Exclusion Method.** The exclusion method is utilized to detect pairs that can never be used when the selections start with a certain pair. The exclusion follows from the fact that if any combination of certain pairs cannot generate a configuration that can be matched by some other pairs, then those other pairs are useless and can be safely ignored. PCP (11) is such an example.

$$\begin{pmatrix} 1 & 0 & 101 \\ 0 & 001 & 1 \end{pmatrix} \tag{11}$$

If we start with *pair 2*, then *pair 3* can be excluded when the instance is further expanded, because all subsequent selections can only be *pair 1* or *pair 2*. The proof can be separated into three steps:

1. Any configuration generated by *pair 1* and *pair 2* is in the bottom.
2. Any combination of the bottom strings of these two pairs cannot have a substring of 101, the top string of *pair 3*. Thus when a configuration generated by these two pairs is at least of a length 3, *pair 3* has no chance to be selected.
3. The only configuration in the bottom with length less than 3 that can be matched by *pair 3* is 10 in the bottom, yet it cannot be generated by selections of *pair 1* and *pair 2*.

Therefore, after selections start at *pair 2*, we only need to deal with an instance consisting of *pair 1* and *pair 2*. This new instance never leads to a solution because of the length of configurations it generates will never decrease. Hence starting at *pair 2* is hopeless. Combined with the fact that this instance has a top mask, we successfully prove it unsolvable.

**Group Method.** If any occurrence of a substring in configurations can only be entirely matched during one selection of pairs, instead of being matched through several selections, we can consider the substring as an undivided entity, or a *group*. In other words, if the first character in a group is matched during one selection, all others in the group will be matched in the same selection. The group method is utilized to detect such groups and help to simplify instances. Consider the following instance, where the substring 10 is undivided.

$$\begin{pmatrix} 011 & 10 & 0 \\ 1 & 0 & 010 \end{pmatrix} \tag{12}$$

The substring 10 can be inserted into configurations through the bottom string in *pair 3*, and then can be matched by 10 in the top string in *pair 2*. If we consider an instance consisting of only *pair 2* and *pair 3*, then it is not difficult to find out that whenever there is a substring 10 occurring in a configuration, this substring must be entirely supplied by a selection of *pair 3* and can only be matched by *pair 2*. Therefore, we can use a new symbol $g$ to represent the group 10, and the instance will be simplified to:

$$\begin{pmatrix} 011 & g & 0 \\ 1 & 0 & 0g \end{pmatrix} \tag{13}$$
$$g = 01$$

If only *pair 2* and *pair 3* are taken into consideration, the configurations they generate will stay in the bottom and have their lengths non-decreasing. So these configurations will lead to no solution. As the new symbol $g$ cannot be matched by 0 or 1, it is easy to see that *pair 1* can be excluded and safely removed when selections start at *pair 3*. Since no other possible starting pair exists, PCP (12) is unsolvable.

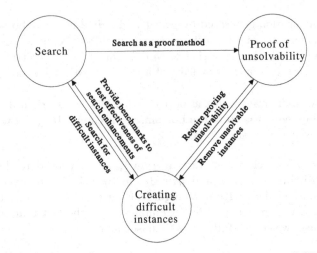

**Fig. 3.** Relations between three research directions in PCP.

## 3.3   Bidirectional Probing

An instance and its reversal are isomorphic to each other in the sense that they share the same solvability, and have the same number of optimal solutions and the same optimal length if solvable. Therefore, solving either one of them is enough. But these two forms may be amazingly different in terms of search difficulties, as shown experimentally in Sect. 5.1.

Hence, we use a probing scheme to decide which search direction is more promising. Initially we set a *comparison depth* (40 in the implementation). Then two search processes are performed for the original instance and its reversal to the comparison depth separately. A comparison of the number of visited nodes in both searches gives a good indication about which direction is easier to explore. The solver then chooses to solve the one with the smaller number of visited nodes. As the branching factor in most instances is quite stable, this scheme worked very well in our experiments.

## 4   Creating Difficult Instances

A strong PCP solver enhanced by the application-dependent methods discussed in the previous section is essential for creating many difficult instances. Besides, the hard instances that we found attracted us to find their solutions efficiently, and those unresolved instances were intriguing for us to come up with new ideas. Thus, the three directions we are working on are interrelated, as shown in Fig. 3.

The task of creating difficult instances can be further categorized into two schemes: random search and systematic search.

## 4.1   Random Search for Hard Instances

A random instance generator plus a PCP solver is a straightforward means of discovering interesting instances, which has been used to create difficult instances in [7]. Statistically, the generator will create a few hard instances with very long optimal lengths sooner or later. However, we can still do much work to increase the chance of finding hard instances. By using several search enhancements and various methods that help to prove instances unsolvable, the program can quickly stop searching hopeless instances and find the optimal solutions to solvable instances faster.

During the search process of an instance, we use three factors as stopping conditions if no solution is found. They are the final depth threshold, the number of visited nodes, and the number of cutoff nodes (nodes pruned by the cache table). Using the number of visited nodes as a stopping condition forces the search process to treat every instance equally, avoiding the situation of frequently getting stuck in instances that have a large branching factor but no solution. Based on the observation of the hard instances we collected, most of those instances do not have a large number of cutoff nodes (the largest is only 28,972). Thus in the implementation, we also use the number of cutoff nodes as one of the stopping conditions.

One method suggested in the literature is the removal of instances that have the *pair purity* feature, that is, a pair consisting wholly of ones or zeroes [7]. This idea is based on the observation that many instances bearing this feature have no solution, but have quite messy search trees that are very costly to explore. However, our experimental results show that this method may cause the failure of discovering some hard instances (described in Sect. 5.3).

## 4.2   Systematic Search for Hard Instances

As the random search scheme randomly chooses instances to consider, the chance of finding difficult instances is still dependent on luck, so a systematic approach seems more convincing to demonstrate the strengths of a PCP solver. If all instances in a specific PCP subclass are examined, they may be completely solved, and lots of hard instances including the hardest one in this subclass will be discovered. Even if we cannot solve all of them, the unresolved instances may be left as incentives for new approaches.

It is not hard to generate all instances in a finite PCP subclass, and there is a related problem about how to remove those isomorphic instances which potentially will cause lots of redundant work. We omit it for brevity, and for further information see [9].

In Lorentz's paper, the subclass $PCP[3, 3]$ was explored, with more than 2000 instances unresolved (most of them have the pair purity feature). We continued the work and rescanned all instances in $PCP[3, 3]$ as well as 9 other subclasses. The results are shown in Sects. 5.2 and 5.3.

**Table 1.** Records of hardest instances known in 5 PCP subclasses.

| subclass | hardest instance known | optimal length | number of optimal solutions |
|---|---|---|---|
| $PCP[3,3]$ | $\begin{pmatrix} 110 & 1 & 0 \\ 1 & 0 & 110 \end{pmatrix}$ | 75 | 2 |
| $PCP[3,4]$ | $\begin{pmatrix} 1101 & 0110 & 1 \\ 1 & 11 & 110 \end{pmatrix}$ | 252 | 1 |
| $PCP[3,5]$ | $\begin{pmatrix} 11101 & 1 & 110 \\ 0110 & 1011 & 1 \end{pmatrix}$ | 240 | 1 |
| $PCP[4,3]$ | $\begin{pmatrix} 111 & 011 & 10 & 0 \\ 110 & 1 & 100 & 11 \end{pmatrix}$ | 302 | 1 |
| $PCP[4,4]$ | $\begin{pmatrix} 1010 & 11 & 0 & 01 \\ 100 & 1011 & 1 & 0 \end{pmatrix}$ | 256 | 1 |

### 4.3   New PCP Records

The random and systematic search schemes for creating difficult instances helped create new records in 4 PCP subclasses. These records are new instances with the longest optimal lengths known in specific PCP subclasses. The records in subclasses $PCP[3,4]$ and $PCP[4,3]$ were found by the systematic search scheme and those in $PCP[3,5]$ and $PCP[4,4]$ were obtained through the random search scheme. Table 1 gives these four records as well as the record in $PCP[3,3]$, which was discovered by R.J. Lorentz and J. Waldmann independently. More information is provided on the websites [13, 14].

## 5   Experimental Results and Analysis

We implemented the methods discussed in Sect. 3 in a depth-first iterative-deepening based search program named PCPSOLVER[3], with the exception of the pattern method and the group method because we could not find a general way to automate them. The program is written in C++ under Linux. All experiments were performed on 200 hard instances. 199 instances have optimal lengths at least 100 and were collected from 4 PCP subclasses through the methods described in Sect. 4; the remaining test case is the hardest instance known in $PCP[3,3]$, as shown in Table 1.

The average branching factor of the 200 test instances is only 1.121 after all enhancements were incorporated. Such a small branching factor makes it feasible to find an optimal solution with length even greater than 300.

---

[3] The source code of PCPSOLVER can be downloaded from http://www.cs.ualberta.ca/~zhao/PCP/PCPSolver/

With other normal search enhancements and programming techniques incorporated, the final version of our PCP solver achieved a search speed of $1.38 \times 10^6$ nodes per second on a machine with a 600MHZ processor and 128MB RAM.

## 5.1   Results of Solving Methods

It is hard to give a quantitative evaluation of the improvements derived from the mask method and the exclusion method, since sometimes they are essential to solve instances. However, we can comment that these two methods help to prune huge search spaces in some cases.

Bidirectional probing is also crucial to solve some hard instances. PCP (14) with an optimal length of 134 is such an example. Up to depth 40, searching this instance directly is more than 15,000 times harder than searching its reversal in terms of the number of visited nodes. Consider that searching to depth 40 has already made such a big difference, if searching to depth 132, the difference will explode exponentially. Thus, it becomes unrealistic to solve this instance when the wrong direction is chosen. This clearly demonstrates how important the bidirectional probing method is.

$$\begin{pmatrix} 110 & 1 & 1 & 0 \\ 0 & 101 & 00 & 11 \end{pmatrix} \tag{14}$$

One nice strength of the above three methods is that they are all done before a deep search is performed and they introduce negligible overhead to the solver.

We implemented two types of admissible heuristic functions to prune hopeless nodes in the forward pruning. The first one is based on the balance of length, and the second is based on the balance of elements, as described under forward pruning in Sect. 3.1. Three separate experiments were conducted on forward pruning, namely, only using the first type of heuristic, only using the second type and using both types. The results show that only using the heuristic on the balance of length gives the best performance.

We tried to compare the improvement achieved by heuristic pruning with the situation when no pruning is used, but it turned out to be too time consuming. PCP (15) is an illustrative example. The solver spent 14,195 seconds to solve this instance when no pruning was used, compared to merely 5.2 seconds when the length balance heuristic was employed. This is a 2730-fold speedup in solving time!

$$\begin{pmatrix} 11011 & 110 & 1 \\ 0110 & 1 & 11011 \end{pmatrix} \tag{15}$$

We also did experiments on some search parameters. For example, the experimental results show that when the depth increment in the iterative deepening is 20, the solving time is minimal over our test set. The result of the experiments on the cache size, however, is a little surprising. It shows that from 8 entries to 65536 entries, the solving time is quite stable. We suspect this phenomenon is largely dependent on the test instances chosen. By default, a large table is employed because it is essential to prove some instances unsolvable.

**Table 2.** Solving results of 7 PCP subclasses.

| PCP subclass | total number | after filter | after mask | after exclusion | solvable instances | unsolvable instances | largest optimal length |
|---|---|---|---|---|---|---|---|
| $PCP[2,2]$ | 76 | 3 | 3 | 3 | 3 | 73 | 2 |
| $PCP[2,3]$ | 2,270 | 51 | 31 | 31 | 31 | 2,239 | 4 |
| $PCP[2,4]$ | 46,514 | 662 | 171 | 166 | 165 | 46,349 | 6 |
| $PCP[2,5]$ | 856,084 | 9,426 | 795 | 761 | 761 | 855,323 | 8 |
| $PCP[2,6]$ | 14,644,876 | 140,034 | 3,404 | 3,129 | 3,104 | 14,641,772 | 10 |
| $PCP[3,2]$ | 574 | 127 | 67 | 61 | 61 | 513 | 5 |
| $PCP[4,2]$ | 3,671 | 1,341 | 812 | 786 | 782 | 2,889 | 5 |

## 5.2   Results of Scanning PCP Subclasses

We scanned seven PCP subclasses that are easy to handle, and all instances in these subclasses have been completely solved. The results are shown in Table 2. All isomorphic instances have been removed.

This table also provides a statistical view on the effectiveness of our unsolvability proof methods, and illustrates how small the percentage of solvable instances in these PCP subclasses is. In addition, the small largest optimal lengths in these subclasses partly explain why they can be completely explored in a small amount of computation.

One special phenomenon we observed is that the hardest instances in PCP subclasses with size 2 can all be represented in the following form[4]:

$$\begin{pmatrix} 1^n0 & 1 \\ 1 & 01^n \end{pmatrix} \tag{16}$$

It is not hard to prove that the optimal length of this kind of instances is $2n$. Lorentz conjectured that such an instance might always be the hardest one in $PCP[2, n + 1]$ in the general case. If the conjecture is true, it will lead to a much simpler proof that $PCP[2]$ is decidable than the existing one [4]. Our experimental results support the conjecture in the cases from $PCP[2,2]$ to $PCP[2,6]$.

We used the systematic method to further examine three PCP subclasses that are much harder to conquer. Table 3 first summarizes the results from $PCP[3,3]$.

In Tables 3 and 4, an instance removed by the exclusion method may still have solutions, but it cannot have a *non-trivial* solution[5]. Since the result of solving such an instance is identical to the combinations of the results from

---

[4] Of the three hardest instances of $PCP[2,2]$, only one instance can be represented in this form.

[5] A solution to an instance is *trivial* if in the solution not all pairs of the instance are used.

**Table 3.** Scanning results of subclass $PCP[3,3]$.

| | |
|---|---|
| Total number | 127,303 |
| After filter | 8,428 |
| After mask | 2,089 |
| After exclusion | 2,002 |
| Solvable instances | 1,961 |
| Unsolvable instances | 40 |
| Unsolved instances | 1 |

instances with smaller sizes, we do not give it any further processing. Similarly, instances removed by the element balance filter (see [7]) may also have trivial solutions, but these solutions are of no interest to us. In addition, 32 instances remained unsolved by our PCP solver, but were proven unsolvable by hand using the methods discussed in Sect. 2. The difference comes from the fact that though some methods can successfully prove several instances unsolvable, some of them are too specific to be generalized, and thus they were not incorporated into our PCP solver.

The only unsolved instance in $PCP[3,3]$ is PCP (17) [6]. Our solver searched to a depth of 300, but still could not find a solution to this instance. Various deduction methods were tried to prove it unsolvable, but also failed. Compared to the fact that the hardest instance in all instances of $PCP[3,3]$ except this one only has an optimal length of 75, we believe that this instance is very likely to have no solution. However, apparently new approaches are needed to deal with it.

$$\begin{pmatrix} 110 & 1 & 0 \\ 1 & 01 & 110 \end{pmatrix} \tag{17}$$

### 5.3 Results of Creating Difficult Instances

We scanned all instances in $PCP[3,4]$ and $PCP[4,3]$ to discover hard instances. The results are summarized in Table 4.

The scanning process took about 30 machine days to finish and resulted in the discovery of 77 hard instances whose optimal lengths are at least 100. At the same time, more than 17,000 instances remain unsolved to the solver, and it becomes impossible to check such a large quantity of instances manually. Although most of these unsolved instances may have no solution, it is still likely that they contain some extremely difficult solvable instances. Thus, these instances are left for future work, waiting for some new search and disproof methods.

Of the 72 hard instances we collected from $PCP[4,3]$, 13 instances (18.1%) have the pair purity feature, and some of them need more than a hundred seconds

---

[6] The problem has recently been proven unsolvable by Mirko Rahn using a generalized pattern method [15].

**Table 4.** Scanning results of subclass $PCP[3, 4]$ and $PCP[4, 3]$.

|  | $PCP[3, 4]$ | $PCP[4, 3]$ |
|---|---|---|
| Total number | 13,603,334 | 5,587,598 |
| After filter | 902,107 | 1,024,909 |
| After mask | 74,881 | 275,389 |
| After exclusion | 65,846 | 266,049 |
| Solvable instances | 61,158 | 249,493 |
| Unsolvable instances | 1,518 | 2,633 |
| Unsolved instances | 3,170 | 13,923 |
| Hard instances | 5 | 72 |

to solve. This evidence suggests that even an instance with the pair purity feature may still have a very long optimal solution (the longest we found is 240), and the quantity of these instances cannot be ignored.

Using the random approach to search for difficult instances, we successfully discovered 21 instances in $PCP[3, 5]$ and 101 instances in $PCP[4, 4]$. Their optimal lengths are all at least 100. The whole process took more than 200 machine days to finish. All of those hard instances and unsolved instances can be found on the website [14].

## 6    Conclusions and Future Work

In this paper, we described some new methods and techniques to tackle PCP instances, including finding optimal solutions quickly, proving instances unsolvable and creating many interesting difficult instances. We successfully solved some instances that were unsolved in Lorentz's paper and scanned all instances in 10 PCP subclasses. Our work resulted in the discovery of 199 difficult instances with optimal length at least 100, and in setting new hardest instance records in 4 PCP subclasses. We also present the empirical results for solving PCP instances and they may serve for investigating some theoretical issues related to this problem.

Although the six new methods described in this paper are all application-dependent, the ideas behind them are not new at all. For example, bidirectional probing uses the idea of performing a shallow search to direct a deep search, and incorporating application-dependent knowledge into general search framework has worked successfully in many domains such as planning and games.

This paper further exploits the ideas discussed in Lorentz's paper, and there is no doubt that the results can be improved. PCP (17) is the only unsolved instance in $PCP[3, 3]$, and more than 17,000 instances in $PCP[3, 4]$ and $PCP[4, 3]$ are unsolved, waiting for new methods to conquer them. Although we implemented most of the methods and techniques discussed in this paper, the group method and pattern method have only been applied by hand to solve some hard instances. We believe that if these methods could be successfully incorporated into our PCP solver, a great portion of unsolved instances would be proven unsolvable.

As PCP instances are closely related to their reversals, bidirectional search can also be applied to solve them. It is also very interesting to evaluate the benefits provided by the complex heuristics mentioned under forward pruning in Sect. 3.1. In this way, PCP can act as a special test bed for general search enhancements.

We anticipate the work to continue to tackle $PCP[3,5]$, $PCP[4,4]$ and more PCP subclasses. If the hardest instances in these subclasses could be found, it may be possible to find some similarities between them and link them to theoretical issues. Nevertheless, identifying more hard instances can provide a better understanding of the complexity of PCP and pave the road for improvement of solving methods.

## Acknowledgment

I would like to thank Dr. Michael Buro and Dr. Lorna Stewart for their instructive advice, and thank Dr. Richard J. Lorentz, Heiko Stamer and Dr. Johannes Waldmann for their generous help. Special thanks are due to my supervisors, Dr. Jonathan Schaeffer and Dr. Martin Müller for their encouragement, patience and meaningful discussions. I am also grateful to the anonymous referees for the valuable comments. This work has been supported by the Alberta Informatics Circle of Research Excellence (iCORE).

## References

1. Post, E.: A variant of a recursively unsolvable problem. Bulletin of the American Mathematical Society **53** (1946) 264–268
2. Hopcroft, J., Ullman, J.: Introduction to Automata Theory, Language, and Computation. Addison–Wesley (1979)
3. Garey, M., Johnson, D.: Computers and Intractability: A Guide to the Theory of *NP*-completeness. W. Freeman (1979)
4. Ehrenfeucht, A., Karhumaki, J., Rozenberg, G.: The (generalized) Post correspondence problem with lists consisting of two words is decidable. Theoretical Computer Science **21** (1982) 119–144
5. Halava, V., Harju, T., Hirvensalo, M.: Binary *generalized* Post correspondence problem. Technical Report 357, TUCS (2000)
6. Matiyasevich, Y., Sénizergues, G.: Decision problems for semi-thue systems with a few rules. In: Proceedings, 11th Annual IEEE Symposium on Logic in Computer Science, IEEE Computer Society Press (1996) 523–531
7. Lorentz, R.: Creating difficult instances of the Post correspondence problem. In Marsland, T., Frank, I., eds.: Second International Conference on Computers and Games (CG 2000). Volume 2063 of Lecture Notes in Computer Science. Springer-Verlag (2001) 145–159
8. Schmidt, M., Stamer, H., Waldmann, J.: Busy beaver PCPs. In: Fifth international workshop on termination (WST '01). (2001)

9. Zhao, L.: Solving and creating difficult instances of Post's correspondence problem. Master's thesis, Department of Computing Science, University of Alberta (2002)
10. Korf, R.: Depth-first iterative-deepening: An optimal admissible tree search. Artificial Intelligence **27** (1985) 97–109
11. Hart, P.E., Nilsson, N.J., Raphael, B.: A formal basis for the heuristic determination of minimum cost paths. IEEE Transactions on Systems Science and Cybernetics **SSC-4(2)** (1968) 100–107
12. Culberson, J., Schaeffer, J.: Searching with pattern databases. In McCalla, G., ed.: Proceedings of CSCSI '96 on Advances in Artificial Intelligence. Volume 1081 of LNAI., Springer-Verlag (1996) 402—416
13. Stamer, H.: PCP at home. `http://www.informatik.uni-leipzig.de/~pcp` (2000—2002)
14. Zhao, L.: PCP homepage. `http://www.cs.ualberta.ca/~zhao/PCP` (2001—2002)
15. Rahn, M.: PCP12 has no solution (unpublished manuscript). `http://www.stud.uni-karlsruhe.de/~uyp0/last/last.ps` (2002)

# Perimeter Search Performance

Carlos Linares López[1]* and Andreas Junghanns[2]

[1] Payload Data Segment – ENVISAT, Ground Segment Engineering Division,
European Space Agency – ESRIN, Frascati
Carlos.Linares.Lopez@esa.int

[2] DaimlerChrysler AG, Research Information and Communication, Berlin
Andreas.Junghanns@dcx.com

**Abstract.** The idea of bidirectional search has fascinated researchers for years: large gains seem intuitively possible because of the exponential growth of search trees. Furthermore, some researchers report significant gains for bidirectional search strategies. This is all the more frustrating for those practitioners that have failed to realize the promised improvements.

We suggest a model for perimeter search performance that, instead of simply counting nodes, counts the execution of important algorithmic subtasks and weights them with their runtime. We then use this model to predict total runtimes of perimeter search algorithms more accurately. Our model conforms to the observation that unidirectional search (IDA*) is a special case of its bidirectional counterpart, perimeter search (BIDA*), with a perimeter depth of 0. Using this insight, we can determine the optimal perimeter depth for BIDA* *a priori*, thus allowing BIDA* to subsume IDA*.

Our model forecasts that applications with expensive heuristic functions have little if anything to gain from perimeter search. Unfortunately, expensive heuristics are often used by high-performance programs. Our experiments show that on the 15-puzzle perimeter search narrowly outperforms its unidirectional counterpart. This finding is consistent with the literature and our model. However, it does not appear that a state-of-the-art implementation of a 15-puzzle solver can benefit from the perimeter search strategy.

## 1  Introduction

Bidirectional search strategies have inspired researchers for many years. Of course, intuitively, the savings can be enormous because of the exponential growth of search trees. But the problems with realizing the promising gains started early and persist to date.

Pure bidirectional search algorithms suffered from different problems. In the case of *front-to-front* bidirectional search (e.g., BHFFA* [1]), it is not easy to guarantee that when both *waves* meet, the optimal solution consists of the paths

---

* Current affiliation: Ground Segment Systems Division, DEIMOS Space S.L., Madrid.
carlos.linares@deimos-space.com

J. Schaeffer et al. (Eds.): CG 2002, LNCS 2883, pp. 345–359, 2003.
© Springer-Verlag Berlin Heidelberg 2003

from both ends to the node where the collision is detected [2]. Therefore, a second version of the BHFFA* algorithm which corrected this error was suggested [3]. On the other hand, *front-to-end* implementations (such as BHPA [4]) seem to be unable to take advantage of the fact that both frontiers are approaching each other. This resulted in a second version of the BHPA algorithm presented some years later, BS* [5]. This algorithm is still no faster than its unidirectional counterpart, the A* search algorithm [6, 7].

These problems, assumed to be at least in part due to the simultaneously performed forward and backward searches, prompted Dillenburg and Nelson [8] as well as Manzini [9] to introduce simultaneously and independently the notion of perimeter search.

Perimeter search algorithms consist of two stages:

**First,** creation of a perimeter (denoted as $P_{p_d}$) around the goal node $t$ which contains all the nodes whose cost for reaching $t$ is equal to a predefined value $p_d$, known as the perimeter depth, and whose children have costs strictly bigger than $p_d$. This perimeter can be created using a brute-force search algorithm such as depth-first search.

**Second,** search from the start node $s$ using a single-agent search algorithm such as IDA* [10] guided by the heuristic $h_{p_d}(\cdot)$, defined as

$$h_{p_d}(n) = \min_{m \in P_{p_d}} \{h(n, m) + h^*(m, t)\}, \tag{1}$$

that forces the forward search to approach the perimeter nodes instead of the target node $t$.

The creators of perimeter search showed that when the original heuristic estimation $h(\cdot)$ is admissible, the newly defined heuristic distance, $h_{p_d}(\cdot)$, is admissible as well. Furthermore, they proved that $h_{p_d}(\cdot)$ is more informed than $h(\cdot)$, meaning that $h_{p_d}(n) \geq h(n)$ for all $n$ in the state space. Moreover, they proved that for the computation of $h_{p_d}(n)$ it is not necessary to compute $h(n, m_j)$ for all the perimeter nodes $m_j$ in $P_{p_d}$. Those nodes which do not exceed the current threshold of their respective parent are so-called *useful* perimeter nodes. Manzini [9] demonstrated that the number of useful perimeter nodes decreases with the depth of the search tree, $d$, explored in the forward search. However, he did not derive a formula for predicting the mean number of useful perimeter nodes.

The authors reported large savings using the perimeter strategy, and others have reported it to be effective, at least in the N-Puzzle domain [2]. However, no indepth theoretical study, as far as we know, on the relative performance of perimeter search algorithms and their unidirectional counterparts has ever been presented in the literature.

Many of the papers in the literature report improvements of bidirectional search over their unidirectional counterparts. Many researchers and practitioners have tried to use those algorithms, but have largely failed to cash in on the promised performance gains.

Most papers trying to compare and/or predict performance of bidirectional search use naive approaches based on counting nodes and measuring total runtimes [5, 2]. However, the time spent per node for bidirectional (here perimeter) searches varies with perimeter depths, as well as between perimeter search and unidirectional search. The only paper we found where more accurate measures than naive node counts are used is by Dillenburg and Nelson [8]: their conclusions, however, are rather simplistic. Superficially, their analysis is similar to ours. However, some rather crude simplifications and assumptions make us question the accuracy of their results. Some examples:

- The fraction of time spent for the evaluation of $h(\cdot)$ ($f$) is not constant throughout the tree in the unidirectional search as assumed by Dillenburg and Nelson: the leaf nodes, a large part of all nodes in the tree, are not expanded, thus less time is spent here as opposed to internal nodes, increasing the relative cost of evaluating $h(\cdot)$ at leaf nodes. Thus, there is no one $f$-value to use in the analysis!
- Furthermore, the heuristic branching factor $b$ is not constant for different perimeter depths $p_d$, because the heuristic improves with increasing $p_d$, which decreases $b$. The heuristic branching factor even depends on the current IDA* threshold!
- Finally, as observed by Dillenburg and Nelson the number of useful perimeter nodes is not constant over the search depth. But they do not take this fact into account for their theoretical analysis.

Our paper takes all of these, in our opinion, important points into account.

Manzini [9] establishes a formula for the number of nodes expanded and the number of heuristic evaluations, but without trying to derive expected runtimes or optimal perimeter depths. The runtime cost per node varies widely and depends on the application, algorithm, implementation, choice of heuristic, hardware, and other factors. This view is shared by Kaindl and Kainz [2]:

...BIDA*'s result here is worse than the data reported by Manzini (1995). This is primarily due to the use of a different machine and a different implementation that is based on the very efficient code of IDA*...

Moreover, in the same paper Kaindl and Kainz state:

...So, we made experiments with increasing perimeter depth in two different domains. The results may seem to be quite surprising. While we cannot yet explain them theoretically, they are important in their own right, and we try to explain them intuitively...

We suggest a model that explains why predictions and reports of some researchers are not necessarily reproducible by others. Our suggested model focuses on runtimes of algorithmic subtasks to capture the increased runtime burden for bidirectional search strategies. The model captures expected gains from such sources as the better informed heuristic in order to predict the overall runtime behaviour. Measuring the runtimes of subtasks such as node expansion

$(t_e)$, heuristic function evaluation $(t_h)$, and collision detection $(t_c)$ takes the subtleties of hardware, implementation, application and choice of heuristic into account during evaluation of our formula. The implementations of these subtasks are largely search algorithm independent and therefore these measured times can serve as a base for the analysis of different search algorithms for specific implementations on specific machines, contrary to rather generic asymptotic analysis.

We believe that this more inclusive approach is better suited than current attempts in predicting search algorithm performance. Several independent reasons are given as to why the high-performance implementations of practitioners are especially hard to beat using bidirectional search methods.

However, we can also show, that under certain conditions, improvements over unidirectional search are possible. Our experimental section illustrates these improvements using the well-known 15-puzzle. These experiments confirm in principle the positive theoretical and practical results reported in the literature.

In the next section we introduce a mathematical model for comparing IDA* and BIDA*. Following that, we compare the predictions of our model with actual performance on the 15-puzzle. Finally, we draw some conclusions.

## 2  Mathematical Model

We introduce a mathematical model for predicting the number of nodes generated and time consumed by perimeter search algorithms, as a function of the perimeter depth employed. Specifically, the algorithms IDA* and BIDA* will be considered. However, the general approach should be applicable to most unidirectional search algorithms and their bidirectional counterparts.

Our model will help us to understand the relationship between unidirectional and bidirectional (perimeter) search. The ultimate goal is to find a way to determine the perimeter depth $p_d$ that minimizes the runtime of the program. The model we develop allows the determination of this optimal $p_d$.

### 2.1  Preconditions

This analysis is restricted to domains that meet the following requirements:

- The cost of all arcs, $c(n, n')$, with $n'$ being a successor of $n$, is equal to 1.
- The only allowed changes of the heuristic function from node $n$ to node $n'$ are $\{-1, +1\}$, when estimating its distance to the same node, $m$.

Examples are domains like the $n$-puzzle when guided by Manhattan distance, or the problem of finding the shortest path in a maze with obstacles when guided by the sum of the difference of the coordinates $x$ and $y$.

These conditions keep the math manageable. However, even though these conditions restrict the applicability of the mathematical analysis that follows, we believe the conclusions apply in principal to a much wider class of applications.

## 2.2   Comparing Unidirectional and Perimeter Search Algorithms

There are two principal reasons why perimeter search algorithms may achieve improved performance over their unidirectional counterparts:

1. the heuristic is better informed, leading to a smaller heuristic branching factor, and
2. the search depth of the forward search is reduced at least by the perimeter search depth $p_d$.

However, these advantages are offset by some more or less obvious computational disadvantages, most notably:

1. the cost of computing the more expensive heuristic function $h_{p_d}(\cdot)$,
2. the cost of computing whether a newly generated node intersects with the perimeter[1], and
3. the cost of generating the perimeter.

It has to be noted that any implementation will perform more tasks than those modeled here. However, we assume that the times for the tasks considered here dominate the overall run times of both search algorithms.

## 2.3   Analysis of the Behaviour of Unidirectional Search Algorithms

$T_u(d)$, the expected runtime of a unidirectional search to depth d, can be computed using Korf and Reid's formula [11]:

$$N_1(b, d, F_h) = \sum_{i=1}^{d+1} b^i F_h(d - i + 1)$$

where $F_h(x)$ is an overall distribution function for the probability of a node randomly selected to have a heuristic distance to the goal node less than or equal to x: $F_h(x) = p(h(n, t) \leq x)$. Likewise, $f_h(x)$, is the overall corresponding density function. On the other hand, though it has been shown that in practice the heuristic branching factor is the same as the brute-force branching factor [12], we consider b to be the former for the sake of completeness.

If $t_h$ and $t_e$ are the times spent by one call to the heuristic function and one node expansion, respectively, the following formula gives the expected runtime of a unidirectional search to depth d in the worst case (goal node found as the last node in the last level of the tree) is:

$$T_u(d) = N_1(b, d, F_h)t_h + N_1(b, d - 1, F_h)t_e \tag{2}$$

Note: $T_u(d)$ is the time spent by a single iteration of IDA*.

---

[1] Adding a node to the perimeter in the generation phase is unavoidable. However, for some implementations, detecting a collision comes nearly for free ($t_c = 0$, find the definition of $t_c$ further down). This is the case if $h_{p_d}$ is guaranteed to return $p_d$ only for nodes on the perimeter: a simple test of the $h$-value of a node delivers the necessary information about containment in the perimeter.

## 2.4   Analysis of the Behaviour of the Perimeter Search Algorithms

In the case of perimeter search algorithms, the computational cost can be decomposed into four parts:

- the time to create and store the perimeter,
- the time to expand the nodes,
- the time consumed for computing $h_{p_d}(n)$, and
- the time spent to check if a newly generated node is in the perimeter.

**Perimeter Creation.** There are two tasks and their respective times to be taken into account when computing the time $T_g(B, p_d)$ for perimeter generation.

- Because the perimeter can be generated with the aid of a brute-force search algorithm, the asymptotic branching factor $B$ is considered instead of the heuristic branching factor $b$, previously employed. The expected number of nodes expanded up to depth $p_d$ by a brute-force search algorithm is:

$$\sum_{i=0}^{p_d-1} B^i = \frac{B^{p_d} - 1}{B - 1}$$

- In addition to the perimeter generation time, the time for inserting all the nodes in the perimeter must be taken into account as well. Perimeters can be implemented with binary trees. Thus, the time complexity for accessing a node in the perimeter is $O(\log n)$, where $n$ is the number of nodes currently stored in the binary tree[2].

If the time for node generation is $t_e$ and the time to access a node in the perimeter is $t_c$, the total time for the perimeter generation is:

$$T_g(B, p_d) \geq t_e \frac{B^{p_d} - 1}{B - 1} + t_c \sum_{i=1}^{B^{p_d}} \log(i)$$

$$\geq t_e \frac{B^{p_d} - 1}{B - 1} + t_c \log\left(B^{p_d}!\right) \tag{3}$$

We should note here that we consider this cost for every problem in our experiments. Thus, we are not reusing the perimeter but generate it every time.

**Node Expansion.** The number of nodes generated or expanded by a perimeter search algorithm will obey a different formula than equation (2), because the heuristic function has changed. The overall density function $f_{p_d}$ is defined to be the likelihood $f_{p_d}(x, p_d) = p(h_{p_d}(n, t) = x)$. $F_{p_d}$ is the corresponding overall distribution function of the perimeter search algorithm.

It follows that the time spent for all the expansions by a perimeter search algorithm, in the worst case, will be equal to:

$$N_1(b, d - p_d - 1, F_{p_d})t_e$$

---

[2] Using hash tables can trade memory for speed, but can only reduce access cost by a constant factor because we have to prevent clobbering.

**Useful Perimeter Nodes.** In order to estimate the time consumed by the heuristic evaluations and the collision detection procedure with a perimeter $P_{p_d}$ generated with depth $p_d$ it is necessary to know how many perimeter nodes are available. The mean number of useful perimeter nodes at any depth $i$ of the search tree developed in the forward search $|\bar{P}_{p_d}^i|$, as a function of the number of useful perimeter nodes at the preceding depth $|\bar{P}_{p_d}^{i-1}|$ is[3]:

$$|\bar{P}_{p_d}^i| = \sum_{j=1}^{|\bar{P}_{p_d}^{i-1}|} p(m_j) \times 1 + \tilde{p}(m_j) \times 0$$

where $p(m_j)$ is the likelihood that a perimeter node $m_j$ that belongs to the useful perimeter $P_{p_d}(n)$ of node $n$ at depth $i-1$ also belongs to the perimeter of the successor $n'$ of $n$, $P_{p_d}(n')$, at depth $i$. $\tilde{p}(m_j)$ is the likelihood of the complementary event: $\tilde{p}(m_j) = 1 - p(m_j)$.

In domains like those considered here, a node $m_j$, can be part of the useful perimeter of node $n$ but not part of the useful perimeter of one of its successors $n'$: this can only happen when the two following conditions are met simultaneously:

- Node $n'$ is estimated to have a larger distance to $m_j$ than that estimated from its parent:

$$h(n', m_j) - h(n, m_j) = 1$$

- The cost of node $n$ using node $m_j$, $f(n, m_j) = g(n) + h_{p_d}(n, m_j) = g(n) + h(n, m_j) + h^*(m_j, t)$, is equal to the threshold currently employed in the forward search, $\eta$:

$$g(n) + h(n, m_j) + h^*(m_j, t) = \eta$$

Summing up both expressions and transforming:

$$g(n) + h(n', m_j) + h^*(m_j, t) = 1 + \eta$$
$$g(n) + h(n', m_j) = 1 + \eta - p_d$$
$$\text{(because } h^*(m_j, t) = p_d)$$
$$h(n', m_j) = 2 + \eta - p_d - i$$
$$\text{(because } c(n, n') = 1,$$
$$\text{it follows that } g(n) = i - 1)$$

where $i$ is the depth of node $n'$.

Assuming the overall density function for node $m_j$ is the same as the overall density function of the target node $t$, it follows that $\tilde{p}(m_j)$ is equal to:

$$\tilde{p}(m_j) = p(h(n', m_j)) = 2 + \eta - p_d - i)$$
$$= f_{p_d}(2 + \eta - p_d - i) \tag{4}$$

---

[3] Note that $|\bar{P}_{p_d}^0|$ (or simply $P_{p_d}$) is defined to be the size of the initial set of perimeter nodes.

Because perimeter search algorithms save at least the last $p_d$ levels of the tree, it follows that $d = \eta - p_d$, in the worst case. Consequently, the mean number of useful perimeter nodes at depth $i$ is:

$$
\begin{aligned}
|\bar{P}^i_{p_d}| &= \sum_{j=1}^{|P^{i-1}_{p_d}|} p(m_j) \times 1 + \tilde{p}(m_j) \times 0 \\
&= \sum_{j=1}^{|P^{i-1}_{p_d}|} (1 - f_{p_d}(2 + d - i)) \\
&= |P^{i-1}_{p_d}|(1 - f_{p_d}(2 + d - i)) \qquad (5)
\end{aligned}
$$

**Heuristic Evaluations.** The number of heuristic evaluations performed by a perimeter search algorithm that has an overall distribution function $F_{p_d}(x)$, at depth $i$ is:

$$
b^i F_{p_d}(1 + d - i)|P^{i-1}_{p_d}|(1 - f_{p_d}(2 + d - i))
$$

where $d$ is the depth of the search tree traversed by the perimeter search algorithm. We define:

$$
N_2(b, d, F) = \sum_{i=1}^{d+1} b^i F(1 + d - i)|P^{i-1}_{p_d}|(1 - f_{p_d}(2 + d - i))
$$

as the expected number of calls to the basic heuristic function $h()$. An interesting effect can be observed here. The number of nodes increases exponentially with tree depth, while the number of useful perimeter nodes decreases exponentially. Which of these two effects dominates is a crucial determinant of the overall performance of perimeter search.

**Collision Detection.** The mean number of accesses to the perimeter to detect collisions at depth $i$ is equal to:

$$
b^i F_{p_d}(1 + d - i) \log(|P^{i-1}_{p_d}|(1 - f_{p_d}(2 + d - i)))
$$

We define:

$$
N_3(b, d, F) = \sum_{i=1}^{d+1} b^i F(1 + d - i) \log(|P_{p_d}|^{i-1}(1 - f_{p_d}(2 + d - i)))
$$

as the expected number of accesses to the perimeter.

## 2.5   Reduction to Unidirectional Case and Optimal $p_d$

Adding all these parts up results in the following total expected runtime for the bidirectional algorithm:

$$T_p(d, p_d) = T_g(B, p_d) +$$
$$N_2(b, d - p_d, F_{p_d})t_h +$$
$$N_1(b, d - p_d - 1, F_{p_d})t_e +$$
$$N_3(b, d - p_d, F_{p_d})t_c \qquad (6)$$

Note, $t_h$ is still the time spent by the heuristic function $h()$, not $h_{p_d}()$.

Setting $p_d$ to 0 reduces Equation 6 to Equation 2 for the unidirectional case as expected. Thus, our model reflects that unidirectional search is a special case ($p_d = 0$) of perimeter search.

Since the number of possible perimeter depths is rather small, we can compute the expected runtimes for all $p_d$ and pick the best. Of course, this best perimeter depth $p_d$ can be 0: in this case unidirectional search outperforms perimeter search.

While we simply measure $t_h$, $t_e$ and $t_c$, we have to sample the function $F_{p_d}$ in order to have all necessary inputs to compute $T_p(d, p_d)$. We do not need to know the solution depth $d$ a priori. We can simply calculate the best perimeter depth for each $d$ and switch $p_d$ if necessary, even from iteration to iteration.

## 2.6   Heavy Evaluation Functions

Search enhancements are not explicitly considered here. However, most search enhancements could be included into either $t_h$ (transposition tables, pattern databases, pattern searches, ...), or $t_e$ (move ordering, ...). Thus, even though not explicitly described here, this style of analysis is also valid for most programs with such search enhancements[4].

However, those expensive, well-informed evaluation functions often used in high performance programs are especially ill-suited to improvement by perimeter search:

- Many additional features that might be added to the naive heuristic functions increase the overall runtime of the heuristic function, making it extremely expensive to call it multiple times per node for a bidirectional search.
- Those additional features cash in on most of the cheap improvements of the heuristic that a better informed bidirectional heuristic could gain; additional improvements are harder and more expensive to come by.
- Creating, maintaining and matching frontiers and/or perimeters adds an additional performance burden that is especially noticeable when considering a sleek, high-performance (unidirectional) implementation. It is especially discouraging that, as illustrated by our model and experiments, these trends increase with larger perimeter depths: the cost of the heuristic increases dramatically while the gains in heuristic quality are moderate and decreasing. The increased infrastructural burden of the larger perimeter adds to the dramatic picture.

---

[4] For a discussion of these search enhancements and others, see [13].

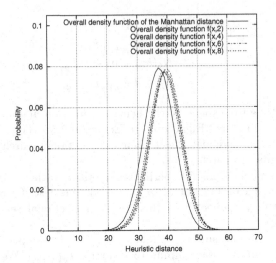

**Fig. 1.** Showing the overall distribution functions $F_h(x) = F_{p_d}(x, 0)$, $F_{p_d}(x, 2)$, $F_{p_d}(x, 4)$, $F_{p_d}(x, 6)$ and $F_{p_d}(x, 8)$.

## 3    Example Domain

This section applies the theory developed in the previous section to the domain of the 15-puzzle.

### 3.1    Heuristic Function Quality

According to Pearl [14], the heuristic $h_{p_d}(n)$ is more informed than $h(n)$, if $h_{p_d}(n) \geq h(n)$, for all $n$. It is easy to derive from this assertion that $F_{p_d}(x, p_d) \leq F_h(x)$, for all $x$. Figure 1 shows the overall density function of the original heuristic function (Manhattan distance) and those obtained with various perimeter depths (namely, 2, 4, 6 and 8).

Figure 1 shows that the largest gain is from $p_d = 0$ to $p_d = 2$. Consequent increases to 4, 6 and 8 show only smaller gains in heuristic function quality. The overall density function for the Manhattan distance was obtained sampling $10^9$ states. Because the other heuristic functions are far more expensive, they were obtained sampling $10^6$ states.

### 3.2    Useful Perimeter Nodes

The following is an attempt to validate the theoretical model using experimental data. Figure 2 shows the small differences between the empirical and predicted values for the mean number of useful perimeter nodes for problem 0 of Korf's test suite, with a perimeter depth equal to 6 units. Other problems show similar graphs. Note that, as Manzini proved [9], the number of useful perimeter nodes decreases with depth.

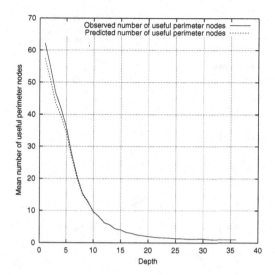

**Fig. 2.** Empirical and predicted values for the mean number of useful perimeter nodes ($p_d$=6).

### 3.3   Example Calculations Using SAL

For problem 0 of Korf's test suite, the times $t_e$, $t_h$ and $t_c$ of an implementation called SAL [15][5] were estimated to be about $1,53 \times 10^{-6}$, $1,25 \times 10^{-6}$ and $1,02 \times 10^{-6}$ seconds, respectively. The heuristic branching factor of this problem turned out to be $b = 2.05$. The asymptotic branching factor for the $n$-puzzle domain derived by Edelkamp and Korf [16] is $B = 2.13$. Figures 3 and 4 illustrate the use of these values in the formulas from the previous analysis.

- Figure 3 shows the theoretical performance of BIDA* relative to its uni-directional version IDA*, in problem 0, for different values of the heuristic branching factor ranging from 1.65 to 2.13 and for four different perimeter depths: 2, 4, 6 and 8.
  The relative performance is computed as $100 \times (T_u(d) - T_p(d, p_d))/T_u(d)$, which is the percentage of time saved by the perimeter search algorithm over its unidirectional counterpart. It can be seen that, the lower the heuristic branching factor, the lower the relative performance of BIDA* will be.
- Figure 4 shows a cross section of the previous three-dimensional graph for the aforementioned heuristic branching factor $b = 2.05$ [6]. Following this theoretical analysis, the best perimeter depth is 6, which is indeed the best perimeter depth observed empirically with SAL [15].

---

[5] SAL is a library written in C++ which implements a wide variety of single-agent search algorithms. Because this library consists of a collection of templates, they can be applied to any domain with any definition of the cost function. SAL can be downloaded from http://savannah.gnu.org/projects/sal.
[6] Computed according to the formula above.

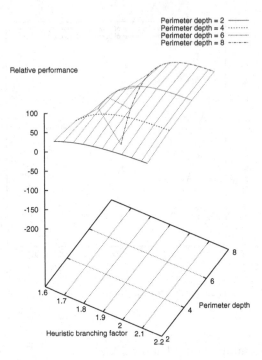

**Fig. 3.** Theoretical relative performance of perimeter search algorithms over their unidirectional version.

## 4   A Real Implementation

In spite of the previous mathematical analysis, practical questions remain unanswered. Because practitioners try to profit as much as possible from the intrinsic properties of the domain under consideration, the resulting implementations are often very well tuned, leaving little room for improvements. However, according to the previous analysis, it should still be possible. Now, a real implementation of BIDA* is presented, which beats its unidirectional version, IDA*, as implemented by Korf.

Using precompiled tables, like those devised by Korf, makes the times $t_e$ and $t_h$ negligible. One cannot expect large savings of time with large perimeter depths. Hence, only perimeter depths equal to 2 and 3 have been tested.

Figure 5 shows the mean time spent by IDA*, BIDA$_2^*$ and BIDA$_3^*$ for solving Korf's test suite consisting of 100 problems [10]. The figure shows that BIDA$_2^*$ is the fastest implementation, saving on average up to 11.74 seconds per problem. BIDA$_2^*$ saves almost 20 minutes over the 68 minutes consumed by IDA*. The runtimes for BIDA$_0^*$ are given to show that our implementation does not improve the original version of the IDA* implementation. Instead, the additional infrastructural burden is reflected in the runtime increase.

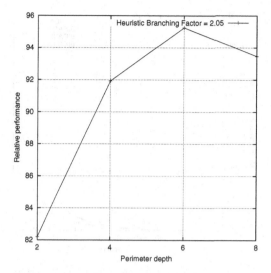

**Fig. 4.** Theoretical relative performance of perimeter search algorithms over their unidirectional version.

## 5    Conclusions

Based on the preceding analysis we can draw several conclusions:

- Perimeter search algorithms profit from: (1) the use of a better informed heuristic function, $h_{p_d}(\cdot)$; (2) saving at least the last $p_d$ levels of the forward search tree (the levels that are the most expensive for unidirectional search). In exchange, perimeter search requires generation of the perimeter $P_{p_d}$, payment of a higher computational cost for computing the heuristic, and checking for collision with the perimeter. Our analysis shows the relations among these cost advantages and disadvantages.
- Unidirectional search is a special case of perimeter search with perimeter depth 0. This insight leads to a unified model which can be used to answer the question of whether to use unidirectional or bidirectional (perimeter) search in a particular situation. Specifically, it is possible to *a priori* estimate the optimal perimeter depth for a given instance, even BIDA*-iteration!. Of course, this is only an estimate: future work is needed to verify how well the model presented here matches practical application.
- The predictions of our model are close to the values observed in practice, at least for the 15-puzzle. We believe this to be a significant indicator of the quality of the suggested model.
  Of course, Korf's code for the 15-puzzle is not the fastest possible, especially given all the search enhancements available for this domain. Still, improving upon the performance of Korf's code using perimeter search indicates that bidirectional search deserves more attention than it currently receives, especially from search practitioners.

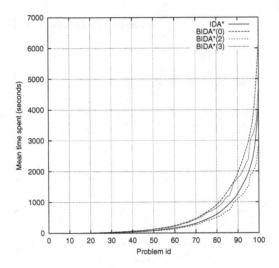

**Fig. 5.** Mean time spent by real implementations of IDA* and BIDA* in the 15-Puzzle.

## 6    Future Work

Several issues remain to be investigated:

- Even though the model developed here can give us overall trends for the behavior of bidirectional and unidirectional algorithms, it would be interesting to specialize the model for certain characteristic run times (e.g., $t_h = 1$, $t_e = t_c = 0$).
- Our model needs to be made as general as possible (removing the restrictions at the beginning of the section "Mathematical Model").
- Furthermore, it would be interesting to resolve the following non-trivial question theoretically and practically: what is the percentage of speedup and cost attributed to first, the reduction in forward search depth $(d - p_d)$ and second, the improved heuristic function quality? It may be that implementations for some applications would benefit from using only the improved heuristic or the collision detection with the perimeter.
- Since $h_{p_d}()$ uses a $min()$-function over all estimates to the perimeter nodes $h(n, m_j)$, one node $m_j$ on that perimeter with bad heuristic estimates can seriously reduce the heuristic function quality of $h_{p_d}()$. These "holes" in the perimeter have a variety of causes. One could try to extend the perimeter strategically in appropriate places in order to "patch" the holes.
- It remains open how much one can save when switching perimeter depths for each IDA* iteration. In such an approach, one could even tune the prediction formulas given here as the search proceeds.

# Acknowledgments

This paper benefited from discussions and interactions with Bart Massey, Manuela Junghanns and Jonathan Schaeffer.

# References

1. de Champeaux, D., Sint, L.: An improved bidirectional heuristic search algorithm. Journal of the Association for Computing Machinery **24** (1977) 177–191
2. Kaindl, H., Kainz, G.: Bidirectional heuristic search reconsidered. Journal of Artificial Intelligence Research **7** (1997) 283–317
3. de Champeaux, D.: Bidirectional heuristic search again. Journal of the Association for Computing Machinery **30** (1983) 22–32
4. Pohl, I.: Bi-directional and heuristic search in path problems. Stanford University. (1969) SLAC Report 104.
5. Kwa, J.: BS*: An admissible bidirectional staged heuristic search algorithm. Artificial Intelligence **38** (1989) 95–109
6. Hart, P., Nilsson, N., Raphael, B.: A formal basis for the heuristic determination of minimum cost paths. IEEE Transactions on Systems Science and Cybernetics **4** (1968) 100–107
7. Hart, P., Nilsson, N., Raphael, B.: Correction to a formed basis for the heuristic determination of minimum cost paths. SIGART Newsletter **37** (1972) 9
8. Dillenburg, J., Nelson, P.: Perimeter search. Artificial Intelligence **65** (1994) 165–178
9. Manzini, G.: BIDA*: An improved perimeter search algorithm. Artificial Intelligence **75** (1995) 347–360
10. Korf, R.: Depth-first iterative-deepening: An optimal admissible tree search. Artificial Intelligence **27** (1985) 97–109
11. Korf, R., Reid, M.: Complexity analysis of admissible heuristic search. In: Fifteenth National Conference of the American Association for Artificial Intelligence (AAAI-98), AAAI Press (1998) 305–310
12. Korf, R., Reid, M., Edelkamp, S.: Time complexity of iterative-deepening-A*. Artificial Intelligence **129** (2001) 199–218
13. Junghanns, A.: Pushing the Limits: New Developments in Single-Agent Search. PhD thesis, University of Alberta (1999)
14. Pearl, J.: Heuristics – Intelligent Search Strategies for Computer Problem Solving. Addison-Wesley Publishing Co., Reading MA (1984)
15. Linares, C.: Caracterización de los modelos de búsqueda de un agente con descripciones generalizadas de los nodos origen y destino. PhD thesis, Facultad de Informática. Universidad Politécnica de Madrid (2001)
16. Edelkamp, S., Korf, R.: The branching factor of regular search spaces. In: Fifteenth National Conference of the American Association for Artificial Intelligence (AAAI-98), AAAI Press (1998) 299–304

# Using Abstraction for Planning in Sokoban

Adi Botea, Martin Müller, and Jonathan Schaeffer

Department of Computing Science University of Alberta, Canada
{adib,mmueller,jonathan}@cs.ualberta.ca

**Abstract.** Heuristic search has been successful for games like chess and checkers, but seems to be of limited value in games such as Go and shogi, and puzzles such as Sokoban. Other techniques are necessary to approach the performance that humans achieve in these hard domains. This paper explores using planning as an alternative problem-solving framework for Sokoban. Previous attempts to express Sokoban as a planning application led to poor performance results. Abstract Sokoban is introduced as a new planning formulation of the domain. The approach abstracts a Sokoban problem into rooms and tunnels. This allows for the decomposition of the hard initial problem into several simpler sub-problems, each of which can be solved efficiently. The experimental results show that the abstraction has the potential for an exponential reduction in the size of the search space explored.

## 1   Introduction

Heuristic search has led to impressive performance in games such as chess and checkers. However, for some two-player games like Go and shogi, or puzzles like Sokoban, approaches based on heuristic search seem to be of limited value. For example, the search effort required to solve Sokoban problems increases exponentially with the difficulty of the problem [1]. Obviously, waiting until computers become 10-fold faster is not the best way to address this problem. New approaches are needed to deal with such hard domains, where humans still perform much better than the best existing programs.

Planning can be a powerful alternative to heuristic search. For example, humans are very good at planning in games, and not quite as good at searching. The last few years have seen major advances in the capabilities of planning systems, in part stimulated by the planning competitions held as part of the AIPS conference [2]. However, there are only a few results in the literature about using planning in a game-playing program [3, 4]. Part of the explanation for this is that the performance-driven aspect of many game-playing research efforts is more conducive to short-term objectives (i.e., what will make an impact in the next tournament), rather than long-term goals. In single-agent search, *macro moves* can be considered as simple plans and are, arguably, the most successful planning idea to make its way into games/puzzle practice. Macro moves are sequences of moves that are treated as a single, more powerful move. They can dramatically reduce the search tree by collapsing a sub-tree into a single move.

J. Schaeffer et al. (Eds.): CG 2002, LNCS 2883, pp. 360–375, 2003.

The idea has been successfully used in the sliding-tile puzzle [5]. Two of the most effective concepts used in the Sokoban solver ROLLING STONE are macro moves (tunnel and goal macros) [1].

Sokoban is an excellent test-bed for planning research, as the domain is recognized as being hard not only for humans, but also for artificial intelligence (AI) applications [1]. In Sokoban, a man in a maze has to push stones from their current location to designated goal locations. The problem is difficult for a computer for several reasons including deadlocks (positions from which no goal state can be reached), the large branching factor (can be over 100), long optimal solutions (can be over 600 moves), and an expensive lower-bound heuristic estimator, which limits search speed. Sokoban problems are especially challenging because they are composed to be as difficult as possible. Many problems are combinations of wonderful and subtle ideas, and finding the solution may require substantial resources — both for humans and computers. Sokoban has been shown to be PSPACE-complete [6]. Junghanns' Sokoban solver ROLLING STONE is able to find solutions for two thirds of the standard 90-problem test suite[1] [1, 7]. There is also a strong Japanese Sokoban community whose best program is written by a researcher who calls himself or herself *deep green* [1].

In this article we introduce a novel planning approach to the game of Sokoban, with the goal of overcoming the limitations of previous approaches based on heuristic search. There are at least two ways to represent Sokoban as a planning domain, and there is a huge difference in terms of efficiency between them. The first, naive approach is to translate all the properties of the domain to a planning representation. We call this the *plain* Sokoban domain representation. For instance, a regular move in Sokoban becomes an action in the planning domain. Previous experiments based on this representation generated poor results [8]. The second approach, which we will use in the paper, is to apply abstraction to the domain, so that the planner need only solve the simpler abstracted problem. We call this the *abstract* Sokoban domain. Our abstract Sokoban representation uses a preprocessing phase to decompose the puzzle into two types of objects: *rooms* and *tunnels*. At the abstract level, the maze is reduced to a graph of rooms linked by tunnels. The rooms are treated as *black boxes*. They have abstract states that model their internal configuration. The planning actions, which are essentially macros of regular Sokoban moves, refer to moving a stone from one object (room or tunnel) to another, rather than simply pushing one stone to an adjacent free square. Planning is done at this high abstraction level, which has a much smaller search space. By splitting the problem into a local component (moves within a room) and a global component (moves between rooms and tunnels), the initial search space is transformed into a hierarchy of smaller spaces, each with a corresponding reduction in the search effort required. The approach is similar to *hierarchical A\** [9].

In our approach, the abstract planning problem is solved using the standard planner TLPLAN [10]. Our initial results are very encouraging, showing a substantial reduction in the planning search effort required. In effect, the search

---

[1] The test suite is available at http://xsokoban.lcs.mit.edu/xsokoban.html

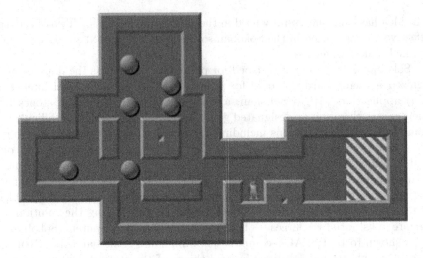

**Fig. 1.** Problem #1 in the standard 90 problem Sokoban test suite. The six goal squares are the marked ones at the right end of the maze.

problem has been split into two much smaller searches: a local preprocessing search and a global search.

The remainder of the paper is structured as follows: In the next section we introduce the domain of Sokoban. Section 3 summarizes our general planning framework for Sokoban and Sect. 4 provides details about our novel abstract representation to the game. Section 5 presents experimental results and Sect. 6 contains conclusions and ideas for further work.

## 2   The Sokoban Domain

Sokoban is a single-player game created in Japan in the early 1980s. The puzzle consists of a maze which has two types of squares: inaccessible *wall squares* and accessible *interior squares*. Several *stones* are initially placed on some of the interior squares. There is also a *man* that can walk around by moving from his current position to any adjacent free interior position. A *free* position is an interior square that is not occupied by either a stone or the man. If there is a stone next to the man and the position behind the stone is free, then the man can push the stone to that free square. The man moves forward to the initial position of the stone. The goal of the game is to push all the stones to some specific marked interior positions called *goal squares*. Figure 1 shows an example of a Sokoban problem.

One of the most interesting features of Sokoban, which contributes to both its hardness and its beauty, is the presence of *deadlocks*. A deadlock refers to an unsolvable position, when there is at least one stone that can never be pushed to a goal square. There are many types of deadlocks in Sokoban, from the simplest, which affect only one stone, to some very subtle blockades, that can involve many stones scattered over the whole puzzle. Humans who create Sokoban levels fully exploit this property of the game to obtain difficult problems. Often, the key issue in solving a problem is to detect some potential deadlocks and develop a strategy that avoids them. In computer programs, the quality of the deadlock-detection algorithm decisively affects the efficiency of the whole program. If a deadlock is detected, then the search tree is pruned and the search effort is significantly reduced. Since there is no goal node in the subtree of a deadlock state, this kind of pruning is always safe. In contrast, if a deadlock exists but is not detected, there is the danger that the corresponding node will be expanded and the search continues in the resulting subtree for a long time, with no results.

In Sokoban, the solution length can be defined in two ways: either the man movements or stone pushes can be counted. As a consequence, there are two types of optimal solutions. However, since the domain is hard enough, we do not require optimal solutions (nor do humans); *any* solution will do. This relaxation allows us to define important equivalence relationships between local configurations of the maze that lead to a simplification of the initial problem. All the equivalences defined in this paper, which support the introduction of the key concept of abstract states, ignore the optimality condition. If desired, non-optimal solutions can be improved in a post-processing phase.

## 3  Planning in Sokoban

While many domains that are used as a test-bed for current planning systems are quite simple, Sokoban is recognized in the planning community as a hard planning domain. General purpose planners cannot deal with the game at a satisfactory level. Junghanns and Schaeffer point out the limited performance that state-of-the-art fully automated planners can achieve in Sokoban [11]. One necessary step in improving the planners' performance in Sokoban is to use additional domain-specific knowledge, such as deadlock detection. This is why we chose to use Fahiem Bacchus' TLPLAN, one of the few planners that allows users to plug-in libraries that contain domain-specific code [10].

The main Sokoban-specific functions that we implemented deal with:

– Deadlock: Since deadlocks affect the search efficiency, we introduced a quick test to detect local deadlock patterns. We use Junghanns' database, which contains all the local deadlock patterns that can occur in a 5x4 area [1]. Although this enhancement was an important gain, the problem of deadlocks was far from being solved.
– Heuristic evaluation function: Since the heuristic function has a big impact on the quality of the search algorithm, we used a custom heuristic, called *Minmatching*, which is also used in ROLLING STONE [1].
– State equivalence: The definition of equivalence of states also has a special importance in Sokoban. For illustration, we provide the following example. Suppose that two states have identical stone configurations but different man positions. Suppose further that the man can walk from one position to the other. The two states are equivalent (unless we seek optimal solutions that

minimize man movements). We want to encode this relationship, as it leads to an impressive reduction of the search space. For this reason TLPLAN was enhanced with functionality such that the planner now supports custom functions to check the equivalence between states.

To enhance system performance, the plain planning representation of Sokoban, which consists of translating the game properties into a planning language such as STRIPS, was also replaced by a partially abstracted representation, called *tunnel* Sokoban. In tunnel Sokoban we identify all the tunnels present in the maze and treat them at a high abstract level (as shown in Sect. 4). All the possible configurations of a tunnel are reduced to a few abstract states and planning actions such as parking a stone inside a tunnel or or pushing a stone across a tunnel are defined.

The above modifications led to a program that is much more efficient than plain translation with no domain-specific functions. However, the system could not deal with even moderately complex puzzles. Only one from the standard test suite of 90 problems can be solved by this approach. The limitations are present because, even though the approach deals with the factors that make Sokoban hard, it is not powerful enough to do it efficiently. The search space is reduced by the tunnel macros, but the reduction is not big enough to achieve reasonable performance. Moreover, although small deadlocks are detected, there are many larger deadlock patterns that still have to be dealt with. We therefore needed a method to further reduce the search space and deal with deadlocks more efficiently. For this reason we introduced a new representation of the domain, *Abstract Sokoban*, which uses abstraction not only for tunnels but also for the rest of the maze.

# 4   Abstraction in Sokoban

*Abstract Sokoban* is a novel and highly abstracted representation of the game that replaces the initial huge search space by several smaller search spaces. The search is decomposed into several local searches and one global search. In other words, the initial problem is transformed into several simpler problems in a divide-and-conquer manner. In this approach planning is done at the global level, in a much simpler abstract space. To obtain the abstraction of the problem, the first step is to perform an initial analysis of the maze that decomposes it into objects of two types: *rooms* and *tunnels*. The next two sub-sections provide details about maze and problem decomposition.

## 4.1   Maze Decomposition

The concept of a tunnel macro was introduced in ROLLING STONE [1]. In the current version of our program, which is called POWER PLAN, tunnels of 7 abstract types are detected, including some trivial ones that have length 0. Figure 2 shows a few examples of tunnels.

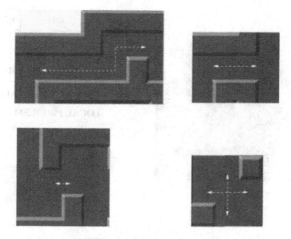

**Fig. 2.** Various types of tunnels.

Tunnels are simple objects that do not need much processing. Their properties can be obtained with little computational effort. We define abstract states for tunnels that characterize their stone configuration. While a tunnel can have many configurations (the longer it is, the larger the number of configurations it can have), the number of possible abstract states is very small. Depending on its type, a tunnel can have between 1 and 3 possible legal abstract states, as we show in the following example. Since the length of the tunnel in the lower-left corner in Fig. 2 is 0 and therefore no stone can be temporarily *parked* inside it, there is only one abstract state that the tunnel can have, corresponding to the situation when it is empty. The *straight* tunnel in the upper-right corner of Fig. 2 can have two abstract states: either the tunnel is empty or there is a stone parked inside it. In the latter case it is not important where exactly the stone is placed — all the configurations are equivalent and they are merged into the same abstract state. The tunnel in the upper-left corner of Fig. 2 has three possible abstract states. One abstract state is defined for the empty tunnel. There is also one state for the situation when a stone is pushed inside through the left end of the tunnel. The last abstract state is for the case when a stone was pushed inside through the right end of the tunnel. These last two abstract states have different properties. In the first case, the stone can be taken out through the left end only, while in the second case the same action can be done through the right end only.

After computing tunnels, all the remaining interior points are grouped together in connected components called *rooms*. Two points belong to the same room if and only if there is a connection between them that does not cross any tunnel. We define abstract states for rooms in the following way: one abstract state represents all the room configurations that can be obtained from each other in such a way that neither any stone nor the man leaves or enters the room (we say that these configurations are *equivalent*). We call a room that contains at

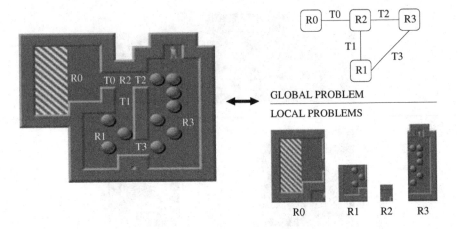

**Fig. 3.** A problem (#6 of the test suite) is decomposed into several abstract sub-problems. There is one global problem as well as one local problem for each room. Rooms and tunnels are denoted by $R$ and $T$, respectively.

least one goal square a *goal room*. We emphasize once more that the definition of abstract states for rooms and tunnels preserves solvability but not optimality of solutions.

## 4.2    Problem Decomposition

Once the maze is split into rooms and tunnels, the initial problem can be decomposed into several smaller ones, as shown in Fig. 3. At the global level, a search problem is transformed into a graph $(R_i, T_j)$, where the nodes $R_i$ represent rooms and the edges $T_j$ represent tunnels. Besides the global planning problem, we also get several local search problems, one for each room. The complexity of a local problem depends on both the size and the shape of a room. The local problem attached to the one-square room $R2$ is much simpler than the one attached to the largest room $R3$. While the complexity of the initial problem increases exponentially with the size of the maze, the complexity of the local problems increase exponentially with the size of the rooms only. Moreover, since the local computation is done only once, its results can be reused many times during the global-level search.

**Local Problems.** The challenge for each room is to compute and provide quick access to information needed at planning time. At the local level, we compute the graph of abstract states of the room. For each abstract state, some properties are also calculated that will be used to check action preconditions at planning time. For instance, for an abstract state we might want to know whether we can push one more stone into the room through a certain tunnel. Another important result of the local computation is that all the states that contain local deadlock patterns are detected and eliminated from the search space.

The steps of the local abstraction are summarized as follows.

---

**Algorithm 1** Rooms Local Processing.

---

1: remove dead squares;
2: build local move graph;
3: mark deadlock configurations;
4: run SCC (strongly connected components) algorithm, find abstract states (AS);
5: determine properties of AS, find abstract moves (actions) for AS.

---

The removal of dead squares is done earlier, during maze preprocessing. Two types of squares are marked as dead: Some squares are completely useless and we remove them from the maze (e.g., tunnels that have one end closed). As done in ROLLING STONE [1], we also mark the *stone-dead* squares, where the man can go but stones cannot be pushed because of deadlock. Next, starting from the initial stone configuration of the room, we compute the *local move graph* of all possible configurations that the room can have. We can use the local move graph to detect all deadlock configurations that can occur locally in the room. We consider a local position to be deadlocked if we cannot clear the room of stones or, equivalently, if there is no path in the local move graph from that position to the empty position. We mark positions using retrograde analysis, starting from the empty position, which is marked as legal. After all the $n$-*stone* configurations have been marked as either legal or deadlock, we go to the next level and mark the $(n + 1)$-*stone* configurations: if there is a path from the current $(n + 1)$-*stone* position to a legal $n$-*stone* or $(n + 1)$-*stone* position, then the current position is legal too. Otherwise, it is deadlocked.

In the next step, to obtain the abstract states of the room, we run the SCC algorithm, which computes the strongly connected components of the local move graph. Each strongly connected component becomes an abstract state. In this way, equivalent configurations are merged together into the same abstract state. All deadlock states are mapped into one abstract deadlock state. To be able to check action preconditions at the planning level, for each abstract state some predicates are also computed (such as "can push one more stone inside the room through entrance X"). When the value of these predicates is $TRUE$, we also compute the resulting abstract states after we perform the corresponding actions, such as pushing one stone, pulling one stone, etc.

For rooms that have up to 15 non-dead squares we are usually able to complete the local processing described above. However, for large rooms it is not feasible to compute all the possible abstract states. Our program currently handles only cases where a large room is directly linked to a goal room. Given an abstract state of the large room, first we check whether we can push a stone to the goal room. If so, we accomplish this action and do not compute preconditions of any other possible actions. This optimization leads to a significant reduction of the local problem complexity. Otherwise, preconditions of other possible actions are computed. To further speed up the computation, we store minimal deadlock patterns and maximal legal patterns. Any position that contains a deadlock pattern is illegal. Any position that is contained in a legal pattern is legal. How to cope with large rooms in general is still an open question and constitutes one of our main directions for future research.

To summarize, the original problem can be abstracted into rooms and tunnels. The internal state of a room can be computed by retrograde analysis as part of the preprocessing stage. Our simplistic approach to computing rooms turns out to be the bottleneck in performance (see Sect. 5). However, it was done this way for simplicity, since it was more important to determine the viability of the abstraction approach than it was to maximize performance. Obviously, not all the states computed in the preprocessing will be needed during a planning search. A better approach would be to compute only subsets of the room's state as needed by the planner.

**Global Problem.** In the global problem, the abstraction is obtained by mapping the maze to a small graph of rooms connected by tunnels. This global problem is solved by planning in abstract Sokoban. To run our experiments, we used TLPLAN enhanced with domain-specific knowledge, as described in Sect. 3. Planning actions now refer to moving one stone from one room or tunnel to another, rather than simply pushing one stone or moving the man by one square. Objects involved in a stone movement change their abstract states after the corresponding action is completed. To be able to move one stone from one room to another, stones in both rooms may have to be re-arranged and the exact way to do it is computed at the local level.

The solutions obtained by our system are correct, but the completeness is not guaranteed. For instance, when moving one stone from one room to another, we currently compute only one possible way to re-arrange stones inside the considered rooms. However, there is a theoretical chance that the way we re-arrange stones is not the one that allows us to get the solution of the problem. To reduce this risk, when accomplishing an action, we do as few local changes as possible and keep open as many room entrances as possible. With this optimization, we have not encountered so far any case when the system missed the solution because of the completeness issue.

Table 1 summarizes the action types in the 3 planning approaches discussed in this paper: plain, tunnel, and abstract Sokoban. While the actions always refer to pushing a stone from one node to another, the meaning of nodes is different: regular squares in plain Sokoban, tunnels and squares in tunnel Sokoban, and rooms and tunnels in abstract Sokoban.

The planning goal is expressed as the conjunction of the conditions that each goal room should obey. The overall goal is reached when all goal rooms are in the abstract state where all its goal squares are occupied by a stone. In Fig. 3, $R0$ is the only goal room of the maze and it contains 8 goal squares.

Compared to plain Sokoban and tunnel Sokoban, the abstract representation shows greater promise for addressing the game as a planning problem. As will be shown in Sect. 5, problems that cannot be solved by the first two approaches

**Table 1.** A comparison of actions in the three Sokoban planning representations.

| Representation Type | Nodes | Actions |
|---|---|---|
| plain Sokoban | squares | $push(node_1, node_2)$ |
| tunnel Sokoban | squares + tunnels | $push(node_1, node_2)$ $cross(tunnel)$ |
| abstract Sokoban | rooms + tunnels (objects) | $push(node_1, node_2)$ |

are easily handled in the abstract one. The two main factors that explain this important improvement are search space size and deadlock detection. In abstract Sokoban the search space is much smaller than in plain and tunnel Sokoban. Both branching factor and solution length are greatly reduced as a result of the abstraction. One abstract move or action is typically composed of several normal Sokoban moves. Planning in abstract Sokoban is also simpler because there are less deadlocks to deal with. All the deadlocks that can occur inside one room are handled by the local analysis. A move never goes into a local deadlock. The only deadlocks that still have to be considered at the planning level are the large ones that involve interactions between several rooms and tunnels.

## 5   Experimental Results

In this section we present experimental results for solving Sokoban using abstraction. Since it is more meaningful to run the tests on real problems (as opposed to using toy problems), we have chosen 10 out of the 90 levels from the standard test suite for our experiments. The 10 problems, which are shown in the Appendix, are solvable by our abstract Sokoban system. Six problems (1, 2, 6, 7, 17, 80) are composed of small rooms and therefore the local preprocessing can be done completely. The other four (3, 4, 5, 9) contain both large and small rooms. Our current framework for large rooms can successfully cope with these problems. These results should be viewed as preliminary, as we scale the system to handle the remaining 80 problems. Our system, called POWER PLAN, uses abstract Sokoban and TLPLAN. To evaluate its performance, we compared POWER PLAN with two other approaches: TLPLAN + tunnel Sokoban (or simply tunnel Sokoban) and ROLLING STONE, which is the best search-based Sokoban solver available.

Figure 4 shows how the solution length is reduced in POWER PLAN and ROLLING STONE. $SP$ represents the number of stone pushes in the solutions found by ROLLING STONE. These are the best values that we know of, and in many cases they are equal to the optimal values. Since ROLLING STONE uses macro moves in its tunnel macros and goal macros, $RS$, which is the length of the solution found by ROLLING STONE, is smaller than $SP$. $AS$, which stands for

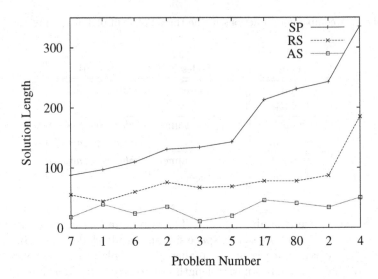

**Fig. 4.** Solution length in POWER PLAN (*AS*) and ROLLING STONE (*RS*). *SP* is close to the number of stone pushes in the optimal solution. Since both POWER PLAN and ROLLING STONE use macro moves, *AS* and *RS* are smaller than *SP*.

abstract Sokoban, is the length of the solution found by POWER PLAN (i.e., the number of planning actions). *AS* is much smaller than *SP*, as one planning action in abstract Sokoban corresponds to several regular moves. Since the solution length reduction is a measure of how the search space is reduced, the graph suggests that our global search space is smaller than the main search space used in ROLLING STONE. This is an important result, as it promises an exponential reduction in the search space.

When using tunnel Sokoban, TLPLAN can seldom solve a problem entirely (in our test subset, only the simplest problem, which has 6 stones, can be solved). For this reason, we solve sub-problems of the initial problem. A sub-problem is obtained by removing from the initial configuration some stones as well as an equal number of goal squares. Figure 5 reveals how the effort for solving sub-problems of Problem #6 evolves for ROLLING STONE (*RS*), tunnel Sokoban (*TS*), and POWER PLAN (*AS*). Here the number of expanded nodes in the main search are plotted (note the logarithmic scale). Tunnel Sokoban is only able to solve sub-problems with 7 or less stones. For the next sub-problem (8 stones) this system didn't find a solution after running for more than 20 CPU hours. Compared to ROLLING STONE, POWER PLAN achieves a reduction by a factor that remains stable over the whole set of sub-problems of Problem #6. This graph also illustrates how dramatically the search effort can be reduced by the choice of problem representation and solving method.

Table 2 presents a more detailed comparison between abstract Sokoban and tunnel Sokoban. The data demonstrates a huge difference in terms of efficiency between the two approaches. Even if the numbers in the *PPN* column seem to

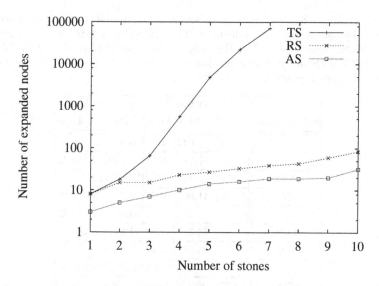

**Fig. 5.** The expanded nodes in the main search for abstract Sokoban (*AS*), tunnel Sokoban (*TS*), and ROLLING STONE (*RS*). The values were obtained for sub-problems of Problem #6.

be large, the preprocessing is fast. The processing cost for a node in the local search space is low, since there is no heuristic function to be computed.

Table 3 shows a comparison between POWER PLAN and ROLLING STONE. As in the case of abstract Sokoban, ROLLING STONE also uses two types of search and, to be able to perform a measurement, we consider a one-to-one correspondence between the search spaces in the two approaches. At the global level ROLLING STONE does the so-called top-level search, whose purpose is to find a goal state. We denote the number of searched nodes at the top level by $TLN$ and compare it to $PlN$ from abstract Sokoban. There is also the pattern search in ROLLING STONE, whose main goal is to determine deadlock patterns and find better bounds for the heuristic function [12] ($PSN$ is the number of expanded nodes in the pattern search). We compare pattern search in ROLLING STONE with local preprocessing in POWER PLAN, as they both are means to simplify the main search.

The good news is that, for many problems, the number of planning nodes $PlN$ is smaller than $TLN$, which supports the claim that our global search space is smaller than the one considered by ROLLING STONE. In contrast, when analyzing $PSN$ and $PPN$, we infer that ROLLING STONE is more efficient from this perspective. Indeed, ROLLING STONE is a finely tuned application, developed over several years of effort, while there is still room to improve our local computation. Although the preprocessing search is very fast, rooms can be big. We therefore need ways to get around the performance bottleneck induced by preprocessing. Some ideas have already been implemented in our system, as shown in Sect. 4. Problems #3, #4, and #9, which contain large rooms, are solved

**Table 2.** Abstract Sokoban vs. tunnel Sokoban. Sub-problem $x(y)$ is obtained from problem $x$ by placing $y$ stones in the maze. The sub-problems listed are the largest that tunnel Sokoban can solve. $PlN$ is the number of expanded nodes during the planning search. $PPN$ is the number of expanded nodes during the preprocessing phase. The time is measured in seconds.

| Sub-problem | Abstract Sokoban | | | Tunnel Sokoban | |
|:---:|---:|---:|---:|---:|---:|
| | PlN | PPN | Time | PlN | Time |
| 1(6) | 71 | 1,044 | 1.57 | 10,589 | 126.24 |
| 2(6) | 24 | 61,113 | 0.93 | 80,740 | 9,490.21 |
| 3(7) | 8 | 482 | 0.12 | 77,919 | 12,248.66 |
| 4(6) | 9 | 41,065 | 0.80 | 27,514 | 3,061.94 |
| 5(6) | 7 | 404 | 0.20 | 53,141 | 11,733.83 |
| 6(7) | 19 | 54,317 | 1.06 | 71,579 | 8,189.77 |
| 7(8) | 13 | 26,011 | 0.75 | 132 | 0.88 |
| 9(6) | 13 | 245 | 0.25 | 35,799 | 4,883.55 |
| 17(5) | 1,047 | 306,224 | 29.63 | 14,189 | 391.42 |
| 80(6) | 10 | 395,583 | 3.02 | 14,266 | 949.98 |

efficiently by our program. However, there are many improvements that can be further added in order to process large rooms efficiently. ROLLING STONE is also faster than our system, with the exception of Problem #4 and Problem #9. The overhead is determined by the local processing as well as the usage of a general purpose planner. The version of TLPLAN that we used in our experiments is often two orders of magnitude slower than the latest experimental version of the planner [13]. On the other hand, our method has the advantage that other planners too can be used to solve the global planning problem, whereas ROLLING STONE is a special purpose system.

Our first results confirm that our problem-solving architecture works. We have already obtained evidence that abstract Sokoban is more efficient than other planning representations of the game. To the best of our knowledge, no previous planning attempts in Sokoban led to solving real problems. However, as the comparison with ROLLING STONE shows, parts of our architecture, such as local computation, are still in an early stage. A few of the ideas about how to improve our system's performance are presented in Sect. 6.

## 6   Conclusions and Future Work

In games such as Sokoban, approaches based on heuristic search seem to be of limited value. In this paper we have proposed an alternative problem-solving architecture based on AI planning. Since the classical representation of Sokoban as a planning domain did not lead to acceptable results, we introduced abstract Sokoban, mapping the puzzle to a highly abstracted planning domain. Our first

**Table 3.** Abstract Sokoban vs. ROLLING STONE. *PlN* is the number of expanded nodes in the planning search, *PPN* is the number of preprocessing nodes, *TLN* is the number of top-level nodes, and *PSN* is the number of nodes in the pattern search. The time is measured in seconds.

| Problem | Abstract Sokoban | | | Rolling Stone | | |
|---|---|---|---|---|---|---|
| | PlN | PPN | Time | TLN | PSN | Time |
| 1 | 71 | 1,044 | 1.57 | 50 | 1,042 | 0.14 |
| 2 | 635 | 62,037 | 16.10 | 80 | 7,530 | 0.63 |
| 3 | 12 | 19,948 | 2.04 | 87 | 12,902 | 0.23 |
| 4 | 128 | 69,511 | 3.20 | 187 | 50,369 | 3.27 |
| 5 | 36 | 297,334 | 23.14 | 202 | 43,294 | 1.72 |
| 6 | 36 | 54,414 | 1.37 | 84 | 5,118 | 0.31 |
| 7 | 54 | 35,813 | 1.57 | 1,392 | 28,460 | 1.37 |
| 9 | 35 | 7,607 | 1.01 | 1,884 | 436,801 | 22.17 |
| 17 | 8,091 | 444,073 | 166.98 | 2,038 | 29,116 | 2.23 |
| 80 | 47 | 877,914 | 4.56 | 165 | 26,943 | 2.25 |

experimental results support the claim that abstract Sokoban outperforms other planning representations of the game. Furthermore, since abstraction leads to a huge reduction of the global search space, we are encouraged to say that planning could be used to overcome the limitations exhibited by heuristic search in Sokoban.

There are many directions that we plan to explore with abstract Sokoban. Our framework does not currently handle all types of large rooms. In addition, the rooms local computation can be further optimized. As a consequence, one of our main future work directions is to complete and improve the local processing for both goal and regular rooms. Another enhancement that we expect to have a great impact on the system performance is a smarter decomposition of the maze into rooms and tunnels. While our heuristic rule that guides this process is quite rigid, it can be replaced by a strategy aiming to optimize several parameters (e.g., minimize the number of rooms and tunnels, minimize the interactions between rooms and tunnels). Moreover, the global planning search space can be further simplified, detecting large deadlocks that involve interactions between several rooms and tunnels. One important future research topic is to try our ideas in real-life planning domains (e.g., robotics related). Automatic abstraction of planning domains can also be an interesting extension of our work.

# References

1. Junghanns, A.: Pushing the Limits: New Developments in Single-Agent Search. PhD thesis, Department of Computing Science, University of Alberta (1999)
2. Bacchus, F.: AIPS'00 planning competition. AI Magazine (2001) 47–56

3. Shapiro, A.: Structured Induction in Expert Systems. Turing Institute Press. Addison-Wesley (1987)
4. Wilkins, D.: Using knowledge to control tree searching. Artificial Intelligence **18** (1982) 1–51
5. Korf, R.: Macro-operators: A weak method for learning. Artificial Intelligence **26(1)** (1985) 35–77
6. Culberson, J.: Sokoban is PSPACE-complete. Technical report, Department of Computing Science, University of Alberta, Edmonton, Alberta, Canada (1997) ftp://ftp.cs.ualberta.ca/pub/TechReports/1997/TR97-02.
7. Junghanns, A., Schaeffer, J.: Sokoban: Enhancing single-agent search using domain knowledge. Artificial Intelligence **129** (2001) 219–251
8. McDermott, D.: Using regression-match graphs to control search in planning. (1997) http://www.cs.yale.edu/HTML/YALE/CS/HyPlans/mcdermott.html.
9. Holte, R., Perez, M., Zimmer, R., MacDonald, A.: Hierarchical A*: Searching abstraction hierarchies efficiently. Technical report, University of Ottawa, TR-95-18 (1995)
10. Bacchus, F., Kabanza, F.: Using temporal logics to express search control knowledge for planning. Artificial Intelligence **16** (2000) 123–191
11. Junghanns, A., Schaeffer, J.: Domain-dependent single-agent search enhancements. In: Sixteenth International Joint Conference on Artificial Intelligence (IJCAI-99), Morgan Kaufmann Publishers (1999) 570–575
12. Junghanns, A., Schaeffer, J.: Single-agent search in the presence of deadlock. In: Fifteenth National Conference of the American Association for Artificial Intelligence (AAAI-98), AAAI Press (1998) 419–424
13. Bacchus, F.: Personal communication (2002)

# Appendix: The 10 Problem Test Suite

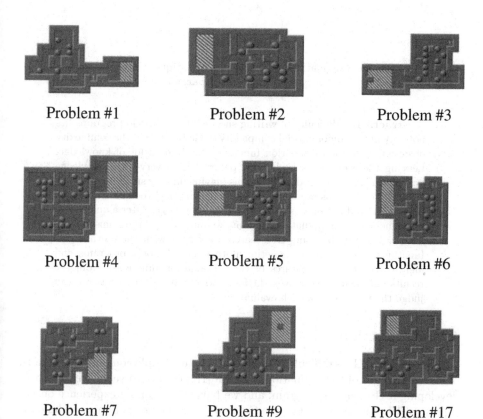

Problem #1          Problem #2          Problem #3

Problem #4          Problem #5          Problem #6

Problem #7          Problem #9          Problem #17

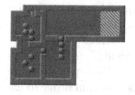

Problem #80

# A Small Go Board Study of Metric and Dimensional Evaluation Functions

Bruno Bouzy

C.R.I.P.5, UFR de mathématiques et d'informatique, Université Paris 5
bouzy@math-info.univ-paris5.fr

**Abstract.** The difficulty of writing successful $19 \times 19$ Go programs lies not only in the combinatorial complexity of Go but also in the complexity of designing a good evaluation function containing a lot of knowledge. Leaving these obstacles aside, this paper defines very-little-knowledge evaluation functions used by programs playing on very small boards. The evaluation functions are based on two mathematical tools, distance and dimension, and not on domain-dependent knowledge. After a qualitative assessment of each evaluation function, we built several programs playing on $4 \times 4$ boards by using tree search associated with these evaluation functions. We set up an experiment to select the best programs and identify the relevant features of these evaluation functions. From the results obtained by these very-little-knowledge-based programs, we can judge the usefulness of each evaluation function.

## 1 Introduction

$19 \times 19$ computer Go is difficult because of tree search explosion, but also because of the complexity of the evaluation function [1]. For several years, we have been developing a Go playing program, and we have accumulated experience on two aspects of $19 \times 19$ computer Go, the Go model and the programming techniques.

With regard to programming techniques, a Go playing program may contain tactical look-ahead, pattern-matching, evaluation function and highly selective global search. In this paper we simplify this technical aspect by reducing the size of the board. Thus, we leave tactical look-ahead, pattern-matching and selectivity and we keep the evaluation function and tree search without selectivity.

Besides, a Go model may contain knowledge about "strings", "liberties", "groups", "eyes", "connections", "territories", "life" and "death" and other useful concepts embedded in an evaluation function. But, in this paper we chose not to take this large domain-dependent knowledge into account, in order to explore the dimensional and metric features of evaluation functions only. We call a model in which the dimension of objects play an important role *dimensional*, and a model in which the distance between objects plays an important role *metric*. Thus, the aim of this paper is to study *dimensional and metric Go evaluation functions* with a system which uses *tree search* on *small boards*.

Section 2 of this paper explains the motivation behind "dimensional" evaluation functions and gives their definitions. Section 3 defines evaluation functions based on a notion of distance. Section 4 shows examples and contains a

J. Schaeffer et al. (Eds.): CG 2002, LNCS 2883, pp. 376–392, 2003.

qualitative assessment of our evaluation functions. Section 5 describes practical experiments to assess the evaluation functions. Before the conclusion, Sect. 6 underlines the results of the experiment. To shorten this presentation, EF stands for evaluation function.

## 2 "Dimensional" Evaluation Functions

This section motivates the dimensional feature and defines dimensional EF.

### 2.1 Motivation

The classical Go evaluation function $E$ corresponds to the sum of the abstract color of each intersection $i$ of the Go board $I$:

$$E = \sum_{i \in I} abstractColor(i) \tag{1}$$

The function $abstractColor(i)$ returns +1 (respectively -1) if the intersection $i$ is *controlled* by Black (respectively White) and returns 0 if the intersection is not controlled at all. The *control* notion can be more or less complex. [2] used an EF with a simplified control. [3] described EF's with complex control including life and death knowledge and morphological operators. In our study, we want to keep the abstractColor function as simple as possible. We do not insert life and death knowledge or morphological knowledge into the *abstractColor* function. The *abstractColor(i)* function returned +1 (respectively -1) if the intersection $i$ was either occupied by Black (respectively White) or empty but surrounded by Black (respectively White) intersections only, otherwise it returns 0. Other ways of writing (1) are possible. The notion of "group" in Go being fundamental, we may render this notion by writing:

$$E = \sum_{g \in G} \sum_{i \in g} abstractColor(i) \tag{2}$$

$G$ is the set of the groups situated on the board, whatever the definition of a group might be. Furthermore, the *abstractColor* function returns the same value for each intersection belonging to the same group. Thus we can define the *abstractColor(g)* of a group $g$ as the constant value of the *abstractColor(i)* in which $i$ is an intersection of $g$.

$$E = \sum_{g \in G} size(g) abstractColor(g) \tag{3}$$

Furthermore, we can define $G_B$ as the subset of $G$ whose groups $g_b$ are black, in other words $abstractColor(g_b) = +1$ and we can define $G_W$ as the subset of $G$ whose groups $g_w$ are white, in other words $abstractColor(g_w) = -1$. Then we can write (3) as follows:

$$E = E_B - E_W \tag{4}$$

$$E_B = \sum_{g \in G_B} size(g) \tag{5}$$

$$E_W = \sum_{g \in G_W} size(g) \tag{6}$$

The size of objects being a basic feature when studying dimensionality of objects, we should bear in mind that the dimensionality of a set $S$ is defined on the basis of measures $M_{d,r}(S)$ such as:

$$M_{d,r}(S) = \sum_{g \in G} size(g)^d \tag{7}$$

Here, $d$ is a dimensional parameter and $G$ is a set of balls $g$ of radius $r$ whose union covers $S$.

$$S \subseteq \bigcup_{g \in G} g \tag{8}$$

These formulas are applied to sets of continuous space to determine their fractal dimension [4]. When the radius $r$ of the ball falls to zero, $M_{d,r}(S)$ reaches a value $M_d(S)$. It is proved that $M_d(S)$ equals either 0 or $+BIG$. The fractal dimension $\delta$ is defined as the unique value for which

$$M_d(S) = 0 \text{ for } d > \delta \text{ and } M_d(S) = +BIG \text{ for } d < \delta \tag{9}$$

Of course, a Go board is not continuous but discrete and finite. Thus, decreasing the radius of ball to zero is nonsense. Nevertheless, we are not aiming at finding any fractal dimension of any object but we may wonder whether the measures defined by (7) provide useful information to a Go program or not. This is the "dimensional" motivation of our paper.

### 2.2    Dimensional Evaluation Functions

With $d$ integer, we can now define the EF $E_d$ by the following formula:

$$E_d = \sum_{g \in G_B} size(g)^d - \sum_{g \in G_W} size(g)^d \tag{10}$$

In our study, we assume that $d \in [0, 2]$. $E_1$ is the classical EF useful for the endgame. $E_0$ is the count of black groups minus white groups. $E_2$ measures the ability of one color to get large groups of this color and small groups of the other color.

## 3    Metric Evaluation Functions

This section defines a "metric" EF. As the connection and the distance notions are important in Go, we may define simple evaluation functions by using simple distance functions like in [5] or in [6].

Formula (11) is a simple way to define an evaluation function for color $c$. When combined with (4), (11) leads to the definition of the "metric" evaluation function. The minus sign stands for increasing the evaluation function of color $c$ when the distance of two intersections is low. The player of color $c$ wants to minimize the distance of color $c$ between the intersections.

$$E_c = - \sum_{i,j \in I} d(c,i,j) \qquad (11)$$

$d(c,i,j)$ is a distance function between two intersections $i$ and $j$ depending on color $c$. It is defined by (12) with a usual distance $d$. $c$ can be equal to 'Black', 'White' or 'Empty', $otherColor(Black)$ equals 'White', $otherColor(White)$ equals 'Black' and $otherColor(Empty)$ equals 'Empty'. $c(i)$ is short for *abstract Color(i)*.

$d(i,j)$ is the distance between two intersections $i$ and $j$ in a metric space. When the number of neighbouring intersections of one intersection is 4, $d$ is the Manhattan distance and we call it $d4$. When the number of neighbouring intersections of one intersection is 8, we call it $d8$. In the following text, $E_{d4}$ (respectively $E_{d8}$) is the evaluation function defined by (4), (11) and (12) by using the distance $d4$ (respectively $d8$).

$$d(c,i,j) = +BIG \text{ if } c(i) = otherColor(c) \text{ or } c(j) = otherColor(c) \qquad (12)$$
$$\text{otherwise, } d(c,i,j) = 0 \text{ if } i,j \in S \text{ connected set and } c(S) = c$$
$$\text{otherwise, } d(c,i,j) = 1 \text{ if } d(i,j) = 1$$
$$\text{otherwise, } d(c,i,j) = \min_{c(k) \neq otherColor(c)} \{d(c,i,k) + d(c,k,j)\}.$$

The first line of (12) means that two intersections are situated at an infinite distance for color $c$ if one of them is of the opposite color of $c$. Otherwise, the second line shows that two intersections belonging to the same connected set S of color $c$ are situated at distance 0 for color $c$. Otherwise, the third line indicates that two distance one intersections for the classical distance are also at distance one for the colored distance. Finally, colored distance $d$ is defined by the fourth line for the other cases.

Formula (12) provides an almost correct definition of a distance, it satisfies two out of three axioms. First, when applied on intersections $x$ and $y$, $d$ of color $c$ is not a distance because $d(c,x,y) = 0 \Rightarrow x = y$ is false. But when aggregating the elements of a connected set into one element, then $d$ is respectful of $d(c,x,y) = 0 \Rightarrow x = y$. Second, $d(c,x,y) = d(c,y,x)$ is true. Third, $d(c,x,y) \leq d(c,x,z) + d(c,z,y)$.

## 4  Qualitative Assessment

This section provides a set of examples of evaluations and several remarks showing that each of these evaluations corresponds to some meaningful concept of Go and recalling the possible downsides of each one.

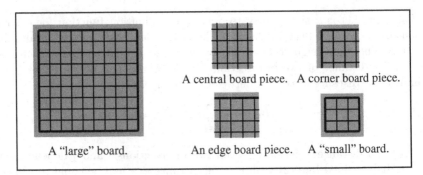

Fig. 1. "Open" and "closed" boards

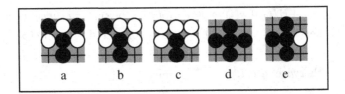

Fig. 2. Five terminal positions

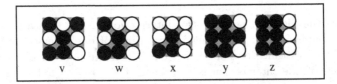

Fig. 3. Nonterminal positions

## 4.1 Evaluation Examples

This subsection gives several examples of position evaluations. But first, we need to distinguish "open" from "closed" boards. Figure 1 gives examples of such boards. Go is always played on "closed" board, for example the $10 \times 10$ board on the left or the $4 \times 4$ board on the right. But, when studying a local position of a large board it is easier to define board pieces. A board piece contains edges that are either "open" or "closed". A closed edge of a board piece corresponds to an actual edge of the initial board. It is drawn with a thick line. An open edge of a board piece is "open" toward other parts of the initial board. It is drawn as if the initial board was cut along this edge. An intersection of an open edge has an unknown number of liberties that depends on the hidden part of the initial board. A board that contains at least one open edge is defined as open, and closed otherwise.

Figures 2, 3, 4 and Tables 1, 2, 3 show evaluations $E_1$, $E_2$, $E_{d4}$ and $E_{d8}$ of some example positions on open boards.

**Table 1.** The evaluation of the terminal positions of Fig. 2. *BIG* value is set to 1024

|        | v     | w     | x      | y      | z      |
|--------|-------|-------|--------|--------|--------|
| $E_0$  | 0     | 0     | 0      | -1     | 0      |
| $E_1$  | +1    | -1    | -3     | +5     | +3     |
| $E_2$  | +5    | -7    | -27    | +47    | +27    |
| $E_{d4}$ | +5120 | -7168 | -27648 | +48128 | +27648 |
| $E_{d8}$ | +9216 | -9216 | -27648 | +48128 | +27648 |

**Table 2.** The evaluation of the non terminal positions of Fig. 3. *BIG* is set to 1024

|        | a      | b      | c      | d      | e      |
|--------|--------|--------|--------|--------|--------|
| $E_0$  | 0      | 0      | 0      | +1     | 0      |
| $E_1$  | +1     | -1     | -3     | +5     | +3     |
| $E_2$  | +3     | -5     | -21    | +25    | +15    |
| $E_{d4}$ | +9208  | -3076  | -33780 | +78784 | +54224 |
| $E_{d8}$ | +11260 | -11260 | -33780 | +78784 | +54224 |

**Table 3.** The evaluation of the terminal positions of Fig. 2. *BIG* value is still set to 1024

|        | f | g      | h      | i      | j      |
|--------|---|--------|--------|--------|--------|
| $E_0$  | 0 | +1     | 0      | -1     | -2     |
| $E_1$  | 0 | +4     | +2     | +1     | -1     |
| $E_2$  | 0 | +16    | +8     | +7     | +1     |
| $E_{d4}$ | 0 | +71612 | +38876 | +11274 | -31640 |
| $E_{d8}$ | 0 | +71612 | +38878 | +23528 | -13304 |

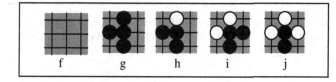

**Fig. 4.** Other positions

Table 4 sums up the final evaluations of perfect play on closed $n \times n$ boards with $n \in \{2, 3, 4\}$.

## 4.2   Remarks

This subsection contains a list of qualitative remarks about the EF's.

*Remark 1.* $E_0$ alone is adapted to the opening of games on large boards. The small board in the upper right of Fig. 5 shows the perfect sequence played by

**Table 4.** The evaluation of the terminal positions of perfect play on small boards

| size | $2 \times 2$ | $3 \times 3$ | $4 \times 4$ |
|------|------|------|------|
| $E_0$ | 0 | +1 | 0 |
| $E_1$ | +1 | +9 | +2 |
| $E_2$ | +3 | +81 | +28 |
| $E_{d4}$ | +5118 | +82944 | +36856 |
| $E_{d8}$ | +5118 | +82944 | +36856 |

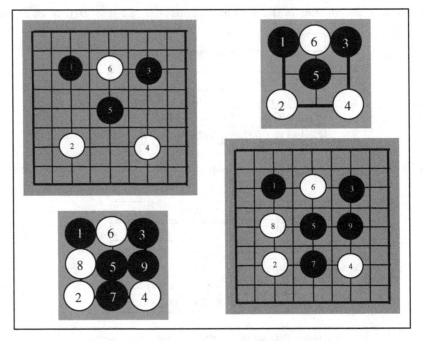

**Fig. 5.** Two openings on a large board obtained by a mapping from the perfect play on a small board by using either $E_0$ or $\lambda E_1 + E_0$ (and no capture rule)

using $E_0$ without the capture rule. The large board in the upper left of Fig. 5 shows the same sequence mapped from the small to the large board by a scaling operator. To some extent, this sequence contains adequate moves of an opening on a large board. The moves are played to occupy big empty points far from friendly stones, which is one of the most important strategies in the opening.

*Remark 2.* Associated with $E_1$, $E_0$ is also adapted to the openings of games on large boards. The small board in the lower left of Fig. 5 shows the perfect game played on a $3 \times 3$ board by using a linear combination evaluation function, $\lambda E_1 + E_0$, $\lambda > 0$, without the capture rule. Again, the board in the lower right shows the same sequence mapped by a scaling operator. This sequence completes

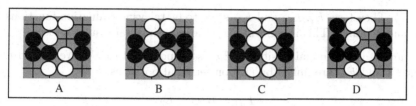

**Fig. 6.** Four open positions. Position A is the starting position, in which B is a good option for Black, C is a good option for White and D is a bad option for Black

the previous one by adding the moves to occupy normal large side extensions, which is another important feature in the opening.

*Remark 3.* $E_1$ is adapted to the endgame. Associated with the *abstractColor* function, this is the classical EF in Go.

*Remark 4.* $E_2$, $E_{d4}$ and $E_{d8}$ are well suited for the middle game. $E_2$ leads the program to grow its own large groups and to reduce or cut the opponent's ones. Figure 6 shows four open positions in which it may be worthwhile to connect or disconnect the stones. Table 5 gives the evaluations of positions of Fig. 6. In the context of the middle game, human Go players will agree that position B is the best option for Black and that C is the best one for White.

**Table 5.** Evaluations of the positions of Fig. 6

|          | A      | B | C      | D      |
|----------|--------|---|--------|--------|
| $E_0$    | 0      | 0 | +1     | 0      |
| $E_1$    | -1     | 0 | -2     | 0      |
| $E_2$    | -5     | 0 | -36    | +2     |
| $E_{d4}$ | -72544 | 0 | -81868 | -51080 |
| $E_{d8}$ | -21564 | 0 | -81868 | -70    |

$E_1$ is dull, evaluating every move from position A the same, with an incentive +1. $E_2$ is more suited to the middle game because, when playing white, option C is far ahead. Unfortunately, when playing black, depth one search using $E_2$ cannot clearly discriminate between the set of moves. Depth one tree search using $E_{d4}$ enables the system to select option B for black because connecting two 4-connected sets of color $c$ into one slightly increases $E_{d4}(c)$. But option C is not far ahead when playing white because adding one element to a connected set does not increase $E_{d4}$. Moreover, 8-connected sets correspond either to the boundary of territories recognized at the end of the game or to the dividers in a fighting position. Therefore, $E_{d8}$ is useful. Depth one tree search using $E_{d8}$ enables the system to select option C for White. Option B is ahead when playing Black, because connecting two 8-connected sets of color $c$ into one increases the

$E_{d8}(c)$. Therefore, $E_2$, $E_{d4}$ and $E_{d8}$ seem to be relevant evaluation functions for the middle game. This is confirmed by the experimental results in Sect. 6.

*Remark 5.* On terminal positions, $E_{d4}$ is proportional to $E_2$. We demonstrate that $E_2$ and $E_{d4}$ are linked by (13) on terminal positions.

$$E_{d4} = BIG \times E_2 \tag{13}$$

Let us assume that the position is terminal. Given the importance of the connected sets, ordering the sum of (11) according to the connected sets $S$ and $S'$ of the position is appropriate. This yields (14).

$$E_c = -\sum_{S,S'} \sum_{i \in S, j \in S'} d(c, i, j) \tag{14}$$

Now, an intersection is either Black or White. Furthermore, $d(c, i, j)$ either equals 0 or $+BIG$. If $i$ and $j$ are of color $c$ and belong to the same connected set, then $d(c, i, j)$ equals 0, otherwise it equals $+BIG$. This gives (15).

$$E_c = -(\sum_{S,S'} \sum_{i \in S, i' \in S'} BIG - \sum_{S:c(S)=c} \sum_{i,j \in S} BIG) \tag{15}$$

Then, we can count the number of elements of these two sums. If $T$ is the number of intersections of the terminal position, the first sum contains $T^2$ elements and the second one contains $E_{2,c}$ elements. $E_{2,c}$ is either $E_{2,B}$ or $E_{2,W}$ (see (4) and (10)). Thus, we simply obtain (16).

$$E_c = -BIG \times (T^2 - E_{2,c}) \tag{16}$$

Finally, the use of (4) and (16) demonstrates (13). We could of course get a similar formula linking $E_{d8}$ and $E_2$ by changing the connection from 4-connection to 8-connection.

*Remark 6.* $E_{d4}$ and $E_{d8}$ are more reliable than $E_1$ or $E_2$ on non-terminal positions. On $19 \times 19$ middle game positions, $E_1$ cannot be used appropriately. Moreover, $E_2$ has the downside of being insensitive to some good moves (see remark 4). In the positions of Fig. 7, depth-one tree search based on $E_{d4}$ or $E_{d8}$ selects the right moves for Black and White.

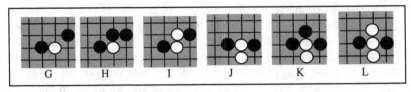

**Fig. 7.** Six open positions corresponding to larger middle game positions. Position G and J are the starting positions in which H and K are "good" options for Black, I and L are "good" options for White

**Table 6.** The evaluations of the positions of Fig. 7

|        | G      | H      | I    | J   | K      | L      |
|--------|--------|--------|------|-----|--------|--------|
| $E_1$  | +1     | +2     | 0    | 0   | +1     | -1     |
| $E_2$  | +1     | +4     | -2   | -2  | -1     | -7     |
| $E_{d4}$ | +29630 | +57438 | -100 | -80 | +27718 | -27810 |
| $E_{d8}$ | +29642 | +57342 | -66  | -30 | +27684 | -27732 |

**Table 7.** The results of $N \times N$ Go in Japanese and Chinese rules, $0 < N < 5$

| Size     | $1 \times 1$ | $2 \times 2$ | $3 \times 3$ | $4 \times 4$ |
|----------|------|--------------|--------------|--------------|
| Japanese | draw | draw         | win          | draw         |
| Chinese  | 0    | $\{+1|-1\}$  | $\{+9|-9\}$  | $\{+2|-2\}$  |

## 5 Practical Experiments

This section describes the two main experiments we carried out: the $4 \times 4$ Go resolution speed-up with $E_2$ instead of $E_1$, and the automatic weight adjustments of a combination of $E_1$, $E_2$, $E_{d4}$, $E_{d8}$ and other parameters by means of an evolving population of $4 \times 4$ Go programs. First, we briefly go over the state of the art of programs playing on small boards to define the test set of the first experiment. Then, we describe the main properties of the tree search algorithm that we used to perform the experiments. Finally, we point out the main features of the evolving population of $4 \times 4$ Go programs which aims at finding an adapted combination of parameters.

### 5.1 Small Go Board Solution

[7] and [8] focused on $2 \times N$ boards while [9] and [2] focused on $N \times N$ boards ($N \leq 4$). [9] provided a solution of $2 \times 2$, $3 \times 3$ and $4 \times 4$ Go by using Japanese rules, little Go knowledge, and alpha-beta search with a transposition table. [2] described retrograde analysis of Go patterns of size $3 \times 3$ or $4 \times 4$ with Chinese rules. Table 7 points out the results of $N \times N$ Go using either Japanese rules or Chinese rules and Fig. 8 shows the optimal sequence for each size of board from $2 \times 2$ up to $4 \times 4$. The sequences apply to both Chinese and Japanese rule sets.

### 5.2 Tree Search Algorithm

Our reference algorithm is alpha-beta with iterative deepening [10]. The time limit at which the last iteration was triggered was 900 seconds on a Pentium 450Mhz with 128MB memory. Iterative deepening enabled the search algorithm to find the correct move quickly. The first three positions of the optimal $4 \times 4$ game could not be played without iterative deepening because of a lack of memory. We used a transposition table [11], [12] with $2^{19}$ entries. In each entry,

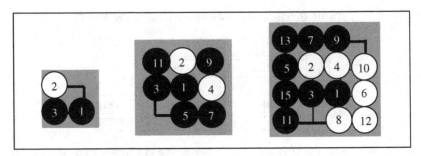

**Fig. 8.** Perfect play on $2 \times 2$, $3 \times 3$ and $4 \times 4$ boards

we stored the Zobrist hash key of the position, the next player to move, whether the last move was a pass or not, the set of moves forbidden by repetition, the depth, the alpha beta bounds and the move found by the previous iteration.

We used the history heuristic [13]: when a move is found to be "sufficient" to create a cut-off somewhere in the tree, the history move value is increased by $2^{depth}$ in the history table. We observed a 22% reduction of visited nodes. Thus, the history heuristic offers a very positive enhancement. Of course, it is not indispensable but it is so easy to implement without any downside that we inserted it into our reference algorithm. We did not use null-move pruning reduction [14] because in Go, null-move is a normal move. We did not use MTD(f) [15] either because the reduction was too small.

Apart from the rules of Go, and the evaluation function that uses a simple abstractColor function, we inserted as little Go knowledge as possible into the move-ordering algorithm. A move has a domain dependent priority that is low near the corners and high in the center of the board. A move has a very low priority when the rules of the game forbid it to the opponent, illustrating the proverb that "my good moves are also my opponent's good moves".

### 5.3  Population of $4 \times 4$ Programs

When starting the experiment, we were looking for a good combination of $E_1$, $E_2$, $E_{d4}$ and $E_{d8}$. Therefore, we used evaluation $E$ defined by (17).

$$E = a_1 E_1 + a_2 E_2 + b_4 E_{d4} + b_8 E_{d8} \qquad (17)$$

We set up a population of eight $4 \times 4$ Go programs, each of them using an instance of the evaluation of (17). In the first stage, $BIG$ was set to 1024. We decided that $a_1$, $a_2$, $b_4$ and $b_8 \in [0, 16]$. The parameters for the first eight programs were selected randomly. One tournament consisted of 56 games in which every program played all other programs twice, once with black and once with white. A win was worth 2 points, a draw one point and a loss no point. After the 56 games, the programs were ranked according to their points total, with the score average as a tie-break. The timeout limit of iterative deepening was set to one second to make the programs play quickly and shorten the time for conducting the experiment.

When a tournament was finished, the first sixth programs plus two new programs entered the next tournament. The first new program was the copy of the winner of the tournament with a random mutation $a_i = a_i \pm 1$ or $b_i = b_i \pm 1$. The second new program was created at random. The tournaments were performed over a "period". After each tournament, a population had an average value that was the average value of the first six programs. We measured the convergence of the population average value with the sum of the square of the deviation of each weight. When a convergence threshold was reached, the period ended.

When a period ended, the space for the next period was modified. When a parameter $a_i$ or $b_i$ converged toward an average value in the previous period, then its new variation interval became centered around this value and its length was reduced for the next period.

This adjustment was performed for several reasons. First, to avoid too slow a convergence. For example, if one parameter, say $a_2$, converged to a fixed value, say 4, and the whole population was in the interval [2,6], then generating a program with one parameter set at random in [0,16] was not appropriate. Therefore, a new set of values for this parameter such as [2, 6] with 17 values was chosen. Second, when one parameter reached one frontier of the interval, the size of the interval was doubled. For example, if one parameter, say $b_4$, reaches the max frontier, say 16, then the new interval was [0, 32]. When a parameter did not converge significantly during the period, this parameter was considered as "noisy" and its interval remained the same for the next period.

We supervised this process "by hand" and stopped it when we considered that either some parameters were sufficiently adjusted or some parameters remained "noisy". This was the end of an "era" — one supervision iteration. At the end of each era, a convergent parameter was fixed to its value, and disappeared from the list of features for the next era, and a "noisy" parameter may give rise to the birth of new features for the next era.

The first era was the life of $a_1$, $a_2$, $b_4$ and $b_8$, at the end of which we observed that the linear combination of Formula 17 could greatly benefit from the tuning of other parameters. The second era marked the adjustment of $BIG$, used by $E_{d4}$ and $E_{d8}$. At the end of this era, no more convergence was foreseeable. However, since the dimensional and the metric evaluation functions are very different by nature, we wondered whether they could be applied in two different stages of the game. Thus, the third era witnessed the introduction and the adjustment of "temporal" parameters reflecting the split of a $4 \times 4$ game into an opening, a middle game and an endgame phase.

## 6   Results of the Experiments

This section provides numerical results from the two experiments assessing the dimensional and metric EF. First, we highlight the node number reduction provided by $E_2$ on $4 \times 4$ resolution. Then, we underline the weights of a linear combination of $E_1$, $E_2$, $E_{d4}$ and $E_{d8}$ obtained by an evolving population of programs playing Go on $4 \times 4$ boards.

## 6.1  $E_2$ Enhancement Assessment

Table 8 illustrates the node number reduction resulting from the classical alpha-beta enhancements such as transposition table, iterative deepening, history heuristic, null move pruning and MTD(f). The measurements were performed on the test set made up of 17 positions of the optimal sequence of $4 \times 4$ Go.

**Table 8.** Results of search enhancements

| TT | ID | HH | null-move | MTD(f) |
|---|---|---|---|---|
| $+\infty$ | $+\infty$ | 22% | 10% | 6% |

Without well-formulated Go knowledge about move ordering, we observed that transposition table and iterative deepening were mandatory. The history heuristic is efficient (22%). Null move reductions are possible (10%). Like in chess, null move pruning on $4 \times 4$ Go actually reduces the number of visited nodes. MTD(f) is a positive but small enhancement (6%).

Instead of $E_1$, we tried to use $E_2$. First, we noticed that the sequences found by the search algorithm using $E_2$ were exactly the same as the sequences found by the normal algorithm. Hence, changing the evaluation function does not change the external behavior of the program and $E_2$ is still able to achieve perfect play on $4 \times 4$ Go. Furthermore, as shown in Fig. 9, the advantage is that, from position number 2 up to the last position of the optimal game, the number of nodes searched by the algorithm using $E_2$ is 21% smaller than with $E_1$.

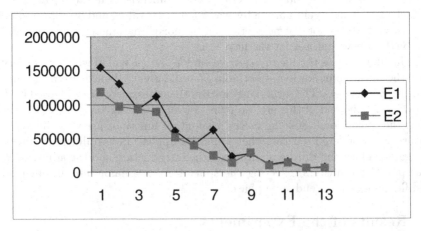

**Fig. 9.** Number of nodes searched by the reference algorithm with $E_1$ or $E_2$

We try to explain this reduction. $E_2$ increases the position evaluations in which the friendly sets are connected. Therefore, the algorithm first explores the moves that increase the size of the connected sets and gives a bad evaluation to

**Table 9.** Results at the end of the first era

| $a_1$ | $a_2$ | $b_4$ | $b_8$ |
|-------|-------|-------|-------|
| [0,8] | [0,8] | [32, 96] | [16, 48] |

positions in which friendly stones are split into parts. This explanation is closely linked to the fact that, unfortunately, $E_2$ based tree search visits slightly more nodes than the normal one on positions 0 and 1. In these two positions, one color is not present. Consequently, $E_2$ cannot benefit from increasing a connected set of this color because this set does not exist. Then, the question is to know how the node number evolves with the "power" $d$ of the evaluation function but we have not carried out this experiment yet.

### 6.2   Assessment of Dimensional versus Metric Evaluation Functions

This subsection deals with results obtained by the evolving population. First, we describe the results of the weight adjustment of Formula 17. This adjustment corresponds to the first era. The first era contained 8 periods and about 400 tournaments at the end of which we observed the results of Table 9:

These results showed the superiority of the metric evaluations over the dimensional ones and also the slowness of the convergence, all parameters remaining "noisy" in their convergence interval.

Because the "best" move of each iteration of iterative deepening was unstable, the decision to set the timeout with a low value lead to almost random play for the first few moves in the openings played in the first tournaments. But the advantage lay in the possibility to explore much space and make the evaluation function weights adapt more quickly.

Given the importance of the metric evaluations and the important weight of $BIG$ within these evaluations, it was urgent to detect the relative importance of the $BIG$ value. Therefore, we added the new feature $BIG$ to the population of programs and started the second era with $BIG \in [0, 2048]$. This era lasted 200 tournaments and 6 periods. Table 10 shows the results.

**Table 10.** Results at the end of the second era

| $a_1$ | $a_2$ | $b_4$ | $b_8$ | $+BIG$ |
|-------|-------|-------|-------|--------|
| [0,8] | [0,1] | [60,68] | [28, 36] | [384, 640] |

Beyond the value of $BIG$, 512, we observed a better convergence in this era than in the previous era: $a_2$ decreased a lot and $b_4$ and $b_8$ converged toward 64 and 32 to some extent. But still, $a_1$ remained noisy. At the end of this era, several observations could be made.

First, the games were not played until their end, as defined by the rules, but stopped before. This was a positive consequence of the weight adjustment. The

**Table 11.** Results at the end of the third era

| depthOpening | endOpening | openEndGame | bOpening | bEndGame |
|:---:|:---:|:---:|:---:|:---:|
| 1 | 4 | 10 | [64, 80] | [8, 16] |

downside was that the good programs stopped early without physically capturing the virtually captured stones. When two programs disagreed, the referee decided who the winner was. Unfortunately, the referee used $E_1$ and did not count the correct evaluation[1] but the naive static one, and the programs playing well had a penalty. Second, we saw that the first iterations of iterative deepening produced a very fluctuant "best" move. Because we wanted the experiment to be finished early enough, we were obliged to set up a short timeout to iterative deepening. Consequently, the first moves of the game (about the first four ones), produced by elapsed timeout iterative deepening were almost random, whereas the middle and endgame moves, produced by deep enough search, were correct. For the opening, we also noticed that the move produced by the first iteration of iterative deepening was surprisingly often a good one, and that the next iterations gave worse moves.

Therefore, for the next era, we required a population of programs whose parameters depended on the stage of the game and also included the maximal depth of iterative deepening as a parameter.

In the third era, we fixed the old features of the playing programs with $a_1 = 1$, $a_2 = 8$, $BIG = 512$ and we correlated $b_4$ and $b_8$ with the formula $b_8 = b_4/2$. We defined five new parameters: *depthOpening*, *bOpening*, *bEndgame*, *openEndGame* and *endOpening*. *endOpening* was the last move number of the opening phase of a game. *openEndGame* was the number of the first move of the endgame. *depthOpening* was the maximal depth of iterative deepening during the opening. *bOpening* was the value of $b_4$ during the opening and the middle game and *bEndgame* was the value of $b_4$ during the endgame. We started the third era with *depthOpening* $\in [1, 6]$, *endOpening* $\in [1, 16]$, *openEndGame* $\in$ [endOpening, 20], *bOpening* $\in [48, 80]$, *bEndgame* $\in [0, 80]$. After 200 tournaments these parameters converged toward the values shown in Table 11.

One result is very surprising: the iterative deepening depth that produces the best result is depth one! We are not able to provide an adequate explanation for it. The other results had no unexpected element. We expected the value of *bEndgame* to decrease to give more importance to the basic $E_1$ evaluation function, relevant to the endgame. This was observed because *bEndGame* converged toward 8, which is the $E_1$ weight. We expected *bOpening* to keep the same value and we actually observed this result. Finally, the two boundaries fixing the opening and endgame phases reached satisfying values: the value 4 marks the end of the opening and the value 10 corresponds the beginning of the endgame.

---

[1] the referee did not perform mini-max search as the players did!

# 7   Conclusion

Here are the contributions of this paper:

- Definition of three "dimensional" Go evaluation functions $E_0$, $E_1$ and $E_2$.
- Definition of two "metric" Go evaluation functions $E_{d4}$ and $E_{d8}$.
- $E_0$ is qualitatively adapted to the opening on large boards.
- $E_1$ is the classical Go evaluation function.
- $E_2$ achieves an experimental speed-up on the solution of $4 \times 4$ Go that is comparable to the classical alpha-beta enhancements.
- $E_{d4}$ and $E_{d8}$ are qualitatively relevant to the middle game on large boards and experimentally adapted to small board Go associated with a depth one search. This constitutes the main contribution of this study.

This paper opens up various perspectives. First, extending the experiment to larger boards until the life and death module becomes necessary, then introducing life and death knowledge into the abstractColor function. Furthermore, it seems worthwhile to integrate the metric evaluation functions $E_{d4}$ and $E_{d8}$ into our $19 \times 19$ program to improve its middle game play, and to integrate the $E_0$ evaluation into the opening evaluation of our $19 \times 19$ playing program.

# References

1. Chen, K.: Computer Go: Knowledge, search, and move decision. International Computer Chess Association Journal **24** (2001) 203–215
2. Bouzy, B.: Go patterns generated by retrograde analysis. In Uiterwijk, J., ed.: The 6th Computer Olympiad Computer-Games Workshop Proceedings. Report CS 01-04 (2001)
3. Bouzy, B., Cazenave, T.: Computer Go: an AI oriented survey. Artificial Intelligence **132** (2001) 39–103
4. Mandelbrot, B.: The fractal geometry of nature. W.H. Freeman and Company, San Francisco (1982)
5. van Rijswijk, J.: Computer Hex: are bees better than fruitflies? Master's thesis, University of Alberta, Edmonton, AB (2000)
6. Enzenberger, M.: The integration of a priori knowledge into a Go playing neural network. http://www.markus-enzenberger.de/neurogo.html (1996)
7. Thorp, E., Walden, W.: A computer assisted study of Go on MxN boards. Information Sciences **4** (1972) 1–33
8. Lorentz, R.: 2xN Go. In: Proceedings of the $4^{th}$ Game Programming Workshop in Japan'97. (1997) 65–74
9. Sei, S., Kawashima, T.: A solution of Go on $4 \times 4$ board by game tree search program. Fujitsu Social Science Laboratory manuscript (2000)
10. Slate, D., Atkin, L.: Chess 4.5 – the Northwestern University chess program. In Frey, P., ed.: Chess Skill in Man and Machine. Springer-Verlag (1977) 82–118
11. Greenblatt, R., Eastlake III, D., Crocker, S.: The Greenblatt chess program. In: Fall Joint Computing Conference 31, (New-York ACM) (1967) 801–810
12. Marsland, T.: A review of game-tree pruning. International Computer Chess Association Journal **9** (1986) 3–19

13. Schaeffer, J.: The history heuristic and alpha-beta search enhancements in practice. IEEE Transactions on Pattern Analysis and Machine Intelligence **11** (1989) 1203–1212

14. Donninger, C.: Null move and deep search: selective-search heuristics for obtuse chess programs. International Computer Chess Association Journal **16** (1993) 137–143

15. Plaat, A., Schaeffer, J., Pijls, W., De Bruin, A.: Best-first fixed-depth minimax algorithms. Artificial Intelligence **87** (1996) 255–293

# Local Move Prediction in Go

Erik van der Werf, Jos W.H.M. Uiterwijk,
Eric Postma, and Jaap van den Herik

Search and Games Group, IKAT, Department of Computer Science,
Universiteit Maastricht
{e.vanderwerf,uiterwijk,postma,herik}@cs.unimaas.nl

**Abstract.** The paper presents a system that learns to predict local
strong expert moves in the game of Go at a level comparable to that
of strong human kyu players. This performance is achieved by four techniques. First, our training algorithm is based on a relative-target approach that avoids needless weight adaptations characteristic of most
neural-network classifiers. Second, we reduce dimensionality through
state-of-the-art feature extraction, and present two new feature-extraction methods, the Move Pair Analysis and the Modified Eigenspace Separation Transform. Third, informed pre-processing is used to reduce state-
space complexity and to focus the feature extraction on important features. Fourth, we introduce and apply second-phase training, i.e., the
retraining of the trained network with an augmented input constituting
all pre-processed features. Experiments suggest that local move prediction will be a significant factor in enhancing the strength of Go programs.

## 1 Introduction

Since the founding years of Artificial Intelligence (AI) computer games have
been used as a test bed for AI algorithms. Many game-playing systems have
now reached an expert level. Go is a notable exception. In the last decades, Go
has received significant attention from AI research [1, 2]. Despite all efforts, the
best computer Go programs are still weak. Many (if not all) of the current top
programs rely on static knowledge bases. As a consequence the programs tend
to become extremely complex and difficult to improve. In principle a learning
system should be able to overcome this problem.

In this paper we investigate methods for building a system that predicts
expert moves based on local observations. For such a prediction we focus on the
ranking that can be made between legal moves which are near to each other.
Ideally, in a local ranking, the move played by the expert should be among the
best. (It should be noted that one would like to predict optimal moves based on
full-board observations. However, due to the complexity of the game and the size
of any reasonable feature space to describe the full board, full-board prediction
is beyond the scope of this paper.)

It is known that many moves in Go conform to some local pattern of play
which is performed almost reflexively by human players. The reflexive nature
of many moves leads us to believe that pattern-recognition techniques, such as

J. Schaeffer et al. (Eds.): CG 2002, LNCS 2883, pp. 393–412, 2003.
© Springer-Verlag Berlin Heidelberg 2003

neural networks, are capable of predicting many of the moves made by human experts. The encouraging results reported by Enderton [3] and Dahl [4] on similar supervised-learning tasks, and by Schraudolph [5] who used neural networks in a reinforcement-learning framework, underline our belief.

The remainder of this paper is organized as follows. In Sect. 2 the local move predictor is introduced. In Sect. 3 we discuss the raw features that are used by the local move predictor for ranking the moves. Section 4 presents feature-extraction and pre-scaling methods for extracting promising features to reduce the dimensionality of the raw-feature space. In Sect. 5 we discuss the option of a second-phase training, performed on the original high-dimensional feature vectors, that can further improve performance in case our feature-extraction method left some important information out. Then Sect. 6 presents experimental results on the performance of the raw features, the feature-extraction, the pre-scaling and the second phase training. In Sect. 7 we assess the quality of our system by comparing it to the performance of human amateurs on a local prediction task, by testing on professional games, and by actually playing against the program GNUGO. Finally, Sect. 8 provides conclusions and suggestions for future research.

## 2    The Predictor

The goal of the move predictor is to rank legal moves, based on a set of local features, in such a way that expert moves are ranked above (most) other moves. In order to rank moves they must be made comparable. This can be done by performing a non-linear mapping of the local feature vector onto a scalar value for each legal move. Such a mapping can be performed by a general function approximator, such as a neural network, which can be trained from examples.

The architecture chosen for our move predictor is the well-known feed-forward Multi Layer Perceptron (MLP). Our MLP has one hidden layer with non-linear transfer functions, is fully connected to all inputs, and has a single linear output predicting the value for ranking the move. Although this local move predictor alone will not suffice to play a strong game, e.g., because it lacks a global perspective, it can be of great value for move ordering and for reducing the number of moves that have to be considered globally.

Functionally the network architecture is identical to the half-networks used in Tesauro's comparison training [6], which were trained by standard back propagation with momentum. Our approach differs from Tesauro's in that it employs a more efficient training scheme and error function especially designed for the task of move ordering.

### 2.1    The Training Algorithm

The MLP must be trained in such a way that expert moves are generally valued higher than other moves. Several training algorithms exist that can be used for this task. In our experiments, the MLP was trained with the resilient propagation algorithm (RPROP) developed by Riedmiller and Braun [7]. RPROP is a gradient-based training procedure that overcomes the disadvantages of gradient-descent techniques (slowness, blurred adaptivity, tuning of learning parameters, etc.). The gradient used by RPROP consists of partial derivatives of each network weight with respect to the (mean-square) error between the actual output values and the desired output values of the network.

In standard pattern-classification tasks the desired output values are usually set to zero or one, depending on the class information. Although for the task of move prediction we also have some kind of class information (expert / non-expert moves) a strict class assignment is not feasible because class membership may change during the game, i.e., moves which may initially be sub-optimal, later on in the game can become expert moves [3]. Another problem, related to the efficiency of fixed targets, is that when the classification or ordering is correct, i.e., the expert move is valued higher than the non-expert move(s), the network needlessly adapts its weights just to get closer to the target values.

To incorporate the relative nature of the move values into the training algorithm, training is performed with move pairs. A pair of moves is selected in such a way that one move is the expert move, and the other is a randomly selected move within a pre-defined maximum distance from the expert move.

With this pair of moves we can devise the following error function

$$E(v_e, v_r) = \begin{cases} (v_r + \epsilon - v_e)^2, & \forall \, v_r + \epsilon > v_e \\ 0, & otherwise \end{cases} \tag{1}$$

in which $v_e$ is the predicted value for the expert move, $v_r$ is the predicted value for the random move and $\epsilon$ is a control parameter that scales the desired minimal distance between the two moves. A positive value for control parameter $\epsilon$ is needed to rule out trivial solutions where all predicted values $v$ would become equal. Although not explicitly formulated in his report [3] Enderton appears to use the same error function with $\epsilon = 0.2$.

Clearly the error function penalizes situations where the expert move is ranked below the non-expert move. In the case of a correct ranking the error can become zero (just by increasing the scale), thus avoiding needless adaptations. The exact value of control parameter $\epsilon$ is not very important, as long as it does not interfere with minimum or maximum step-sizes for the weight updates. (Typical settings for $\epsilon$ were tested in the range $[0.1, 1]$ in combination with standard RPROP settings.)

From (1) the gradient can be calculated by repeated application of the chain rule, using standard back propagation. A nice property of the error function is that no gradient needs to be calculated when the error signal is zero (which practically never happens for the standard fixed target approach). As the performance grows, this significantly reduces the time between weight updates.

The quality of the weight updates strongly depends on the generalization of the calculated gradient. Therefore, all training is performed in batch, i.e., the gradient is averaged over all training examples before performing the RPROP weight update. To avoid overfitting, training is terminated when the performance on an independent validation set does not improve for a pre-defined number of weight updates. (In our experiments this number is set to 100.)

# 3   Choice of Raw Features

Features are important for prediction. In this section an overview of raw input features is presented which can be used as inputs to the local move predictor. The list is by no means complete and could be extended with a (possibly huge) number of Go-specific features. Our selection comprises a simple set of locally computable features that are common in most Go representations [3, 4] used for similar learning tasks. In a tournament program our representation can readily be enriched with extra features which may be obtained by more extensive (full-board) analysis or by specific goal-directed searches.

Our selection consists of the following eight features: stones, edge, ko, liberties, liberties after, captures, nearest stones and the last move.

- **Stones**

  The most fundamental set of features to describe a board configuration is the positioning of black and white stones. Local stone configurations can be represented by two bitmaps. One bitmap represents the positions of the black stones, the other represents the white stones.

  Since we are interested in local move prediction, it seems natural to define a local region of interest (ROI) centered on the move under consideration. The points inside the ROI are then selected from the two bitmaps, and concatenated into one binary vector, points outside the ROI are discarded. A question which arises when defining the ROI is: what are good shapes and sizes? For this we tested differently sized square, diamond, and circular shapes, shown in Fig. 1, all centered around the free point considered for play. For simplicity (and for saving training time) we only included the edge and the ko features. In Fig. 2 the percentages of incorrectly ordered move-pairs are plotted for the ROIs. These results were obtained from several runs with training sets of 100,000 raw feature vectors. Performances were measured on independent test sets. The standard deviations (not shown) were less than 0.5 percent.

  The results do not reveal significant differences between the three shapes. In contrast, the size of the ROI does affect performance considerably. Figure 2 clearly shows that initially, as the size of the ROI grows, the error decreases. However, at a size of about 30 the error starts to increase with size. The increase of classification error with the size (or dimensionality) of the input is known as the "peaking phenomenon" [8] which is caused by the "curse of dimensionality" [9] to be discussed in Sect. 4.

  Since the shape of the ROI is not critical for performance, in all further experiments we employ a fixed shape, i.e., the diamond. The optimal size of the ROI cannot yet be determined at this point since it depends on other factors such as the number of training patterns, the number of other raw features and the performance of the feature-extraction methods that will be discussed in the next section.

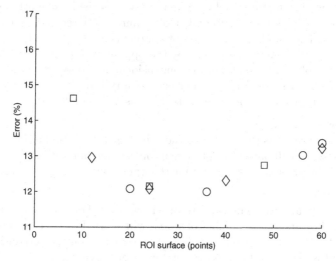

**Fig. 1.** Shapes and sizes of the ROI.

**Fig. 2.** Performance for different ROIs.

– **Edge**

In Go, the edge has a huge impact on the game. The edge can be encoded in two ways, 1) by including the coordinates of the move that is considered for play, or 2) by a binary representation (board=0, edge=1) using a 9-bit string vector along the horizontal and the vertical line from the move towards the closest edges. Preliminary experiments showed a slightly better performance for the binary representation (around 0.75%). Therefore we implemented the binary representation. (Note that since the integer representation can be obtained by a linear combination of the larger binary representation, there is no need to implement both. Furthermore, since on square boards the binary representation can directly be transformed to a full binary ROI by a logical OR it is not useful to include more edge features.)

– **Ko**

The ko-rule, which forbids returning to previous board states, has a significant impact on the game. In a ko-fight the ko-rule forces players to play threatening moves, elsewhere on the board, which have to be answered locally by the opponent. Such moves often are only good in a ko-fight since they otherwise just reduce the player's number of ko-threats. For the experiments presented here only two ko-features were included. The first one is a binary feature which says if there is a ko or not. The second one is the distance from the point considered for play to the point which is illegal due to the ko-rule.

In a tournament program it may be wise to include more information, such as (an estimate of) the value of the ko (number of points associated with winning or losing the ko), assuming this information is available.

– **Liberties**
An important feature used by human players is the number of liberties. The number of liberties is the number of unique free intersections connected to a stone. The number of liberties of a stone is a lower bound on the number of moves which must be played by the opponent to capture that stone.
In our implementation for each stone inside a diamond-shaped ROI the number of liberties is calculated. For each empty intersection only the number of directly neighboring free intersections is calculated.

– **Liberties after**
This feature, directly related to the previous, is the number of liberties that a new stone will have after placement on the position that is considered. The same feature is also calculated for the opponent moving here first.

– **Captures**
Capturing and avoiding captures are important. For this we include two features. The first is the number of stones which are directly captured (removed from the board) after placing a stone on the position that is considered. The second is the same feature for the opponent moving first.

– **Nearest stones**
In Go stones can have long-range effects which are not visible inside the ROI. To detect such long-range effects a number of features are incorporated characterizing stones outside the ROI.
Since we do not want to incorporate all stones outside the ROI, features are calculated only for a limited set of stones near the point considered for play. These nearest stones are found by searching in eight directions (2 horizontal, 2 vertical and 4 diagonal) starting from the points just outside the ROI. In Fig. 3 the two principal orderings for these searches are shown.

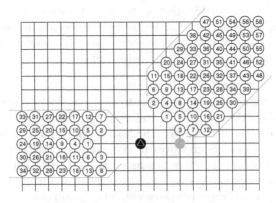

**Fig. 3.** Ordering of nearest stones.

The point considered for play is represented by the marked stone, directly surrounded by the diamond-shaped ROI. Outside the ROI a horizontal beam and a diagonal beam of numbered stones are shown representing the order in which stones are searched. For each stone found we store the following four features: 1) color, 2) Manhattan distance to the proposed move, 3) offset perpendicular to the beam direction, and 4) number of liberties.

In our implementation only the first stone found in each direction is used. However, at least in principle, a more detailed set of features might improve the performance. This can be done by searching for more than one stone per direction or by using more (narrow) beams (at the cost of increased dimensionality).

- **Last move**
The last feature in our local representation is the Manhattan distance to the last move played by the opponent. This feature is often a powerful cue to know where the action is. In the experiments performed by Enderton [3] (where bad moves were selected randomly from all legal moves) this feature was even considered harmful since it dominated all others and made the program play all moves directly adjacent to the last move played. However, since in our experiments both moves are selected from a local region we do not expect such a dramatic result.

A standard technique to simplify the learning task, that is applied before calculating the raw feature vector, is canonical ordering and color reversal. The game of Go is played on a square board which contains eight symmetries. Furthermore, positions with Black to move are equal to positions where White is to move if all stones reverse color. Rotating the viewpoint of the move under consideration to a canonical region in one corner, and reversing colors so that only one side is always to move, effectively reduces the state space by a factor approaching 16.

## 4   Feature Extraction and Pre-scaling

The local board representation, presented in the previous section, can grow quite large as more information is added (either by simply increasing the ROIs or by adding extra Go-specific knowledge). The high-dimensional feature-vectors significantly slow down training and can even decrease the performance (see Fig. 2). As stated in Sect. 3 the negative effect on performance is known as the "curse of dimensionality" [9] and refers to the exponential growth of hyper-volume of space as a function of its dimensionality. The "curse of dimensionality" leads to the paradoxical "peaking phenomenon" where adding more information decreases classification performance [10, 8]. Efficiency and performance may be reduced even further due to the fact that some of the raw features are correlated or redundant.

Feature-extraction methods deal with peaking phenomena by reducing the dimensionality of the raw-feature space. In pattern recognition there is a wide range of methods for feature extraction such as principal component analysis [11], discriminant analysis [12], independent component analysis [13], Kohonen's mapping [14], Sammon's mapping [15] and auto-associative diabolo networks [16]. In this section we will focus on simple linear feature-extraction methods which can efficiently be used in combination with feed-forward networks.

The most well-known feature-extraction method is principal component analysis (PCA). PCA is an unsupervised feature-extraction method that approximates the data by a linear subspace using the mean-square-error criterion. Mathematically, PCA finds an orthonormal linear projection that maximizes preserved variance. Since the amount of variance is equal to the trace of the covariance matrix, an orthonormal projection that maximizes the trace for the extracted feature space is considered optimal. Such a projection is constructed by selecting the eigenvectors with the largest eigenvalues of the covariance matrix. The PCA technique is of main importance since all other feature-extraction methods discussed here rely on trace maximization. The only difference is that for other mappings other (covariance-like) matrices are used.

PCA works well in a large number of domains where there is no class information available. However, in domains where class information is available supervised feature-extraction methods usually work better. The task of learning to predict moves from examples is a supervised-learning task. A commonly used supervised feature-extraction method is linear discriminant analysis (LDA). In LDA, inter-class separation is emphasized by replacing the covariance matrix in PCA by a general separability measure known as the Fisher criterion, which results in finding the eigenvectors of $S_w^{-1}S_b$, the product of the inverse of the within-class scatter matrix, $S_w$, and the between-class scatter matrix $S_b$ [12].

The Fisher criterion is known to be optimal for linearly separable Gaussian class distributions only. However, in the case of move prediction, the two classes of random and expert moves, are not likely to be so easily separable for at least two reasons. First, the random moves can become expert moves later on in the game. Second, sometimes the random moves may even be just as good as (or better than) the expert moves. Therefore, standard LDA may not be well suited for the job.

### 4.1   Move-Pair Scatter

As an alternative to LDA we propose a new measure, the move-pair scatter, which may be better suited to emphasize the differences between good and bad moves. The move-pair scatter matrix $S_{mp}$ is given by

$$S_{mp} = E[(x_e - x_r)(x_e - x_r)^T] \qquad (2)$$

in which $E$ is the expectation operator, $x_e$ is the expert vector and $x_r$ the associated random vector. It should be noted that the trace of $S_{mp}$ is the expected quadratic distance between a pair of vectors.

The move-pair scatter matrix $S_{mp}$ can be used to replace the between scatter matrix in the Fisher criterion. Alternatively a mapping that directly maximizes move-pair separation is obtained by replacing the covariance matrix in PCA by $S_{mp}$. We call this mapping, on the largest eigenvectors of $S_{mp}$, Move Pair Analysis (MPA).

Although MPA linearly maximizes the preserved distances between the move pairs, which is generally a good idea for separability, it has one serious flaw. Since the mapping aggressively tries to find features which separate the move pairs, it can miss some features which are also relevant but of a more global and static nature. An example is the binary ko-feature. In Go the ko-rule can significantly alter the ordering of moves, i.e., a move which is good in a ko-fight can be bad in a position without a ko. However, since this ko-feature is globally set for both expert and random moves the preservable distance will always be zero, and thus the feature is regarded as uninteresting.

To overcome this problem the mapping has to be balanced with a mapping that preserves global structure such as PCA. This can be done by extracting a set of features, preserving global structure, using standard PCA followed by extracting a set of features using MPA on the subspace orthogonal to the PCA features. In the experimental section we will refer to this type of balanced feature extraction as PCAMPA. Naturally balanced extraction can also be performed in reversed order starting with MPA, followed by PCA performed on the subspace orthogonal to the MPA features, to which we will refer as MPAPCA. Another approach, the construction of a balanced scatter matrix by averaging a weighted sum of scatter and covariance matrices, will not be explored in this paper.

## 4.2 Eigenspace Separation Transforms

Recently an interesting supervised feature-extraction method, called the Eigenspace Separation Transform (EST), was proposed by Torrieri [17]. EST aims at maximizing the difference in average length of vectors in two classes, measured by the absolute value of the trace of the correlation difference matrix. For the task of local move prediction the correlation difference matrix $M$ is defined by

$$M = E[x_e x_e^T] - E[x_r x_r^T] \tag{3}$$

subtracting the correlation matrix of random move vectors from the correlation matrix of expert move vectors. (Notice that the trace of $M$ is the expected quadratic length of expert vectors minus the expected quadratic length of random vectors.) From $M$ the eigenvectors and eigenvalues are calculated. If the sum of positive eigenvalues is larger than the absolute sum of negative eigenvalues, the positive eigenvectors are used for the projection. Otherwise, if the absolute sum of negative eigenvalues is larger, the negative eigenvectors are used. In the unlikely event of an equal sum, the smallest set of eigenvectors is selected.

Effectively EST tries to project one class close to the origin while having the other as far away as possible. The choice for either the positive or the negative eigenvectors is directly related to the choice which of the two classes will be close to the origin and which will be far away. EST was experimentally shown to perform well especially in combination with radial basis networks (on an outlier-sensitive classification task). However in general the choice for only positive or negative eigenvectors seems questionable.

In principle it should not matter which of the two classes are close to the origin, along a certain axis, as long as the classes are well separable. Therefore, we modify the EST by taking eigenvectors with large eigenvalues regardless of their sign. This feature-extraction method, which we call Modified Eigenspace Separation Transform (MEST), is easily implemented by taking the absolute value of the eigenvalues before ordering the eigenvectors of $M$.

### 4.3   Pre-scaling the Raw Feature Vector

Just like standard PCA, all feature-extraction methods discussed here (except LDA) are sensitive to scaling of the raw input features. Therefore features have to be scaled to an appropriate range. The standard solution for this is to subtract the mean and divide by the standard deviation for each individual feature. Another simple solution is to scale the minimum and maximum values of the features to a fixed range. The latter method however can give bad results in the case of outliers. In the absence of a priori knowledge of the data, such uninformed scalings usually works well. However, for the task of local move prediction, we have knowledge on what our raw features represent and what features are likely to be important. Therefore we may be able to use this extra knowledge for finding an informed scaling that emphasizes relevant information.

Since moves are more likely to be influenced by stones that are close than stones that are far away, it is often good to preserve most information from the central region of the ROI. This is done by scaling the variance of the individual features of the ROI inversely proportional to the distance to the center point. In combination with the previously described feature-extraction methods this biases our extracted features towards representing local differences, while still keeping a relatively large field of view.

## 5   Second-Phase Training

A direct consequence of applying the feature-extraction methods discussed above is that potentially important information may not be available for training the predictor. Although the associated loss in performance may be compensated by the gain in performance resulting from the reduced dimensionality, it can imply that we have not used all raw features to their full potential. Fortunately, there is a simple way to have the best of both worlds by making the raw representation available to the network, and improve the performance even further.

Since we used a linear feature extractor, the mapping from the raw features onto the extracted features is a linear mapping. In the network the mapping from input to hidden layer is linear too. As a consequence both mappings can be combined into one (simply by multiplying the matrices). The result is a network that takes the raw feature vectors as input, with the performance obtained on the extracted features.

Second-phase training entails the (re)training of the network formed by the combined mapping. Since all raw features are directly available, the extra free parameters (i.e.,weights) give way for further improvement. (Naturally the validation set prevents the performance from getting worse.)

# 6   Experimental Results

In this section we present experimental results obtained with our approach on predicting the moves of strong human players. Most games used for the experiments presented here were played on IGS. Only games from rated players were used. Although we mainly used dan-level games, a small number of games played by low(er)-ranked players was incorporated in the training set too. The reason was our belief that the network should be exposed to positions somewhat less regular, which are likely to appear in the games of weaker players.

The raw feature vectors for the move pairs were obtained by replaying the games from the beginning to the end, selecting for each move the actual move that was played by the expert together with a second move selected randomly from all other free positions within a Manhattan distance of 3 from the expert move.

The dataset was split up into three subsets. One for training, one for validation (deciding when to stop training), and one for testing. Due to time constraints most experiments were performed with a relatively small data set. Unless stated otherwise, the training set contained 25,000 examples (12,500 move pairs), the validation and test set contained 5,000 and 20,000 (independent) examples, respectively. The predictor had one hidden layer of 100 neurons with hyperbolic tangent transfer functions.

The rest of this section is organized as follows. First we investigate the relative contribution of individual feature types. Then in Sect. 6.2 we present experimental results for different feature-extraction and pre-scaling methods. In Sect. 6.3 we show the gain in performance by the second-phase training.

## 6.1   Relative Contribution of Individual Feature Types

A strong predictor requires good features. Therefore, an important question for building a strong predictor is: how good are the features? The answer to this question can be found experimentally by measuring the performance of different configurations of feature types. The results can be influenced by the number of examples, peaking phenomena, and a possible bias of the predictor towards certain distributions. Although our experiments are not exhaustive, they give a reasonable indication of the relative contribution of the feature types.

We performed two experiments. In the first experiment we trained the predictor with only one feature type as input. Naturally, the performance of most single feature types is not expected to be high. The added performance (compared to 50% for pure guessing) is shown in the column headed "Individual" of table 1. The first experiment shows that individually, the stones, liberties and the last move are the strongest features.

**Table 1.** Added performance in percents of raw-feature types.

|  | Individual | Leave one out |
|---|---|---|
| Stones | +32.9 | -4.8 |
| Edge | +9.5 | -0.9 |
| Ko | +0.3 | -0.1 |
| Liberties | +24.8 | 0.0 |
| Liberties after | +12.1 | -0.1 |
| Captures | +5.6 | -0.8 |
| Nearest stones | +6.2 | +0.3 |
| Last move | +21.5 | -0.8 |

For the second experiment, we trained the predictor on all feature types except one, and compared the performance to a predictor using all feature types. The results, shown in the last column of table 1, show that again the stones are the best feature type. (It should be noted that negative values indicate good performance of the feature type that is left out.) The edge, captures and the last move also yield a small gain in performance. For the other features there seems to be a fair degree of redundancy, and possibly some of them are better left out. However, it may be that the added value of these features is only in the combination with other features. The liberties might benefit from reducing the size of their ROI. The nearest stones performed poorly (4 out of 5 times), possibly due to peaking effects. Unfortunately the standard deviations, which were around 0.5%, do not allow hard conclusions.

## 6.2 Results for Different Feature-Extraction and Pre-scaling Methods

Feature extraction reduces the dimensionality, while preserving relevant information, to overcome harmful effects related to the curse of dimensionality. In Sect. 4 a number of methods for feature extraction and pre-scaling of the data were discussed. Here we present empirical results on the performance of the feature-extraction methods discussed in combination with the three techniques for pre-scaling the raw feature vectors, discussed in Sect. 4.3.

Table 2 lists the results for the different pre-scaling and feature-extraction methods. In the first row the pre-scaling is shown. The three types of pre-scaling are (from left to right): normalized mean and standard deviation ($[\mu, \sigma]$), fixed-range pre-scaling ($[\min, \max]$) and ROI-scaled mean and standard deviation ($[\mu, \sigma]$ , $\sigma^2 \sim 1/d$ in ROI). The second row shows the (reduced) dimensionality, as a percentage of the dimensionality of the original raw-feature space. Rows 3 to 11 show the percentages of correctly ranked move-pairs, measured on the independent test set, for the nine different feature-extraction methods. Though the performances shown are averages of only a small number of experiments, all standard deviations were less than 1%. It should be noted that LDA* was obtained by replacing $S_b$ with $S_{mp}$. Both LDA and LDA* used a small regularisation term (to avoid invertability and singularity problems). The balanced mappings, PCAMPA and MPAPCA, both used 50% PCA and 50% MPA.

**Table 2.** Performance of extractors for different dimensionalities.

| pre-scaling | $[\mu, \sigma]$ | | [min, max] | | $[\mu, \sigma]$ , $\sigma^2 \sim 1/d$ in ROI | | | |
|---|---|---|---|---|---|---|---|---|
| Dimensionality | 10% | 25% | 10% | 25% | 10% | 25% | 50% | 90% |
| PCA | 78.7 | 80.8 | 74.4 | 80.4 | 83.9 | 85.8 | 84.9 | 84.5 |
| LDA | 75.2 | 74.9 | 75.2 | 75.5 | 75.3 | 75.4 | 75.8 | 76.9 |
| LDA* | 72.5 | 73.8 | 70.0 | 72.1 | 72.6 | 74.0 | 75.9 | 76.7 |
| MPA | 73.7 | 76.7 | 70.3 | 73.8 | 80.7 | 84.6 | 85.5 | 84.3 |
| PCAMPA | 77.3 | 80.6 | 73.5 | 80.5 | 84.4 | 85.9 | 84.1 | 83.7 |
| MPAPCA | 77.2 | 80.6 | 74.0 | 80.1 | 83.6 | 85.6 | 84.4 | 84.3 |
| EST | 80.7 | 79.3 | 78.1 | 78.9 | 83.5 | 81.1 | 79.2 | 79.5 |
| MEST | 84.3 | 82.4 | 82.6 | 82.2 | 86.0 | 84.6 | 83.9 | 84.4 |
| MESTMPA | 83.8 | 82.5 | 82.6 | 82.2 | 86.8 | 85.6 | 84.5 | 84.5 |

Table 2 reveals seven important findings.

- *A priori* knowledge of the feature space is useful for pre-scaling the data as is evident from the overall higher scores obtained with the scaled ROIs.
- In the absence of a priori knowledge it is better to scale by the mean and standard deviation than by the minimum and maximum values as follows from a comparison of the results for both pre-scaling methods.
- PCA performs quite well despite the fact that it does not use class information.
- LDA performs poorly. Replacing the between scatter matrix in LDA with our move-pair scatter matrix (i.e., LDA*) degrades rather than upgrades the performance. We suspect that the main reason for this is that minimization of the within scatter matrix, which is very similar to the successful covariance matrix used by PCA, is extremely harmful to the performance.
- MPA performs reasonable, but is inferior to PCA. Presumably this is due to the level of global information in MPA.
- The balanced mappings PCAMPA and MPAPCA are competitive and sometimes even outperform PCA.
- MEST is the clear winner. It outperforms both PCA and the balanced mappings.

Our modification of the Eigenspace Separation Transform (MEST) significantly outperforms standard EST. Unfortunately, MEST does not seem to be very effective at the higher dimensionalities. It may therefore be useful to balance this mapping with one or possibly two other mappings such as MPA or PCA. One equally balanced combination of MEST and MPA is shown in the last row. Other possible combinations are left for future study.

### 6.3   Second-Phase Training

After having trained the predictor on the extracted features we now turn to second-phase training. Table 3 displays the performances for both training phases performed on a training set of 200,000 examples. The first column shows the size of the ROIs, first for the stones and second for the liberties. In the second column the dimensionality of the extracted-feature space is shown, as a percentage of the raw-feature space. The rest of the table shows the performances and duration of the first- and second-phase training experiments.

**Table 3.** First- and second-phase training statistics.

| ROI | dim. (%) | Phase 1 | | Phase 2 | |
|---|---|---|---|---|---|
| | | perf.(%) | time (h) | perf.(%) | time (h) |
| 40,40 | 10 | 87.3 | 14.9 | 89.9 | 33.0 |
| 40,40 | 15 | 88.8 | 12.0 | 90.5 | 35.9 |
| 40,12 | 20 | 88.4 | 11.6 | 90.1 | 18.5 |
| 60,60 | 15 | 89.0 | 10.2 | 90.4 | 42.5 |
| 60,24 | 20 | 89.2 | 16.9 | 90.7 | 31.0 |
| 60,12 | 20 | 88.8 | 11.8 | 90.2 | 26.1 |

As is evident from the results in table 3, second-phase training boosts the performance obtained with the first-phase training. The extra performance comes at the price of increased training time, though it should be noted that in these experiments up to 60% of the time was spent on the stopping criterion, which can be reduced by setting a lower threshold.

## 7   Assessing the Quality of the Local Move Predictor

Below we assess the quality of the local move predictor. First we compare human performance on the task of local move prediction to the performance of the local move predictor. Then in Sect. 7.2 the local move predictor is tested on professional games. Finally, in Sect. 7.3 our system is tested by actual play against the program GNUGO.

### 7.1   Human Performance with Full-Board Information

We compared the performance of our (best) predictor with that of human performance. For this we selected three games played by strong players (3d*-5d* IGS). All three games were replayed by a number of human amateur Go players, all playing black. The task faced by the human was identical to that of the neural predictor, the main difference being that the human had access to complete full-board information. At each move the human player was instructed to choose

**Table 4.** Human and computer (lmp) performance on local move prediction.

|        | game 1 | game 2 | game 3 | average |
|--------|--------|--------|--------|---------|
| 3 dan  | 96.7   | 91.5   | 89.5   | 92.4    |
| 2 dan  | 95.8   | 95.0   | 97.0   | 95.9    |
| 2 kyu  | 95.0   | 91.5   | 92.5   | 92.9    |
| lmp    | 90.0   | 89.4   | 89.5   | 89.6    |
| 2 kyu  | 87.5   | 90.8   | n.a.   | 89.3    |
| 5 kyu  | 87.5   | 84.4   | 85.0   | 85.5    |
| 8 kyu  | 87.5   | 85.1   | 86.5   | 86.3    |
| 13 kyu | 83.3   | 75.2   | 82.7   | 80.2    |
| 14 kyu | 76.7   | 83.0   | 80.5   | 80.2    |
| 15 kyu | 80.0   | 73.8   | 82.0   | 78.4    |

between two moves: one of the two moves was the expert move, the other was a move randomly selected within a Manhattan distance of 3 from the expert move.

Table 4 shows the results achieved by the human players. The players are ordered according to their (Dutch) rating, shown in the first column. It should be noted that some of the ratings may be off by one or two grades, which is especially true for the low-ranked kyu players (5-15k). Only of the dan-level ratings we can be reasonably sure, since they are regulated by official tournament results. The next three columns contain the scores of the human players on the three games, and the last column contains their average scores over all moves. (Naturally all humans were given the exact same set of choices, and were not exposed to these games before.)

From table 4 we estimate that dan-level performance lies somewhere around 95%. Clearly there is still a significant variation most likely related to some inherent freedom for choosing between moves that are (almost) equally good. Strong kyu level is somewhere around 90%, and as players get weaker we see performance dropping even below 80%. On the three games our local move predictor (lmp) scored an average of 89.6% correct predictions thus placing it in the region of strong kyu-level players.

## 7.2   Testing on Professional Games

### 19×19 Games

The performance of our predictor was tested on 52 professional games played for the title matches of recent Kisei, Meijin and Honinbo Tournaments. The Kisei, Meijin and Honinbo are the most prestigious titles in Japan with a total first-prize money of about US$ 600,000. The 52 games contained 11,460 positions with 248 legal moves on average (excluding the pass move). For each position our system was used to rank all legal moves. We calculated the probability that the professional move was among the first $n$ moves (cumulative performance). In Fig. 4 the cumulative performance of the ranking is shown for the full board as well as for the local neighborhoods (within a Manhattan distance of 3 from the professional move).

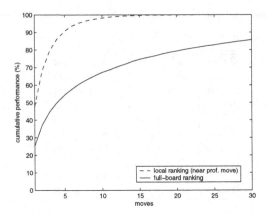

**Fig. 4.** Ranking professional moves on 19×19.

In local neighborhoods the predictor ranked 48% of the professional moves first. On the full board the predictor ranked 25% of the professional moves first, 45% within the best three, and 80% in the top 20. The last 20% was in a long tailed distribution reaching 99% at about 100 moves.

In an experiment performed ten years ago by Müller and reported in [18] the program EXPLORER ranked one third of the moves played by professionals among its top three choices. Another third of the moves was ranked between 4 and 20, and the remaining third was either ranked below the top twenty or not considered at all. Though the comparison may be somewhat unfair, due to the fact that EXPLORER was not optimized for predicting professional moves, it still seems that significant progress has been made.

**9×9 Games**

For fast games, and for teaching beginners, the game of Go is often played on the 9x9 board. It has been pointed out by Chen and Chen [19] that for the *current* Go programs, despite of the reduced state space, branching factor and length of the game, 9x9 Go is just as difficult as 19x19 Go. We tested the performance of our predictor on 17 professional 9x9 games. The games contained 862 positions with 56 legal moves on average (excluding the pass move). Figure 5 shows the cumulative performance of the ranking for the full board as well as for the local neighborhoods (again within a Manhattan distance of 3 from the professional move). On the full 9x9 board the predictor ranked 37% of the profesional moves first and over 99% of the professional moves in the top 25.

**7.3   Testing by Actual Play**

To assess the strength of our predictor in practice we tested it against GNUGO version 3.2. This was done by always playing the first-ranked move. Despite of many good local moves in the opening and middle-game our system lost all games. Hence our belief was confirmed that the local move predictor in itself

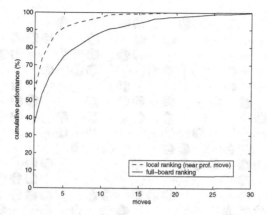

**Fig. 5.** Ranking professional moves on 9×9.

is not sufficient to play a strong game. The main handicap of the local move predictor was that it did not understand (some of) the tactical fights. This occasionally resulted in the loss of large groups, and poor play in the endgame. Another handicap was that always playing the first-ranked move often turned out to be too passive. (The program followed GNuGo's moves and seldomly took initiative in other regions of the board.)

We believe that a combination of the move predictor with a decent tactical search procedure would have a significant impact on the performance, and especially on the quality of the fights. Since we did not have a reasonable search procedure (and evaluation function) available, we were not yet able to explore this idea. As an alternative we tested some games where the first author (a strong kyu level player) selected moves from the first $n$ candidate moves ranked by the local move predictor. Playing with the first ten moves we were able to defeat GNuGo even when it played with up to five handicap stones. With the first twenty moves the strength of our combination increased even further. In Fig. 6 a game is shown where GNuGo played black with nine handicap stones against the first author selecting from the first twenty moves ranked by the local move predictor. The game is shown up to move 234, where White is clearly ahead. After some small endgame fights White convincingly won the game with 57.5 points.

The results indicate that, at least against other Go programs, a relatively small set of high-ranked moves is sufficient to play a strong game.

## 8   Conclusions and Future Research

We have presented a system that learns to predict moves in the game of Go from observing strong human play. The performance of our system on the task of local move prediction is comparable to that of strong kyu-level players.

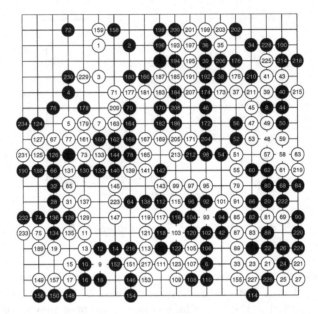

**Fig. 6.** Nine stone handicap game against GNUGO (white 29 at 136).

The training algorithm presented in this paper is more efficient than standard fixed-target implementations. This is mainly due to the avoidance of needless weight adaptation when rankings are correct. As an extra bonus, our training method reduces the number of gradient calculations as performance grows, thus speeding up training. A major contribution to the performance is the use of feature-extraction methods. Feature extraction reduces the training time while increasing the quality of the predictor. Together with a sensible scaling of the original features and an optional second-phase training, superior performance over direct-training schemes can be obtained.

Our conclusion is that it is possible to train a learning system to predict good local moves with a performance at least comparable to strong kyu-level players. The performance was obtained from a simple set of locally computable features, thus ignoring a significant amount of information which can be obtained by more extensive (full-board) analysis or by specific goal-directed searches. Therefore, we believe that there is still significant room for improving the performance, possibly even into the strong dan-level region.

The local move predictor can be used for move ordering and forward pruning in a full-board search. The performance obtained on ranking professional moves indicates that a large fraction of the legal moves may be pruned directly without any significant risk.

On a modern GHz PC our system evaluates moves with a speed in the order of 5000 moves per second. This translates to around 0.05 seconds for a full-board ranking. As a consequence our approach may not be directly applicable for deep searches. The speed can however be increased greatly by parallel computation.

Trade-offs between speed and predictive power are also possible since the number of hidden units and the dimensionality of the raw feature vector both scale linearly with computation time.

## 8.1 Future Research

Experiments showed that our system performs well on the prediction of moves which are played in strong human games. The downside however is that our system cannot be trusted (yet) in odd positions which do not show up in (strong) human games. Future work should therefore focus on ensuring reliability of the predictor in more odd regions of the Go state space.

We are considering to train our predictor further through some type of Q-learning. In principle Q-learning works regardless of who is selecting the moves, consequently training should work both by self-play (in odd positions) and by replaying human games. Furthermore, since Q-learning does not rely on the assumption of optimality of human moves, it may be able to solve possible inconsistencies in its current knowledge (due to the fact that some human moves were bad).

Finally we intend to investigate the application of our predictor for move ordering and forward pruning in full-board search. We expect that our predictor can greatly improve the search especially if we can combine it with a sensible full-board evaluation function.

## Acknowledgments

We are grateful to all Go players that helped us perform the experiments reported in Sect. 7.1.

## References

1. Bouzy, B., Cazenave, T.: Computer Go: An AI oriented survey. Artificial Intelligence **132** (2001) 39–102
2. Müller, M.: Computer Go. Artificial Intelligence **134** (2002) 145–179
3. Enderton, H.: The GOLEM Go program. Technical Report CMU-CS-92-101, School of Computer Science, Carnegie-Mellon University (1991)
4. Dahl, F.: HONTE, a Go-playing program using neural nets, 16th International Conference on Machine Learning (1999)
5. Schraudolph, N., Dayan, P., Sejnowski, T.: Temporal difference learning of position evaluation in the game of Go. In Cowan, J., Tesauro, G., Alspector, J., eds.: Advances in Neural Information Processing 6. Morgan Kaufmann (1994) 817–824
6. Tesauro, G.: Connectionist learning of expert preferences by comparison training. In Touretzky, D., ed.: Advances in Neural Information Processing Systems 1 (NIPS-88). Morgan Kaufmann (1989) 99–106
7. Riedmiller, M., Braun, H.: A direct adaptive method for faster backpropagation: the RPROP algorithm. In: IEEE Int. Conf. on Neural Networks (ICNN). (1993) 586–591

8. Jain, A., Chandrasekaran, B.: Dimensionality and sample size considerations in pattern recognition practice. In Krishnaiah, P., Kanal, L., eds.: Handbook of Statistics. Volume 2. North-Holland (1982) 835–855
9. Bellman, R.: Adaptive Control Processes: A Guided Tour. Princeton University Press (1961)
10. Bishop, C.: Neural Networks for Pattern Recognition. Clarendon Press (1995)
11. Jollife, I.: Principal Component Analysis. Springer-Verlag (1986)
12. Fukunaga, K.: Introduction to Statistical Pattern Recognition, Second Edition. Academic Press (1990)
13. Karhunen, J., Oja, E., Wang, L., Vigario, R., Joutsensalo, J.: A class of neural networks for independant component analysis. IEEE Transactions on Neural Networks 8 (1997) 486–504
14. Kohonen, T.: Self-organising maps. Springer-Verlag (1995)
15. Sammon Jr., J.: A non-linear mapping for data structure analysis. IEEE Transactions on Computers 18 (1969) 401–409
16. van der Werf, E.: Non-linear target based feature extraction by diabolo networks. Master's thesis, Pattern Recognition Group, Department of Applied Physics, Faculty of Applied Sciences, Delft University of Technology (1999)
17. Torrieri, D.: The eigenspace separation transform for neural-network classifiers. Neural Networks 12 (1999) 419–427
18. Müller, M.: Computer Go as a sum of local games: An application of combinatorial game theory. PhD thesis, ETH Zürich (1995) Diss. ETH No. 11.006.
19. Chen, K., Chen, Z.: Static analysis of life and death in the game of Go. Information Sciences 121 (1999) 113–134

# Evaluating Kos in a Neutral Threat Environment: Preliminary Results

William L. Spight

Oakland, California
BillSpight@aol.com

**Abstract.** The idea of a komaster made it possible to use thermography to analyze loopy games, called kos, in Go. But neither player may be komaster. In that case a neutral threat environment permits the construction of thermographs for kos. Using neutral threat environments some hyperactive kos are evaluated and general results are derived for approach kos.

## 1 Introduction

### 1.1 Combinatorial Games

Combinatorial games are 2-person games of perfect information [1]. The players, called Left and Right (Black and White, respectively, in Go), alternate play until one player has no move, and therefore loses. Numbers are games, and the play may end when a number, or score, is reached. Scores in favor of Left are positive. Independent games may be combined in a sum. Play alternates in the sum, but not necessarily in each component of the sum. Thus, which player has the move is not part of the definition of a combinatorial game.

Equivalent games need not have the same form, but each game has a simplest, or *canonical*, form, and equivalent games have the same canonical form. The canonical form for 0, for instance, is $\{|\}$. The vertical slash indicates the root of the game tree. Options for Left are to the left of the slash, and Right's options to thr right. Here, neither player has a move. Zero is a win for the second player. A parent node has more slashes than its children, or *followers*. In the game $G = \{3|1|| - 4\}$, Right can play to a score of -4 and Left can play to the game $H = \{3|1\}$. To find the negative of a game, flip it from left to right and negate any scores or other subgames. The negative of $G$ is $\{4|| - 1| - 3\}$.

### 1.2 Mean Value and Temperature

What does Left gain from a play from $G$ to $H$? Strictly speaking, it is their difference, $H - G$. But we may also speak of the average gain in terms of the mean values of $G$ and $H$. The mean value of $G$ is not apparent, but $H = \{3|1\}$ has an obvious mean value of 2. Each player gains 1 point on average by a play in $H$, and this gain indicates the urgency of a play in $H$. This urgency is called

J. Schaeffer et al. (Eds.): CG 2002, LNCS 2883, pp. 413–428, 2003.

the *temperature* of $H$. The mean value of a game is not always the average of the mean values of two of its followers, nor is its temperature always the difference between it and the mean value of an immediate follower.

## 1.3   Thermography

Thermography is a method for finding the mean value and temperature of a combinatorial game [1]. The thermograph of a game, $G$, represents the results, $v$, of optimal play (which may be no play at all) in that game for each temperature $t$. By convention $t$ is plotted on the vertical axis with positive values above the origin, and $v$ is plotted on the horizontal axis with positive values to the left of the origin. $t$ may be considered a tax on each play. The imposition of the tax is called *cooling*. For each $t$, the left wall of the thermograph, $LW(G)$, indicates the result with optimal play if Left (Black) plays first, the right wall, $RW(G)$, the result with optimal play if Right (White) plays first. If the tax is high enough, the player will choose not to make a play. At the top of each thermograph neither player can afford to play, and the walls coincide in a vertical mast, $v = m(G)$. $m(G)$ is called the mast value of the game. (For classical combinatorial games the mean value and mast value are the same, but that may not be the case for some Go positions involving loopy games called *kos*.) The temperature at the bottom of this vertical mast is called the temperature of the game, $t(G)$.

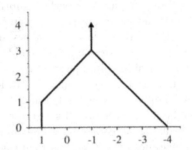

**Fig. 1.** Thermograph of $\{3|1|| - 4\}$.

Figure 1 shows the thermograph of $G$. $RW(G) = -4 + t$, when $0 \le t \le 3$. When $0 \le t \le 1$, $LW(G)$ is vertical with $v = 1$, indicating that Right will reply if Left plays to $\{3|1\}$, and is the line $v = 2 - t$ when $1 \le t \le 3$. The two walls form a mast at temperature 3. $t(G) = 3$ and $m(G) = -1$.

Note that below $t(G)$, $LW(G) = RW(H) - t$; similarly, $RW(G) = LW(-4) + t$ below $t(G)$. (The left wall of a number is simply its mast.) The thermograph of a non-ko game is derived directly from the thermographs of its followers. The left scaffold of $G$, $LS(G)$, is the maximum of the right walls of the left followers of $G$ minus $t$. Likewise, RS(G) is the minimum of the left walls of the right followers of $G$ plus $t$. The intersection or coincidence of the scaffolds determines $t(G)$ and $m(G)$.

## 1.4   Komaster

In Go there are loopy games called kos or superkos, which repeat a previous board position. (Superkos repeat a board position after a cycle longer than 2 plies. In this paper "ko" will refer to either a ko or superko.) Ko rules require a player to make another play before repeating a position in a ko. Because of this ko ban, for a single ko each player may be able to escape the loop favorably when his opponent is barred from playing in the ko. Escaping favorably from the loop is called *winning the ko*. The players may engage in a *ko fight* over winning the ko.

In a ko fight when a player is barred from playing in a ko, he may make another play that carries a sizable threat. Such a play is called a *ko threat*. If his opponent answers the threat, he is then allowed to play in the ko. Then his opponent may make a ko threat, and the fight may continue until someone wins the ko.

Because of their dependence on ko threats, how to construct a thermograph for a ko is not obvious. Berlekamp extended thermography to kos with the idea of *komaster* [2]. The komaster of a ko is able to win it, but is not able to reduce the temperature of alternative plays before doing so. The mast values of some kos, called *placid*, remain the same regardless of who is komaster, but the mast values of other kos, dubbed *hyperactive*, depend on who is komaster. Often, however, neither player is komaster. In the summer of 2000 Berlekamp proposed a way of evaluating kos which favors neither player [3]. The ko is summed with a neutral threat environment (NTE).

## 1.5   Neutral Threat Environment

A neutral threat environment is a sum of pairs of opposing ko threats of equal size, such that in a ko fight each player has a sufficient number of ko threats of any required size to guarantee an equitable outcome of the fight. The value of a ko plus an NTE depends on the structure of the ko threats. Here we shall utilize basic ko threats, of the form $\{2\theta|0||\ \}$ or its negative. The size of the threat is the temperature, $\theta$, of its follower. Each basic ko threat evaluates to zero as a combinatorial game.

We may construct an NTE, $E_{n,s}$, in similar fashion to a universal enriched environment [2]. Let

$$\delta_n = 1/lcm\{1, 2, ..., n\}, \tag{1}$$

where $lcm$ is the least common multiple of a set of numbers. Let $E_{n,s}$ consist of $n$ pairs of opposing threats of each size $\delta_n, 2\delta_n, \ldots, s$, the maximum size of a threat. To evaluate a given ko in an NTE, a sufficient number of threat pairs of certain sizes must exist. We may make $n$ and $s$ large enough so that they do. Since each required size is rational, a sufficiently large $n$ will make the denominator of $\delta_n$ large enough to divide it.

## 1.6    Example: Simple Ko

Figure 2 shows a simple ko. Black may play at $a$ to capture the 2 marked White stones for a score of 5 points, 1 point for each captured stone plus 1 point for each point of surrounded territory. If White takes the circled Black stone with a play at $b$, the resulting White stone there could be recaptured by Black except for the ko ban. This take-and-take-back situation is a simple ko. After taking the ko at $b$, White may win the ko by filling it. Then the Black stone marked with a triangle is killed and counts as a point for White. White gets 4 points for winning the ko, for 2 captured stones and 2 points of territory. (In all Go diagrams in this paper the framing stones connected to the center of the board are considered to be immortal.) Figure 3 shows the game graph for this ko. A indicates the initial position, B the position after White takes the ko. The curved line between them indicates the ko. The White score is negative.

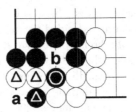

**Fig. 2.** Simple ko.

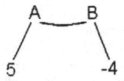

**Fig. 3.** Simple ko game graph.

What is the thermograph of this simple ko, A, in an NTE of basic threats? The left scaffold of A, $LS(A) = 5 - t$. Let us find the right scaffold. For any given rational temperature, $0 \le t < 3$, let the NTE contain an opposing pair of threats of size $\theta = (9 + t)/4$. (Note: $\theta > t$, or the threat is not large enough to play effectively.) In the ko fight if White takes the ko, let Black play her threat of size $\theta$. If White wins the ko, Black completes her threat. The result is $2\theta - 4 = (1 + t)/2$. Since each player has made the same number of plays, the result cooled by $t$ remains the same. If White answers Black's ko threat, Black takes the ko back. Now White plays his threat of size $\theta$. This time Black wins the ko and White plays the second leg of his threat. The result is $5 - 2\theta = (1 - t)/2$. Since White has made one more play than Black, the cooled value is $(1 - t)/2 + t = (1 + t)/2$. That is the same as the cooled value if Black wins the ko. In fact, $\theta$ was chosen to make that the case. $\theta$ is the *critical size* of the ko threat, and $(1 + t)/2$ is the *equilibrium value* of the ko fight.

## 1.7 Equilibrium Strategy

Each player can assure the equilibrium value by playing the *equilibrium strategy*. The player makes minimum threats no smaller than the critical size. Then if the opponent wins the ko the player is guaranteed at least the equilibrium value. The player also answers threats larger that the critical size, allowing the opponent to win the ko only if the player gets at least the equilibrium value. As the NTE is finite, eventually the opponent will run out of larger threats, and must win the ko or make a threat no larger than the critical size, which is not answered.

Even if the opponent has a few extra ko threats larger than the critical size, the number of ko threats of each size in the NTE will have been made large enough to exhaust the opponent of threats larger than the equilibrium value while the player still has threats at least that large. Say that White has $k$ large threats. The number, $n$, of pairs of threats of each size in the NTE is made large enough so that $n > k$. Now suppose that White does not answer any Black threat and plays threats larger than $\theta$, which Black answers. This continues until we reach the point where White has $n$ threats of size $\theta$ plus $k$ large threats, and Black has $n$ threats of size $s$, the maximum size of threats in the NTE. Now White plays the large threats, leaving Black with $n - k$ threats of size $s$. At worst it is Black's turn and he makes a threat of size $s$, which White answers. White must now play a threat of size $\theta$, which Black ignores.

Returning to our example, when White takes the ko the cooled result is $(1 + t)/2$, and $RS(A) = (1 + t)/2$. At temperature 3 the two walls meet at the mast value, $m(A) = 2$, and so $t(A) = 3$. Figure 4 shows the NTE thermograph of the ko in the center, flanked by the komaster thermographs on either side.

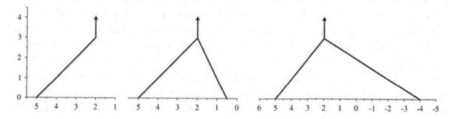

**Fig. 4.** Simple ko thermographs: Black komaster, NTE, and White komaster.

Since this ko is placid, all three thermographs have the same temperature and mast value. On the left, the right wall of the Black komaster thermograph coincides with the left wall, indicating that Black wins the ko no matter who plays first. On the right, the right wall of the White komaster thermograph indicates that White can win the ko in two net plays in the ko if he plays first.

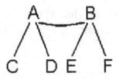

**Fig. 5.** General one-step ko.

(The inverse of the slope of the wall indicates the net number of local plays.) The right wall of the NTE thermograph indicates the intermediate result if White takes the ko.

What does it take to be komaster of this ko? If White has $w$ ko threats, Black requires $w + 1$ *primary* threats to be komaster. (A primary threat is one so large that the opponent must answer it.) Then she can exhaust White's ko threats and have a threat left to win the ko fight without giving up anything by ignoring a White threat. If Black has $b$ threats, White requires $b$ primary threats to be komaster.

## 2    General Method

Figure 5 shows a general one-step ko. C, D, E, and F stand, not for single nodes, but for sets of followers of A and B. For any given temperature, $t$, $LS(A)$ is the maximum result cooled by $t$ of Black's choice among the options of C, given subsequent optimal play, or simply $\max(C)$. Similarly, $RS(A) = \min(D, B)$.

Let us unfold the play when White takes the ko, playing from A to B. Let us assume for the moment that the temperature of B is at least $t$. Black will therefore respond. Black is banned from taking the ko back immediately, but may make a play to one of the options in E or play a ko threat of critical size, $\theta$. In that case White may win the ko, achieving $\min(F + \{2\theta|0\})$. (After White's play to an option in F, the uncompleted ko threat is still on the board, so we are concerned with their sum.) If White answers the ko threat, Black can play to E or return to A. We can ignore Black's play to E. If it were correct she could have made it before. After Black takes the ko back, White can play to D or continue the ko fight. In a real game, White might play to D, satisfied with having removed one of Black's ko threats before making the play. But in the NTE one threat does not matter. We can ignore White's play to D as we ignored Black's play to E. So we let White play a threat of critical size, $\theta$, and Black win the ko, for $\max(C + \{0| - 2\theta\})$. Unless $\min(F) > \max(C)$, we have for some $\theta > 0$

$$\max(C + \{0| - 2\theta\}) = \min(F + \{2\theta|0\}), \tag{2}$$

for after one player wins the ko the other can complete the threat. By completing his threat White can achieve $LS(C_0) - 2\theta$, where $C_0$ is the option of C that maximizes its left wall at temperature, $t$, and, similarly, by completing her threat Black can achieve $RS(F_0) + 2\theta$. For $\theta$ large enough equality can be achieved.

Completing the threat may not be correct play, but in that case the loser of the ko can do even better, and still achieve at least equality. If $\min(F) > \max(C)$, no ko threat is necessary, and there will be no ko fight at temperature $t$. For some kos, $\min(F) > \max(C)$ for any non-negative temperature. We may evaluate such unusual kos, but that is not the concern of this paper.

Equation (2) determines the equilibrium value of the ko fight, or $K$. If White takes the ko, Black can get $\max(E, K)$. Then $RS(A) = \min(D, \max(E, K))$.

## 2.1   Finding the Equilibrium Value

We may solve (2) by a process of successive refinement, as indicated by the partial Prolog program shown in Fig. 6.

Each player has a set of strategies for play in each subgame being evaluated. It will be convenient to use behavior strategies, assuming optimal play at unspecified game nodes. Given a strategy for the first player, $S_0$, and an estimate for $\theta$, $T_0$, we determine the best strategy for the second player, $S_1$. From $S_0$ and $S_1$ we now derive a new estimate for $\theta$, $T_1$, and for $K$, $K_1$. For $S_1$ and $T_1$ we determine the best strategy for the first player, $S_2$. If the results for $S_1$, $S_2$, and $T_1$ equal $K_1$, then $\theta = T_1$ and $K = K_1$, and we are done. Otherwise we derive new estimates for $\theta$ and $K$ from $S_1$ and $S_2$ and continue. If a strategy is not the best response to the best response to it, we ignore it in further calculations. We may obtain our original estimates by assuming that each player, upon losing the ko, completes the ko threat. That yields

$$LS(C_0) - 2T_0 = RS(F_0) + 2T_0, \tag{3}$$

and, hence,

$$T_0 = (LS(C_0) - RS(F_0))/4, \tag{4}$$

and

$$K_0 = (LS(C_0) + RS(F_0))/2. \tag{5}$$

We need not use these initial estimates, of course. For instance, calculating the komaster thermographs may have suggested other strategies.

## 3   Approach Ko

An approach ko is hyperactive. In Fig. 7 White can win the ko immediately by a play at $a$, capturing Black's 5 stones. If Black takes the ko at $b$ and then fills it White can still capture her stones with a play at $a$. To win the ko Black must take at $b$, then make an approach move at $c$ to fill a liberty of the White group, and then win the subsequent ko by capturing the White stones.

Figure 8 shows the game graph of this ko. (We ignore the *dame*, or neutral point at $c$ if White wins the ko. As a combinatorial game it is an infinitesimal called *star*. It does not affect the Go score.) If White wins the ko the local score is -11; if Black wins it is 14.

```
% values(PlayerStrats,OppStrats,CritSize,Value)
% finds the critical size and equilibrium value
values([PStrat|PSs],[OStrat0|OSs0],T,K) :-
    estimates(PStrat,OStrat0,T0,K0),
    best_response(PStrat,[OStrat0|OSs0],T0,OStrat,OSs),
    (   OStrat = OStrat0
        -> T1 = T0, K1 = K0
        ;   estimates(PStrat,OStrat,T1,K1)
    ),
    values1(OStrat,OSs,[PStrat|PSs],T1,K1,T,K).

% values1(BestResp,Rs,[S|Ss],SizeEst,ValueEst,Size,Value)
% BestResp is the best response to strategy S.
% Rs are other responses.
values1(PStrat,PSs,[OStrat0|OSs0],T0,K0,T,K) :-
    best_response(PStrat,[OStrat0|OSs0],T0,OStrat,OSs),
    (   OStrat = OStrat0
        -> T = T0, K = K0
        ;   estimates(PStrat,OStrat,T1,K1),
            values1(OStrat,OSs,[PStrat|PSs],T1,K1,T,K)
    ).

% estimates(PlayerStrat,OppStrat,CritSize,Value)
% estimates the critical size and equilibrium
% value, given the players' strategies

% best_response(PStrat,[OStrat0|OSs0],Size,OStrat,OSs)
% OStrat is the first optimal response to PStrat found
% in [OStrat0|OSs0], given critical size, Size.
% OSs is [OStrat0|OSs0] minus OStrat.
```

**Fig. 6.** Pseudo-code for finding the equilibrium value using successive refinement.

Let us derive the thermograph of A. Since there are no options to escape a ko fight without winning it, the left wall will be formed by the equilibrium value of the fight. If White wins the ko fight Black will complete her ko threat, $\{2\theta_0|0||\ \}$, and, since Black has made one more play than White, the result will be $-11 + 2\theta_0 - t$. (Since the game tree has two kos, one between A and B and the other between C and D, there will be two critical sizes for threats. The subscript indicates the number of approach moves.) If Black makes the approach move, she reaches $C + \{0| - 2\theta_0\}$. To get our first estimate of $\theta_0$, $T_0$, we let White complete the threat. At this point each player has made the same number of plays, so the result will be $LW(C) - 2\theta_0$. Since C is a simple ko we find that $LW(C) = 14 - t$ when $0 \le t < 8\frac{1}{3}$, and $5\frac{2}{3}$ when $t \ge 8\frac{1}{3}$. Solving for $T_0$ we find that $T_0 = 6\frac{1}{4}$ when $0 \le t < 8\frac{1}{3}$, which is effectively $0 \le t < 6\frac{1}{4}$, since $\theta_0 > t$. In that case our first estimate for the value of the ko fight, $K_0 = 6\frac{1}{4} - t$. When

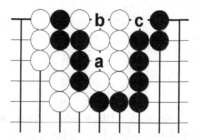

**Fig. 7.** Approach ko.

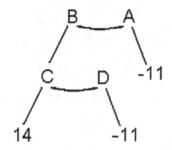

**Fig. 8.** Approach ko game graph.

$t > 8\frac{1}{3}$, $T_0 = 4\frac{1}{6} + t/4$, but then $T_0 < t$, and there is no ko fight. So when White completes his threat, Black wins the second ko.

Assuming that $\theta_0 = 6\frac{1}{4}$, can White do better after Black ignores his threat and plays to C by taking the ko rather than completing his threat? To evaluate this ko fight we find our first estimate, $U_0$, for its critical size. If Black wins the ko we let White complete the ko threat for the second ko and then Black will take the first ko threat back. The result is $14 - 2U_0 - t$. If White wins the ko we let Black complete her ko threat and then White completes his earlier ko threat. The result is $-11 + 2U_0 - 13$, or $-24 + 2U_0$. Equating the two results we find that $U_0 = 9\frac{1}{2} + t/4$, and an estimated value for the second ko fight, $L_0 = -5 - t/2$. Since $0 \le t < 6\frac{1}{4}$, that is better for White than completing his first threat. And since $U_0 > T_0$, it is correct to complete or answer the second threat before the first.

For our second estimates, we take our current play strategies and solve for critical sizes and equilibrium values. For the second ko fight we have

$$14 - 2U_1 - t = -11 + 2U_1 - 2T_1. \tag{6}$$

Since the equilibrium value of the second ko fight is equal to the equilibrium value of the first, we also have

$$14 - 2U_1 - t = -11 + 2T_1 - t. \tag{7}$$

Solving these two equations we find

$$T_1 = 4\frac{1}{6} + \frac{t}{6},\qquad(8)$$

$$U_1 = 8\frac{1}{3} - \frac{t}{6}\qquad(9)$$

There is no second strategy for Black when White wins the first ko, nor is there a third strategy for White when Black plays an approach move to C. Therefore these are the exact values for the critical sizes. Solving for the equilibrium value we find

$$LS(A) = -2\frac{2}{3} - \frac{2}{3}t.\qquad(10)$$

The right scaffold is formed by White's playing to -11.

$$RS(A) = -11 + t.\qquad(11)$$

The two scaffolds intersect at temperature 5 to yield

$$t(A) = 5\qquad(12)$$

and

$$m(A) = -6.\qquad(13)$$

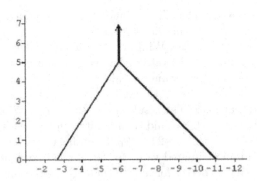

**Fig. 9.** Approach ko thermograph.

Figure 9 shows the thermograph.

## 4   N-Move Approach Kos

What is the value of an $n$-move approach ko in a basic NTE, where White to play can win the ko? Let us add a constant to the scores so that the local score if White wins the ko is 0 and the local score if Black wins is $x$.

With the 1-move approach ko we saw that when one player makes an approach move, rather than completing the ko threat the opponent should take the ko, starting a new ko fight. In this fight the players ignore the uncompleted threat until after the ko is won, playing threats at a new, hotter critical value. For n-move approach kos we start by assuming that each approach move ushers in a new ko fight with a hotter critical size in which previously played threats are ignored until the ko is won. Then after the ko is won, any remaining threats are played in order from the most recent, hottest one down to the earliest, coolest one. Then we shall see that such play guarantees each player the equilibrium value of the ko fight.

Suppose that Black takes the ko and then White wins the ko after $m$ approach moves. At this point each player will have made the same number of plays and $m$ uncompleted White ko threats will remain plus one uncompleted Black threat, $\{2\theta_m|0\}$. Black will complete this threat, achieving $2\theta_m - W - t$, where $W$ is the result of play in the sum of White threats with White playing first at temperature, $t$. Let us compare that result with the one when White wins the ko after $m + 2$ approach moves. Now there will be $m + 3$ uncompleted threats, $\{2\theta_{m+2}|0\}$, $\{0|-2\theta_{m+1}\}$, $\{0|-2\theta_m\}$, and the same uncompleted White ko threats as after $m$ approach moves. Black to play will achieve $2\theta_{m+2} - 2\theta_{m+1} - W - t$. Setting these results to be equal we get

$$2\theta_{m+2} - 2\theta_{m+1} - W - t = 2\theta_m - W - t, \tag{14}$$

from which we derive

$$\theta_{m+2} = \theta_{m+1} + \theta_m \tag{15}$$

for $0 \leq m \leq n - 2$.

If White wins the ko after $n - 1$ approach moves and then Black completes her threat, the cooled result is $2\theta_{n-1} - W - t$, where $W$ is the result of play in the sum of the White threats with White playing first at temperature $t$. If Black wins the ko fight after $n$ approach moves and then White completes his threat, the cooled result is $x - 2\theta_n - W - t$. Setting these equal we get

$$x - 2\theta_n - W - t = 2\theta_{n-1} - W - t \tag{16}$$

from which we derive

$$x/2 = \theta_n + \theta_{n-1}. \tag{17}$$

From this and (15) it follows by induction that

$$x/2 = F(n + 1)\theta_1 + F(n)\theta_0, \tag{18}$$

where $F(n)$ is the $n$-th Fibonacci number.

When White wins the ko fight after zero approach moves Black completes her ko threat for a cooled result of $2\theta_0 - t$. When White wins the ko fight after one approach move Black completes her threat and then White completes his earlier threat, for a cooled result of $2\theta_1 - 2\theta_0$. Setting these equal we have

$$2\theta_1 - 2\theta_0 = 2\theta_0 - t, \tag{19}$$

from which we derive

$$\theta_1 = 2\theta_0 - t/2. \tag{20}$$

Substituting in (18) we find

$$x/2 = F(n+1)(2\theta_0 - t/2) + F(n)\theta_0, \tag{21}$$

from which we derive

$$x/2 = F(n+3)\theta_0 - F(n+1)t/2 \tag{22}$$

and

$$\theta_0 = (x + F(n+1)t)/2F(n+3). \tag{23}$$

The equilibrium value $K$ is

$$K = 2\theta_0 - t. \tag{24}$$

Substituting from (23) we find

$$K = (x - F(n+2)t)/F(n+3). \tag{25}$$

Taking this as the left scaffold of the approach ko, A, with

$$RS(A) = t, \tag{26}$$

the scaffolds intersect when

$$t = (x - F(n+2)t)/F(n+3), \tag{27}$$

from which we find

$$t(A) = x/F(n+4) \tag{28}$$

and

$$m(A) = x/F(n+4). \tag{29}$$

The players can try to improve on this equilibrium value by playing the uncompleted ko threats in a different order or by not taking the ko after an approach move. For each critical size $\theta_m$, $0 < m \le n$, if $\theta_m \ge \theta_{m-1}$, then neither stratagem will work. Obviously, it is correct to play the hottest uncompleted threat first.

Suppose that, after the $m$-th approach move White does not take the ko but completes his threat of size $\theta_{m-1}$. Black then makes another approach move or wins the ko. Play then continues along one of the lines as with the proposed strategy. After the ko fight is won a sum $S$ of uncompleted ko threats remains. Playing the proposed strategy with the ko fight won at the same place, White will have played his threat of size $\theta_m$, which Black will have ignored. Otherwise the play will have been the same, and the sum $R$ of uncompleted ko threats will include all of the uncompleted threats in $S$ plus $\{0| -2\theta_m\}$ and $\{0| -2\theta_{m-1}\}$. $S$

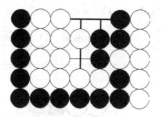

**Fig. 10.** Throw-in 10,000 year ko.

will also include the score $-2\theta_{m-1}$ from having completed that threat. Taking the difference,

$$R - S = \{0| - 2\theta_m\} + \{2\theta_{m-1}|0\}. \tag{30}$$

(The other uncompleted threats cancel and the second game is equal to $\{0| - 2\theta_{m-1}\} - (-2\theta_{m-1})$.) This difference favors White. If Black plays first she will play to 0 in the first game and White will reply to 0 in the second. If White plays first she will also play in the first game, with the result $2\theta_{m-1} - 2\theta_m$, which is non-positive. The fact that this difference favors White means that $R$ is at least as good for White as $S$.

It remains to show that $\theta_m \geq \theta_{m-1}$. Since $\theta_{m+2} = \theta_{m+1} + \theta_m$, it suffices to show that $\theta_1 \geq \theta_0$ in the range of the ko fights. As $\theta_1 = 2\theta_0 - t/2$, this is so when $t = 0$. If $\theta_1 \geq \theta_0$ then

$$2\theta_0 - t/2 \geq \theta_0 \tag{31}$$

and

$$\theta_0 \geq t/2 \tag{32}$$

Substituting from (23), we get

$$(x + F(n+1)t)/2F(n+3) \geq t/2, \tag{33}$$

from which we derive

$$x \geq F(n+2)t. \tag{34}$$

The ko fight ranges up to $t = x/F(n+4)$. Then

$$x \geq F(n+2)x/F(n+4), \tag{35}$$

which is the case.

## 5 Various Results

The following examples are hyperactive ko positions from [4]. I shall give their figure numbers there along with their mast values and temperatures when Black is komaster, when White is komaster, and in the basic NTE.

Figure 10 shows Fig. 67 of [4]. Either player can throw in to make a ko or can prevent the throw-in to make *seki*, a standoff position with a score of 0. Such

**Table 1.** Throw-in 10,000 year ko.

|             | Black komaster | White komaster | NTE |
|-------------|----------------|----------------|-----|
| Mast value  | $7\frac{2}{3}$ | 1              | $4\frac{1}{3}$ |
| Temperature | $7\frac{2}{3}$ | 11             | $4\frac{1}{3}$ |

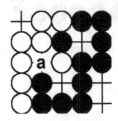

**Fig. 11.** Rogue ko.

positions are called 10,000 year kos because of the reluctance of each player to make the ko. However, the player with fewer stones at risk usually has less to lose by making the ko. In the NTE Black will be inclined to make the ko while White will be inclined to avoid it.

If Black throws in to make the ko she threatens to garner 23 points in 2 moves (1 to make the ko and 1 to win it), while White can score 10 points in 1 net move. The left scaffold from the throw-in is therefore $v = 6\frac{1}{2} - t/2$. This is better for Black than playing for a seki. White prefers seki, however, and the right scaffold is $v = t$. The two scaffolds intersect at $t = 4\frac{1}{3}$.

Figure 11 is Fig. 76 in [4]. This is one of the first hyperactive kos to be analyzed [5]. Black can play at $a$, heating up the ko by threatening to capture the White stone. It acts like a 1-step approach ko for Black[1].

Figure 12 is Fig. 40 in [4]. This ko was misidentified in [4] as a simple ko. It is actually a rogue ko that acts like a 2-move approach ko. Black can play approach moves in the ko fight, first a move 3, ignoring White's first threat, and then at move 9, after White takes the ko back, Black plays a threat that White answers, and then White plays a second threat that Black ignores. After another cycle Black can ignore White's third threat and win the ko at move 15, leaving behind a $\frac{1}{3}$ point ko at $b$. The local count will then be $8\frac{1}{3}$. If White fills the ko at 1 the result is -2. As a 2-move approach ko the temperature is $(8\frac{1}{3} - (-2))/8 = 1\frac{7}{24}$, yielding a mast value of $-\frac{17}{24}$.

Each player has reasonable alternatives. After B 3 White might fill at $b$, and then fill again at 15 when Black continues at 9. That line of play leaves a $\frac{1}{3}$ point ko at $a$. The resulting position is worth $\frac{2}{3}$. If that line of play determined the temperature and mast value, they would be $1\frac{1}{3}$ and $-\frac{2}{3}$, respectively. In fact, that is White's preferred line of play when Black is komaster, but in the NTE White can do slightly better ($\frac{1}{24}$ point on average).

---

[1] Rogue kos are hyperactive, but do not fall into any traditional category. Here Black can win the ko immediately or raise its temperature with another play. In all known rogue kos one player has this kind of choice at some point.

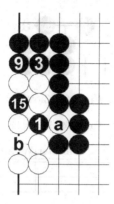

**Fig. 12.** 2-move approach rogue ko.

**Table 2.** Rogue ko.

|  | Black komaster | White komaster | NTE |
|---|---|---|---|
| Mast value | $\frac{7}{9}$ | $\frac{1}{3}$ | $\frac{5}{9}$ |
| Temperature | $1\frac{1}{9}$ | $1\frac{1}{3}$ | $\frac{8}{9}$ |

**Table 3.** 2-move approach ko.

|  | Black komaster | White komaster | NTE |
|---|---|---|---|
| Mast value | $-\frac{2}{3}$ | -1 | $-\frac{17}{24}$ |
| Temperature | $1\frac{1}{3}$ | 1 | $1\frac{7}{24}$ |

At move 3 Black might fill the ko at $a$ for a local score of 1. That is correct when White is komaster, yielding a temperature of 1 and a mast value of 1. But Black does almost $\frac{1}{3}$ point better on average in the NTE.

Figure 13 shows Fig. 79 in [4]. This rogue ko, analyzed by Kao Kuo-Yuan, is one of the smallest hyperactive positions. After taking the ko White threatens to play at $a$. The mast value is the same in the NTE as when White is komaster. In such a case White is called the dogmatic komaster.

## 6    Future Research

There are as yet few results for NTEs. The thermographs for non-ko positions and for ko thermographs under komaster conditions may be constructed from the thermographs of their followers. Finding optimal results for a ko fight in an NTE is rather more difficult. Perhaps we will find easier and more practical methods.

Already the NTE raises new questions. Under what conditions is one player dogmatic komaster? Sometimes the temperature of a ko in an NTE falls between the temperatures when Black is komaster and when White is komaster, but often it is lower than in both komaster conditions. When is that so? Can the

**Fig. 13.** Kao's ko.

**Table 4.** Kao's ko.

|  | Black komaster | White komaster | NTE |
|---|---|---|---|
| Mast value | $\frac{1}{2}$ | $\frac{1}{3}$ | $\frac{1}{3}$ |
| Temperature | $\frac{1}{2}$ | $\frac{2}{3}$ | $\frac{2}{3}$ |

temperature in an NTE ever be higher than under both komaster conditions? (Conjecture: No.)

We are just beginning to scratch the surface of NTE analysis. We have one general result for approach kos. What nuggets remain to be uncovered?

# References

1. Berlekamp, E., Conway, J., Guy, R.: Winning Ways. Academic Press (1982)
2. Berlekamp, E.: The economist's view of combinatorial games. In Nowakowski, R., ed.: Games of No Chance. Cambridge University Press (1996) 365–405
3. Berlekamp, E.: (2000) Personal communication.
4. Müller, M., Berlekamp, E., Spight, W.: Generalized thermography: Algorithms, implementation, and application to Go endgames. Technical Report 96-030, ICSI Berkeley (1996)
5. Berlekamp, E., Wolfe, D.: Mathematical Go: Chilling Gets the Last Point. A K Peters (1994)

# Author Index

Althöfer, Ingo   142
Avetisyan, Henry   123

Berlekamp, Elwyn   213
Botea, Adi   360
Bouzy, Bruno   376
Buro, Michael   280

Cazenave, Tristan   75

Demaine, Erik D.   188
Demaine, Martin L.   188

Eladhari, Mirjam   292

Fang, Haw-ren   264
Fleischer, Rudolf   188
Fraenkel, Aviezri S.   201
Fuchs, Gil   42

Grimbergen, Reijer   171

Herik, Jaap van den   61, 154, 393
Hsu, Shun-chin   264
Hsu, Tsan-sheng   264

Junghanns, Andreas   345

Kendall, Graham   29
Kocsis, Levente   154

Levinson, Robert   42
Liao, Simon   230
Linares López, Carlos   345

Lincke, Thomas R.   249
Lindley, Craig A.   292
Lorentz, Richard J.   123

Müller, Martin   88, 360

Nakamura, Teigo   213

Pawlak, Miroslaw   230
Postma, Eric   154, 393

Rahat, Ofer   201
Rollason, Jeff   171

Schaeffer, Jonathan   360
Shapiro, Ari   42
Shaw, Stephen   29
Snatzke, Raymond Georg   142
Spight, William L.   413
Sturtevant, Nathan   108

Tournavitis, Konstantinos   11

Uiterwijk, Jos W.H.M.   61, 154, 393

Verbrugge, Clark   308

Werf, Erik van der   393
Winands, Mark H.M.   61
Wolfe, David   1

Yang, Jing   230

Zhao, Ling   326

# Game Index

15-puzzle   345

Amazons   123
Awari   249

Blackjack   1

Cellular Automata Games   201
Chess   142, 154, 249
Chinese checkers   108
Chinese chess   264
Clobber   188
Cribbage   29

Diplomacy   42

Go   75, 142, 213, 376, 393, 413

Hearts   108
Hex   230

Lines of Action   61

Othello   11

PCP   326

RPG   292
RTS   280

Shogi   171
Sokoban   360
Spades   108

# Lecture Notes in Computer Science

For information about Vols. 1–2812
please contact your bookseller or Springer-Verlag

Vol. 2770: E. Becker, W. Buhse, D. Günnewig, N. Rump (Eds.), Digital Rights Management. XI, 805 pages. 2003.

Vol. 2813: I.-Y. Song, S.W. Liddle, T.W. Ling, P. Scheuermann (Eds.), Conceptual Modeling – ER 2003. Proceedings, 2003. XIX, 584 pages. 2003.

Vol. 2814: M.A. Jeusfeld, Ó. Pastor (Eds.), Conceptual Modeling for Novel Application Domains. Proceedings, 2003. XVI, 410 pages. 2003.

Vol. 2815: Y. Lindell, Composition of Secure Multi-Party Protocols. XVI, 192 pages. 2003.

Vol. 2816: B. Stiller, G. Carle, M. Karsten, P. Reichl (Eds.), Group Communications and Charges. Proceedings, 2003. XIII, 354 pages. 2003.

Vol. 2817: D. Konstantas, M. Leonard, Y. Pigneur, S. Patel (Eds.), Object-Oriented Information Systems. Proceedings, 2003. XII, 426 pages. 2003.

Vol. 2818: H. Blanken, T. Grabs, H.-J. Schek, R. Schenkel, G. Weikum (Eds.), Intelligent Search on XML Data. XVII, 319 pages. 2003.

Vol. 2819: B. Benatallah, M.-C. Shan (Eds.), Technologies for E-Services. Proceedings, 2003. X, 203 pages. 2003.

Vol. 2820: G. Vigna, E. Jonsson, C. Kruegel (Eds.), Recent Advances in Intrusion Detection. Proceedings, 2003. X, 239 pages. 2003.

Vol. 2821: A. Günter, R. Kruse, B. Neumann (Eds.), KI 2003: Advances in Artificial Intelligence. Proceedings, 2003. XII, 662 pages. 2003. (Subseries LNAI).

Vol. 2822: N. Bianchi-Berthouze (Ed.), Databases in Networked Information Systems. Proceedings, 2003. X, 271 pages. 2003.

Vol. 2823: A. Omondi, S. Sedukhin (Eds.), Advances in Computer Systems Architecture. Proceedings, 2003. XIII, 409 pages. 2003.

Vol. 2824: Z. Bellahsène, A.B. Chaudhri, E. Rahm, M. Rys, R. Unland (Eds.), Database and XML Technologies. Proceedings, 2003. X, 283 pages. 2003.

Vol. 2825: W. Kuhn, M. Worboys, S. Timpf (Eds.), Spatial Information Theory. Proceedings, 2003. XI, 399 pages. 2003.

Vol. 2826: A. Krall (Ed.), Software and Compilers for Embedded Systems. Proceedings, 2003. XI, 403 pages. 2003.

Vol. 2827: A. Albrecht, K. Steinhöfel (Eds.), Stochastic Algorithms: Foundations and Applications. Proceedings, 2003. VIII, 167 pages. 2003.

Vol. 2828: A. Lioy, D. Mazzocchi (Eds.), Communications and Multimedia Security. Proceedings, 2003. VIII, 265 pages. 2003.

Vol. 2829: A. Cappelli, F. Turini (Eds.), AI*IA 2003: Advances in Artificial Intelligence. Proceedings, 2003. XIV, 552 pages. 2003. (Subseries LNAI).

Vol. 2830: F. Pfenning, Y. Smaragdakis (Eds.), Generative Programming and Component Engineering. Proceedings, 2003. IX, 397 pages. 2003.

Vol. 2831: M. Schillo, M. Klusch, J. Müller, H. Tianfield (Eds.), Multiagent System Technologies. Proceedings, 2003. X, 229 pages. 2003. (Subseries LNAI).

Vol. 2832: G. Di Battista, U. Zwick (Eds.), Algorithms – ESA 2003. Proceedings, 2003. XIV, 790 pages. 2003.

Vol. 2833: F. Rossi (Ed.), Principles and Practice of Constraint Programming – CP 2003. Proceedings, 2003. XIX, 1005 pages. 2003.

Vol. 2834: X. Zhou, S. Jähnichen, M. Xu, J. Cao (Eds.), Advanced Parallel Processing Technologies. Proceedings, 2003. XIV, 679 pages. 2003.

Vol. 2835: T. Horváth, A. Yamamoto (Eds.), Inductive Logic Programming. Proceedings, 2003. X, 401 pages. 2003. (Subseries LNAI).

Vol. 2836: S. Qing, D. Gollmann, J. Zhou (Eds.), Information and Communications Security. Proceedings, 2003. XI, 416 pages. 2003.

Vol. 2837: N. Lavrač, D. Gamberger, H. Blockeel, L. Todorovski (Eds.), Machine Learning: ECML 2003. Proceedings, 2003. XVI, 504 pages. 2003. (Subseries LNAI).

Vol. 2838: N. Lavrač, D. Gamberger, L. Todorovski, H. Blockeel (Eds.), Knowledge Discovery in Databases: PKDD 2003. Proceedings, 2003. XVI, 508 pages. 2003. (Subseries LNAI).

Vol. 2839: A. Marshall, N. Agoulmine (Eds.), Management of Multimedia Networks and Services. Proceedings, 2003. XIV, 532 pages. 2003.

Vol. 2840: J. Dongarra, D. Laforenza, S. Orlando (Eds.), Recent Advances in Parallel Virtual Machine and Message Passing Interface. Proceedings, 2003. XVIII, 693 pages. 2003.

Vol. 2841: C. Blundo, C. Laneve (Eds.), Theoretical Computer Science. Proceedings, 2003. XI, 397 pages. 2003.

Vol. 2842: R. Gavaldà, K.P. Jantke, E. Takimoto (Eds.), Algorithmic Learning Theory. Proceedings, 2003. XI, 313 pages. 2003. (Subseries LNAI).

Vol. 2843: G. Grieser, Y. Tanaka, A. Yamamoto (Eds.), Discovery Science. Proceedings, 2003. XII, 504 pages. 2003. (Subseries LNAI).

Vol. 2844: J.A. Jorge, N.J. Nunes, J.F. e Cunha (Eds.), Interactive Systems. Proceedings, 2003. XIII, 429 pages. 2003.

Vol. 2846: J. Zhou, M. Yung, Y. Han (Eds.), Applied Cryptography and Network Security. Proceedings, 2003. XI, 436 pages. 2003.

Vol. 2847: R. de Lemos, T.S. Weber, J.B. Camargo Jr. (Eds.), Dependable Computing. Proceedings, 2003. XIV, 371 pages. 2003.

Vol. 2848: F.E. Fich (Ed.), Distributed Computing. Proceedings, 2003. X, 367 pages. 2003.

Vol. 2849: N. García, J.M. Martínez, L. Salgado (Eds.), Visual Content Processing and Representation. Proceedings, 2003. XII, 352 pages. 2003.

Vol. 2850: M.Y. Vardi, A. Voronkov (Eds.), Logic for Programming, Artificial Intelligence, and Reasoning. Proceedings, 2003. XIII, 437 pages. 2003. (Subseries LNAI)

Vol. 2851: C. Boyd, W. Mao (Eds.), Information Security. Proceedings, 2003. XI, 443 pages. 2003.

Vol. 2852: F.S. de Boer, M.M. Bonsangue, S. Graf, W.-P. de Roever (Eds.), Formal Methods for Components and Objects. Proceedings, 2003. VIII, 509 pages. 2003.

Vol. 2853: M. Jeckle, L.-J. Zhang (Eds.), Web Services – ICWS-Europe 2003. Proceedings, 2003. VIII, 227 pages. 2003.

Vol. 2854: J. Hoffmann, Utilizing Problem Structure in Planning. XIII, 251 pages. 2003. (Subseries LNAI)

Vol. 2855: R. Alur, I. Lee (Eds.), Embedded Software. Proceedings, 2003. X, 373 pages. 2003.

Vol. 2856: M. Smirnov, E. Biersack, C. Blondia, O. Bonaventure, O. Casals, G. Karlsson, George Pavlou, B. Quoitin, J. Roberts, I. Stavrakakis, B. Stiller, P. Trimintzios, P. Van Mieghem (Eds.), Quality of Future Internet Services. IX, 293 pages. 2003.

Vol. 2857: M.A. Nascimento, E.S. de Moura, A.L. Oliveira (Eds.), String Processing and Information Retrieval. Proceedings, 2003. XI, 379 pages. 2003.

Vol. 2858: A. Veidenbaum, K. Joe, H. Amano, H. Aiso (Eds.), High Performance Computing. Proceedings, 2003. XV, 566 pages. 2003.

Vol. 2859: B. Apolloni, M. Marinaro, R. Tagliaferri (Eds.), Neural Nets. Proceedings, 2003. X, 376 pages. 2003.

Vol. 2860: D. Geist, E. Tronci (Eds.), Correct Hardware Design and Verification Methods. Proceedings, 2003. XII, 426 pages. 2003.

Vol. 2861: C. Bliek, C. Jermann, A. Neumaier (Eds.), Global Optimization and Constraint Satisfaction. Proceedings, 2002. XII, 239 pages. 2003.

Vol. 2862: D. Feitelson, L. Rudolph, U. Schwiegelshohn (Eds.), Job Scheduling Strategies for Parallel Processing. Proceedings, 2003. VII, 269 pages. 2003.

Vol. 2863: P. Stevens, J. Whittle, G. Booch (Eds.), «UML» 2003 – The Unified Modeling Language. Proceedings, 2003. XIV, 415 pages. 2003.

Vol. 2864: A.K. Dey, A. Schmidt, J.F. McCarthy (Eds.), UbiComp 2003: Ubiquitous Computing. Proceedings, 2003. XVII, 368 pages. 2003.

Vol. 2865: S. Pierre, M. Barbeau, E. Kranakis (Eds.), Ad-Hoc, Mobile, and Wireless Networks. Proceedings, 2003. X, 293 pages. 2003.

Vol. 2867: M. Brunner, A. Keller (Eds.), Self-Managing Distributed Systems. Proceedings, 2003. XIII, 274 pages. 2003.

Vol. 2868: P. Perner, R. Brause, H.-G. Holzhütter (Eds.), Medical Data Analysis. Proceedings, 2003. VIII, 127 pages. 2003.

Vol. 2869: A. Yazici, C. Şener (Eds.), Computer and Information Sciences – ISCIS 2003. Proceedings, 2003. XIX, 1110 pages. 2003.

Vol. 2870: D. Fensel, K. Sycara, J. Mylopoulos (Eds.), The Semantic Web - ISWC 2003. Proceedings, 2003. XV, 931 pages. 2003.

Vol. 2871: N. Zhong, Z.W. Raś, S. Tsumoto, E. Suzuki (Eds.), Foundations of Intelligent Systems. Proceedings, 2003. XV, 697 pages. 2003. (Subseries LNAI)

Vol. 2873: J. Lawry, J. Shanahan, A. Ralescu (Eds.), Modelling with Words. XIII, 229 pages. 2003. (Subseries LNAI)

Vol. 2875: E. Aarts, R. Collier, E. van Loenen, B. de Ruyter (Eds.), Ambient Intelligence. Proceedings, 2003. XI, 432 pages. 2003.

Vol. 2876: M. Schroeder, G. Wagner (Eds.), Rules and Rule Markup Languages for the Semantic Web. Proceedings, 2003. VII, 173 pages. 2003.

Vol. 2877: T. Böhme, G. Heyer, H. Unger (Eds.), Innovative Internet Community Systems. Proceedings, 2003. VIII, 263 pages. 2003.

Vol. 2878: R.E. Ellis, T.M. Peters (Eds.), Medical Image Computing and Computer-Assisted Intervention - MICCAI 2003. Part I. Proceedings, 2003. XXXIII, 819 pages. 2003.

Vol. 2879: R.E. Ellis, T.M. Peters (Eds.), Medical Image Computing and Computer-Assisted Intervention - MICCAI 2003. Part II. Proceedings, 2003. XXXIV, 1003 pages. 2003.

Vol. 2880: H.L. Bodlaender (Ed.), Graph-Theoretic Concepts in Computer Science. Proceedings, 2003. XI, 386 pages. 2003.

Vol. 2881: E. Horlait, T. Magedanz, R.H. Glitho (Eds.), Mobile Agents for Telecommunication Applications. Proceedings, 2003. IX, 297 pages. 2003.

Vol. 2883: J. Schaeffer, M. Müller, Y. Björnsson (Eds.), Computers and Games. Proceedings, 2002. XI, 431 pages. 2003.

Vol. 2884: E. Najm, U. Nestmann, P. Stevens (Eds.), Formal Methods for Open Object-Based Distributed Systems. Proceedings, 2003. X, 293 pages. 2003.

Vol. 2885: J.S. Dong, J. Woodcock (Eds.), Formal Methods and Software Engineering. Proceedings, 2003. XI, 683 pages. 2003.

Vol. 2886: I. Nyström, G. Sanniti di Baja, S. Svensson (Eds.), Discrete Geometry for Computer Imagery. Proceedings, 2003. XII, 556 pages. 2003.

Vol. 2887: T. Johansson (Ed.), Fast Software Encryption. Proceedings, 2003. IX, 397 pages. 2003.

Vol. 2888: R. Meersman, Zahir Tari, D.C. Schmidt et al. (Eds.), On The Move to Meaningful Internet Systems 2003: CoopIS, DOA, and ODBASE. Proceedings, 2003. XXI, 1546 pages. 2003.

Vol. 2889: Robert Meersman, Zahir Tari et al. (Eds.), On The Move to Meaningful Internet Systems 2003: OTM 2003 Workshops. Proceedings, 2003. XXI, 1096 pages. 2003.

Vol. 2891: J. Lee, M. Barley (Eds.), Intelligent Agents and Multi-Agent Systems. Proceedings, 2003. X, 215 pages. 2003. (Subseries LNAI)

Vol. 2893: J.-B. Stefani, I. Demeure, D. Hagimont (Eds.), Distributed Applications and Interoperable Systems. Proceedings, 2003. XIII, 311 pages. 2003.

Vol. 2895: A. Ohori (Ed.), Programming Languages and Systems. Proceedings, 2003. XIII, 427 pages. 2003.

Vol. 2897: O. Balet, G. Subsol, P. Torguet (Eds.), Virtual Storytelling. Proceedings, 2003. XI, 240 pages. 2003.

Vol. 2899: G. Ventre, R. Canonico (Eds.), Interactive Multimedia on Next Generation Networks. Proceedings, 2003. XIV, 420 pages. 2003.